Materials Selection for Sulfuric Acid

Materials Selection for Sulfuric Acid

Michael Davies
CARIAD Consultants

MS-1: Sulfuric Acid

Second Edition

Publication No. MS-1

Materials Technology Institute of the Chemical Process Industries, Inc.

Printed and bound in the United States of America

ISBN:1-57698-035-9

Published by
Materials Technology Institute
www.mti-global.org

Table of Contents

List of Figures

List of Tables

1

Foreword

This book is the second edition of MS-1 in the MTI series on materials selection for manufacture, handling, storage, and shipment of critical hazardous chemicals. The first edition of this volume was prepared and edited by C. P. Dillon and W. I. Pollock based on "Chem. Cor 1, Concentrated Sulfuric Acid and Oleum" by C. P. Dillon. This second edition has been completely reviewed, updated, and rewritten to include weak and intermediate-strength acid by M. Davies of CARIAD Consultants based on the first edition and a first draft of a section on weak sulfuric acid. Most of the figures have been prepared by P.J.B. Scott, also of CARIAD Consultants.

Technical input and comments have been provided by a number of people, including Jim Alexander and Russell Schnell (DuPont Dow Elastomer); Paul Crook (Haynes International); Jim McCoy (Special Metals); Randy Scheel (Wah Chang); Göran Sjoqvist (Edmeston); John Rodda (McMaster University); and Doug Shaw, George Miller, and Jim Thomson (Chemetics).

Information is provided in this monograph on the properties of all strengths of sulfuric acid, production methods, health and safety issues, forms of corrosion specific to sulfuric acid, definitions, and relevant specifications for materials of construction, as well as pertinent laboratory and field corrosion data. Alloys are identified by their UNS number together with their generic or trade name where appropriate.

The materials of construction preferred for manufacturing, storing, transporting, and handling sulfuric acid as a commercial product are described in detail.

2

Introduction

Sulfuric acid, H_2SO_4, is the largest-tonnage heavy industrial chemical manufactured. It is a major feedstock for many inorganic and organic products. In Western Europe in 1997 over 19 million tonnes were produced, the total production worldwide being estimated at around 150 million tonnes. About half of this output is produced in North America, Western Europe, and Japan.[1] U.S. imports and exports for 2000 were approximately 1,420,000 and 192,000 metric tons, respectively.[2] In 2002 the figures were 1,058,055 metric tons imported into the United States and 147,452 metric tons exported.[3]

The output of sulfuric acid at base metal smelters today represents about 20% of all acid production. Whereas in 1991 smelter acid production amounted to 27.98 million tonnes, it was calculated that the output in the decade following would grow to 44.97 million tonnes, and that smelter acid would comprise more than 25% of world sulfuric acid production by 2001, compared with 18% in 1991. Sulfuric acid is used directly or indirectly in nearly all industries and is a vital commodity in any national economy. In fact, sulfuric acid is so widely used that its consumption rate, like steel production or electric power, can be used to indicate a nation's prosperity.[4]

Concentrated sulfuric acid (i.e., from about 70% to about 120%) is primarily manufactured and shipped in North America as either:

- 93% acid, or
- nominally 98.5% acid,

although there are other concentrations (e.g., High Stage Concentrator [HSC] Make at about 90.5%). The Chemical Manufacturers Association also recognizes two other acid concentrations, 58° Baumé (74.36%) and 60° Baumé (77.67%), where:

$$°\text{Baumé} = 145 - 145/\text{Sp. Gr.}$$

This relationship becomes somewhat indeterminate above 93% concentration, maximum specific gravity being in the 97–98% range.

The monohydrate (1 mol SO_3 / 1 mol water) is 100% acid; the dihydrate (1 mol SO_3 / 2 mols water) is 84.5%; the trihydrate is 73%. Acid of 70% concentration is 55° Baumé, while strengths of 93% and above are 66° Baumé.

Oleum is fuming sulfuric acid, containing an excess of uncombined sulfur trioxide. It is characterized as 20–65% oleum for commercial handling. It may also be designated as the concentration of sulfuric acid *equivalent* to the contained H_2SO_4 *plus* the acid content that would form if the SO_3 were reacted with water to form additional acid. By such a calculation, for example, 20% oleum is designated as 104.5% H_2SO_4. This relationship is primarily of value in understanding and interpreting graphical data over the 70 to 120% range.

References

1. Sulphur 241, November–December 1995, p. 35 and No. 258, September–October 1998, p. 54, in "Production of Sulphuric Acid," Vol. 3 (Brussels, Belgium: EFMA, European Fertilizer Manufacturers' Association, 2000): 68 pp.
2. Anon, "Strong Inorganic Acid Mists Containing Sulfuric Acid," from 10th report on carcinogens, U.S. Department of Health and Human Services (2003), http://ntp-server.niehs.nih.gov/NewHomeRoc/AboutRoC.html.
3. Anon, "U.S. Trade Quick-Reference Tables," Office of Trade and Economic Analysis (OTEA), U.S. Department of Commerce (2003), http://www.ita.doc.gov/td/industry/otea/Trade-Detail.
4. Anon, "Production of Sulphuric Acid," Vol. 3 (Brussels, Belgium: EFMA, European Fertilizer Manufacturers' Association, 2000): 68 pp.

3

Properties of Sulfuric Acid

Sulfuric acid (Chemical Abstraction Service; CAS# 7664-93-9) is a clear, colorless, oily, and odorless liquid. It is also known as sulphine acid, battery acid, dihydrogen sulfate, mattling acid, oil of vitriol, spirit of sulfur, and hydrogen sulfate. More sulfuric acid is produced in the United States than any other chemical. Its main use is in phosphate fertilizer production, but it is also used to manufacture other acids, explosives, dyestuffs, parchment paper, glue, wood preservatives, and lead-acid batteries for vehicles. It is used in the purification of petroleum, pickling of metals, electroplating baths, nonferrous metallurgy, and production of rayon and film, and as a laboratory reagent.[1]

The boiling points and freezing points for the complete range of sulfuric acid and oleum are shown in Figures 3.1[2] and 3.2,[3] respectively. Details of other physical properties are discussed below.

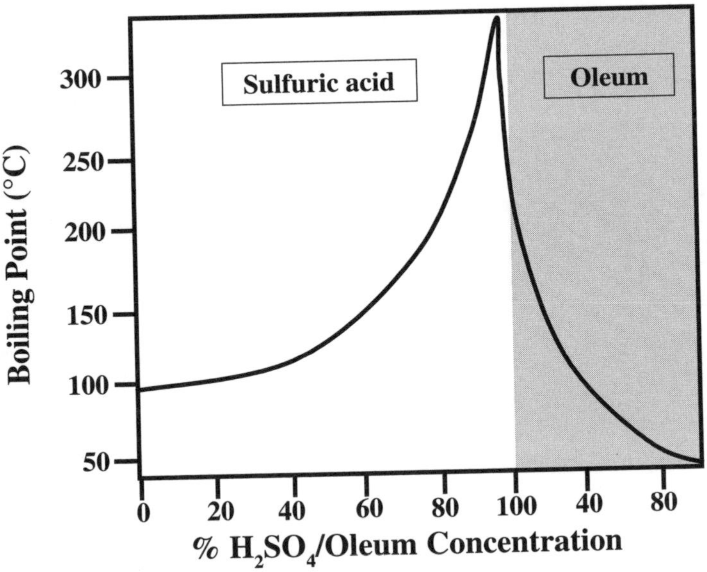

Figure 3.1 Boiling Point of Sulfuric Acid and Oleum

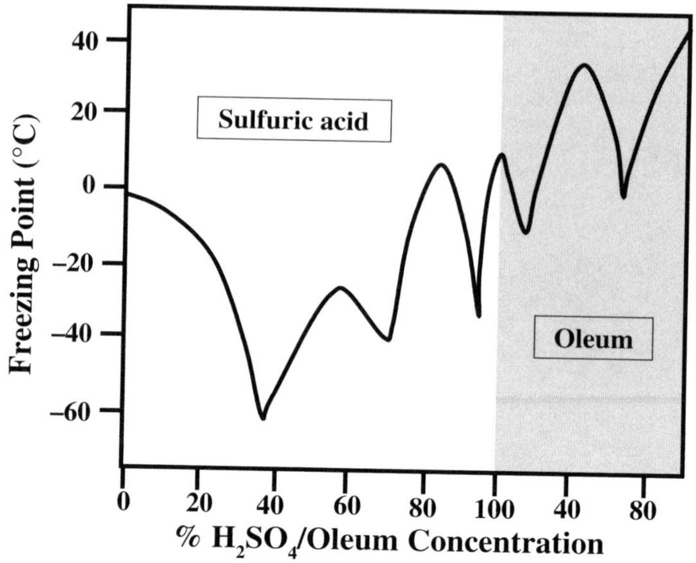

Figure 3.2 Freezing Point of Sulfuric Acid and Oleum

Physical Properties of Concentrated Sulfuric Acid

The physical properties of sulfuric acid from 70 to 100% are given in Table 3.1.[4,5]

Table 3.1 Physical Properties of Concentrated Sulfuric Acid

% H_2SO_4	Degrees Baumé[a]	Freezing Point °C	Boiling Point[b] °C	Sp. Gr.[c]	Gm/L	Lb/Gal
70	55.0	–43	156	1.611	1127	9.41
71	55.6	—	—	1.622	1152	9.61
72	56.2	—	—	1.634	1176	9.82
73[d]	56.9	–40	167	1.646	1201	10.02
74	57.5	—	—	1.657	1226	10.24
75	58.1	–30	—	1.669	1252	10.45
76	58.7	—	—	1.681	1278	10.66
77	59.3	–15	—	1.693	1303	10.88
78	59.9	–12	193	1.704	1329	11.09
79	60.5	—	—	1.716	1355	11.31
80	61.1	–4	198	1.727	1382	11.53
81	61.6	—	—	1.738	1408	11.75
82	62.1	—	—	1.749	1434	11.97

(continued)

Table 3.1 Physical Properties of Concentrated Sulfuric Acid *(continued)*

% H_2SO_4	Degrees Baumé[a]	Freezing Point °C	Boiling Point[b] °C	Sp. Gr.[c]	Gm/L	Lb/Gal
83	62.6	8	—	1.759	1460	12.19
84	63.0	—	—	1.769	1486	12.40
84.5[e]	63.3	8	224	1.774	1499	12.51
85	63.5	—	—	1.779	1512	12.62
86	63.9	—	—	1.787	1537	12.83
87	64.2	—	—	1.795	1562	13.03
88	64.5	—	—	1.802	1586	13.23
89	64.8		—	1.809	1610	13.42
90	65.1	–6	271	1.814	1622	13.63
91	65.3	—	—	1.820	1656	13.82
92	65.5	–19	—	1.824	1678	14.00
93	65.7	–28	280	1.828	1700	14.19
94	65.8	–32	—	1.831	1721	14.36
95	65.9	–21	304	1.834	1742	14.54
96	66.0	–14	308	1.836	1762	14.70
97	66.0	–6	317	1.836	1781	14.87
98[f]	66.0	–1	327	1.836	1781	14.87
99	65.9	5	310	1.834	1816	15.15
100[g]	65.8	10	274	1.831	1831	15.28

[a] degrees Baumé (Bé) = 145–145/Sp. Gr.
[b] at atmospheric pressure
[c] specific gravity @ 20°/4°C (determined @ 20°C and corrected to 4°C)
[d] trihydrate (1 mol SO_3 / 3 mols H_2O)
[e] dihydrate (1 mol SO_3 / 2 mols H_2O)
[f] Constant Boiling Mixture (CBM)
[g] monohydrate (1 mol SO_3 / 1 mol H_2O)

Chemical Properties

Sulfuric acid can be found in the air as small droplets or it can be attached to other small particles in the air. When concentrated sulfuric acid is mixed with water there is a violent, exothermic reaction. Concentrated sulfuric acid can catch fire or explode when it comes into contact with many chemicals, including acetone, alcohols, and metals. When heated, it emits highly toxic fumes that include sulfur trioxide. It is capable of igniting finely divided combustible materials. It is incompatible or reactive with organic materials, chlorates, carbides, fulminates, water, and powdered metals. It is soluble in water and ethyl alcohol.[1] It is corrosive to metals.

Commercial Grades of Concentrated Sulfuric Acid

There is a wide range of grades based on acid strength and purity depending on final application. Typical grades include commercial, electrolyte (high purity), textile (low organic content), and chemically pure or reagent grades. Many of these grades are available in different acid strengths.[6]

93% Sulfuric Acid

93% sulfuric acid (66° Baumé) is the common commercial strength in North American practice, convenient for shipping and storage because of its low freezing point and low corrosivity toward steel. It is, however, virtually unknown in the United Kingdom and elsewhere in Europe. This concentration is also known as oil of vitriol. The iron content is sometimes specified to be 50 ppm maximum (e.g., if the acid is used for rayon manufacture).

A typical analysis of some commercial grades[7] is shown in Table 3.2 together with similar data for two high purity grades.

The water-white and battery-quality grades are clear, colorless liquids while the commercial grades may be turbid and off-white in appearance. The battery-quality grade is produced by chemical treatment and filtration of the water-white grade. Specifications further limit such contaminants as ammonium ions, antimony, copper, manganese, nickel, selenium, sulfurous acid, and zinc. There are also specifications for tests to determine light transmission, color, organic matter, platinum, and sedi-

Table 3.2 Typical Analyses of Various Grades of Strong Sulfuric Acid

	66o Bé 93%				
Property	Commercial	Water-White	Battery Quality	Commercial 98/99%	Commercial 60° Bé 77%
Minimum acidity	93.19	93.2	93.2	98.0	76.0
Gravity @60°F (15.6°C)	—	66.03	66.1	—	—
Iron ppm max	50	20	<20	50	50
Nitrogen oxide (NO_3) ppm max	10	5	<4	10	10
Sulfur dioxide (SO_2) ppm max	50	—	—	50	50
Permanganate oxidizable matter (ml 0.02 N $KMnO_4$) max	5	—	—	—	—
Pb ppm max	—	<1	—	—	—
Chlorides ppm max	—	<1	—	—	—

ment. These grades are intended to meet both federal specification FS 0-S-801C and the food chemical codex specification.

93% acid concentration is used in drying air in the contact process and in drying sulfur dioxide feed-gas. It is also used in process applications—e.g., synthesis of bis-methylphenylethane at 8–20°C (45–70°F)—by reacting toluene and acetylene in the presence of mercuric sulfate catalyst, and (with nitric acid) in the manufacture of trinitrotoluene (TNT).

Acid of 93% concentration is produced:

- as an interim product from the contact process (see Chapter 5);
- in drying the air feed with stronger acid;
- as a final product from combustion of sulfidic ores; or
- by reconcentration of spent acid from chemical or petrochemical processes (e.g., in vacuum concentrators or drum concentrators).[8]

There are both dry and wet processes used to produce either 93% or 98% acid. Dry processes, as described in Chapter 5 and illustrated in Figure 5.1, involve the direct combustion of either sulfur or sulfidic ores as a source of sulfur dioxide. Wet processes, which involve potentially more corrosive conditions ahead of the drying tower, include both ore combustion (with subsequent effluent gas scrubbing) and the combustion (for reconcentration) of dilute or intermediate concentrations of acid.

95–96% Sulfuric Acid

This strength is supplied as a reagent or Chemically Pure (C.P.) grade at 95.5% acid, produced under very special conditions and shipped in small volumes in glass bottles or carboys.

96–99% Sulfuric Acid

Acid in the 96–99% concentration range is the normal, final product in the Contact Process for manufacturing sulfuric acid; see Chapter 5. Known as Nominal 98.5% acid in North America, this product is widely stored and distributed as a bulk product for use in inorganic processes (e.g., manufacture of ammonium sulfate), as a powerful desiccant (e.g., for drying chlorine, methylene chloride, etc.), and in organic syntheses. Because of its high freezing temperature range of approximately –14 to +5°C (6 to 40°F), this concentration range may require heated tanks and well-insulated tank cars. In Europe, the practice is for the 96% to be the normal commodity grade, with the nominal 98.5% acid kept at 99% concentration and routinely protected against low-temperature exposure.

Other Concentrations

Concentrated sulfuric acid is loosely defined as 70–100% concentration. In addition to 93% and 98.5% acid, 96% acid is a common concentration in Europe. 90–91%, known as High Stage Concentrator (HSC) Make is encountered in some countries (e.g., the United Kingdom), while various dilutions (still above 70%) of the stronger acids may be used in chemical operations. In the United States, HSC Make may be as high as 93%, depending on temperatures and pressures in the concentrators.

Industrial strength acids in the 70% (i.e., 68–72%), 73% (60° Bé), 74% (58° Bé), and nominal 78% range are sometimes encountered. 80–85% acid is formed during sulfonation of alkylated benzene with strong sulfuric acid or oleum. Concentrations of less than 100% sulfuric acid are usually interpreted as mass percent *in water*, the solutions free of other major constituents. However, a "mixed acid" containing 80% sulfuric acid, 10% nitric acid, and 10% water is not the same chemical as 80% or 90% sulfuric acid. In the absorption of ethylene in strong sulfuric acid, to produce ethanol via the intermediate diethyl sulfate (DES), the absorber product of approximately 40% DES / 55% sulfuric has characteristics more like strong sulfuric than 55–60% acid.

Besides the commercial products described above, there are three other concentrations commonly encountered:

- Nitration Spent Acid (68–72%; nominal 70%),
- Alkylation Spent Acid (88–89%), and
- Sulfonation Spent Acid (70–90%).

These usually contain uncontrolled amounts of contaminants from nitration, sulfation, sulfonation, or alkylation reactions.

Despite the uncertainties raised by contaminants, these intermediate concentrations within strong acid parameters may constitute a feedstock for reconcentration.

Physical Properties of Oleum

Service experience indicates that oleum below about 20% (104.5% H_2SO_4), and especially below 14% (103.15% H_2SO_4), is more aggressive than higher concentrations. However, some producers consider anything over 10% oleum (102.25% H_2SO_4) to be satisfactory in steel at ambient temperatures. Physical properties of oleum are given in Table 3.3.[5,9]

Table 3.3 Physical Properties of Oleum

% Oleum	FP °C	BP[a] °C	Sp. Gr.[b]	Lb/Gal	Equivalent H_2SO_4 [c]
0	11	275	1.839	15.33	100.00
5	6	205	1.862	15.52	101.13
10	−2	173	1.880	15.67	102.25
15	−9	155	1.889	15.82	103.38
20	1	142	1.915	15.96	104.50
25	14	131	1.934	16.12	105.62
30	22	121	1.952	16.28	106.75
35	29	104	1.968	16.40	107.87

(continued)

Table 3.3 Physical Properties of Oleum *(continued)*

% Oleum	FP °C	BP[a] °C	Sp. Gr.[b]	Lb/Gal	Equivalent H_2SO_4 [c]
40	34	95	1.983	16.54	109.00
45	35	85	1.993	16.62	110.13
50	34	75	2.001	16.68	111.25
55	26	66	2.051	17.11	112.38
60	17	63	2.102	17.53	113.50
65	2	58	1.992	16.60	114.63
70	—	55	1.982	16.50	115.75
80	—	51	1.949	16.25	118.00
90	—	47	1.911	15.92	120.25
100	17	43	1.857	15.50	122.50

[a] at atmospheric pressure
[b] specific gravity @ 20°/4°C (determined @ 20°C and corrected to 4°C)
[c] equivalent H_2SO_4: see equation in section "Commercial Grades of Oleum"

20% Oleum

Oleum should contain at least 20–22% "free" sulfur trioxide (equivalent to approximately 105% H_2SO_4) to minimize corrosion of carbon steel equipment. With a freezing range of 1 to 6°C (34 to 44°F), it also may require thermal control of tanks and tank cars. In the United Kingdom, neither heating nor insulation is employed, and one industrial plant has handled 17–18% oleum (freezing point [FP]: –4°C) without problems for about 20 years.

35% Oleum

35% oleum is the usual product from the contact process. It has a freezing point of 29°C (84°F) and needs careful attention paid to heating for shipment and storage. Small amounts of strong nitric acid may be added as antifreeze, although this is not done when low NO_X specifications apply. It is good practice to maintain temperatures at least 15°C (27°F) *above* the freezing point. Even vapor lines and vents must be kept at 90°C (205°F) minimum to prevent blockage with solid sulfur trioxide.

Commercial Grades of Oleum

Oleum comprises sulfuric acid concentrations nominally in excess of 100%. The Chemical Manufacturers Association recognizes 20%, 30%, 40%, and 65% oleum as standard industrial grades. Pure sulfuric acid with an excess of sulfur trioxide results in concentrations greater than 100%. However, the nomenclature is somewhat confusing. The term "equivalent H_2SO_4" is used:

$$\text{Equivalent } H_2SO_4 = (\%SO_3 \times 98/80) + (100 - \%SO_3)$$

The product designated as 20% oleum is 20% by weight SO_3 dissolved in 80% by weight pure H_2SO_4. If the value for SO_3 is converted to equivalent H_2SO_4 (i.e., 20% × 98/80) and added to the 80% figure, then the equivalent sulfuric acid = 80 + (20 × 98/80) = 104.5% H_2SO_4.

Both types of nomenclature are used. Oleum is described either as 20 to 65% oleum or as the equivalent sulfuric acid (100.225 to 114.625%). Tables of physical properties usually give both types of nomenclature or the excess SO_3 value in lieu of "percent oleum." Published corrosion data tables usually show "% oleum," while graphical data usually show the equivalent percent acid in order to extend the X-axis to values over 100% concentration.

Concentrations of 60–65% oleum, preferred for lower freezing points than the 35% grade, are made by adding sulfur trioxide to the 20–36% product. An oleum/nitric blend acid is also sometimes supplied.

Physical Properties of Weak and Intermediate Grades

Some of the physical properties of weak and intermediate concentrations of sulfuric acid are shown in Table 3.4. [5,10,11]

Table 3.4 Physical Properties of Weak and Intermediate Grades of Sulfuric Acid

% H_2SO_4	Degrees Baumé[a]	Freezing Point °C	Boiling Point[b] °C	Sp. Gr.[c]	Gm/L	Lb/Gal
1	0.7	–0.42	100	1.0051	10.05	0.0839
5	4.5	–2.05	101	1.0317	51.59	0.4305
10	9.0	–4.64	102	1.0661	106.6	0.8897
20	17.7	–13.6	104	1.1394	227.9	1.902
30	26.0	–33.5	108	1.2185	365.6	3.051
40	33.7	–55.2	114	1.3028	521.1	4.349
50	41.1	–36.5	123	1.3951	697.6	5.821
60	48.2	–28.9	140	1.4983	899.0	7.502
65	51.7	–36.6	151	1.5533	1,010	8.426
70	55.0	–43.0	165	1.6105	1,127	9.408

[a] degrees Baumé (Bé) = 145 – 145/Sp. Gr
[b] at atmospheric pressure
[c] specific gravity @ 20°/4°C (determined @ 20°C and corrected to 4°C)

Commercial Grades of Weak Acid

There are a number of commercially available acid grades within the 0–70% range. Data for three of them, 10N, 5N, and 0.1N, are shown in Table 3.5.[12]

Table 3.5 Commercial Weak and Intermediate Grades of Sulfuric Acid

Grade	10N H_2SO_4 (UN1760)	5N H_2SO_4 (UN130)	0.1N H_2SO_4
Nominal % H_2SO_4	49	24.5	0.49
Comments	Compounded from 93% H_2SO_4 and demineralized water	DILUT-IT® Analytical Concentrate	Commercially available for particular applications (e.g., pH control)
Characteristics	Clear, colorless, odorless liquid. Corrodes Al ~1450 mpy (37 mm/y); steel ~ 7900 mpy (200mm/y). Keep closed; store in a cool, dry place.	Clear, colorless, odorless. Keep closed; store in a cool, dry place.	The product is a clear, colorless, odorless liquid.
SG	1.270	—	—
pH	<0.5	—	—

References

1. Anon, "Sulfuric Acid Chemical Backgrounder," NSC, National Safety Council (2003), http://www.nsc.org/library/chemical/sulfuric.htm.
2. F. C. Zeisberg (1922) in anon, "Sulphuric Acid" (Montreal, QC, Canada: CIL Inc., 1980): p. 19.
3. C. M. Gabie et al. (1950) in anon, "Sulphuric Acid" (Montreal, QC, Canada: CIL Inc., 1980): p. 19.
4. R. C. Weast, ed., "CRC Handbook of Chemistry and Physics," 63rd edition (Boca Raton, FL: CRC Press LLC, 1982): p. F-7.
5. Anon, "The Corrosion Resistance of Nickel-Containing Alloys in Sulfuric Acid and Related Compounds," CEB-1 (Suffern, NY: The International Nickel Co. Inc., 1983): 90 pp.
6. Anon, "Strong Inorganic Acid Mists Containing Sulfuric Acid," from 10th report on carcinogens, U.S. Department of Health and Human Services (2003), http://ntp-server.niehs.nih.gov/NewHomeRoc/AboutRoC.html.
7. Anon, "Sulfuric Acid" (North York, ON, Canada: Marsulex, 2003), http://www.marsulex.com/PDF_MLX/specs/SulfuricAcid.specs.pdf.
8. G. M. Smith, E. Mantius, "The Concentration of Sulfuric Acid," Chem. Eng. Prog. (September 1978).
9. R. C. Weast, ed., "CRC Handbook of Chemistry and Physics," 63rd edition (Boca Raton, FL: CRC Press LLC, 1982): p. F-8.
10. R. C. Weast, ed., "CRC Handbook of Chemistry and Physics," 63rd edition (Boca Raton, FL: CRC Press LLC, 1982): pp. D-271, F-7.
11. Anon, "Sulphuric Acid—Properties," DKL Engineering (2003), http://members.rogers.com/acidmanual3/properties_acid_properties.htm.
12. Anon, "Material Safety Data Sheets" (Phillipsburg, NJ: J.T. Baker Co., 2002).

4

Safety and Health Considerations

Sulfuric acid is an intrinsically hazardous substance and care should be taken with all aspects of handling it. The hazards involved and the correct procedures to avoid them are well known and information is widely available from producers, suppliers, and others.

Concentrated Sulfuric Acid and Oleum

Concentrated sulfuric acid is a hazardous chemical, and fuming acid and oleum pose special problems regarding safety and health.[1] Even a small oleum leak can produce copious amounts of a dense and irritating fog. The combination of sulfur trioxide and relative humidity (moisture) in the atmosphere forms submicron sulfuric acid particles ("white smoke"), which are a hazard to personnel.

Sulfuric acid is very corrosive and irritating and can cause direct effects on the skin, eyes, and respiratory and gastrointestinal tracts when there is direct exposure to sufficient concentrations. It can cause blindness by direct contact with eyes. Drinking concentrated sulfuric acid can burn the mouth and throat, erode a hole in the stomach, and possibly cause death. Breathing sulfuric acid mists can result in tooth erosion and respiratory tract irritation.

Breathing can be difficult in polluted air that contains small droplets of sulfuric acid. Breathing large amounts of sulfuric acid droplets will also decrease the ability of the respiratory tract to remove other small particles in the respiratory tract.

Exposure through inhalation, ingestion, or contact with the skin can cause pulmonary edema, bronchitis, emphysema, conjunctivitis, stomatis, tracheobronchitis, and dermatitis.

Various regulatory bodies set exposure limits intended to protect humans that might come into contact with sulfuric acid. Typical limits are as follows:[2]

IDLH: 15 mg/m^3 (NIOSH, 1997)
TLV TWA: 3 mg/m^3
TLV STEL: 3 mg/m^3 (ACGIH, 1999)
ERPG-1: 2 mg/m^3 (AIHA, 1999)
ERPG-2: 10 mg/m^3 (AIHA, 1999)
ERPG-3: 30 mg/m^3 (AIHA, 1999)
NIOSH REL: TWA 1 mg/m^3
OSHA PEL: 1 mg/m^3 TWA

Acute Toxicity

Skin contact results in severe damage, with immediate charring of the flesh. Eye contact results in severe damage followed by loss of sight. Inhalation of fumes causes damage to the upper respiratory tract and even to lung tissue (1-hour LC_{50}: 347 ppm in rats). Swallowing will cause severe injury and possibly death (oral LDS_{50}: 2140 mg/kg in rats).[3,4]

Chronic Toxicity

Repeated contact with sulfuric acid mist may cause skin irritation. Repeated inhalation of smoke or mist may cause inflammation of the upper and lower respiratory tract. Concentrations as low as 3 mg/m^3 are disagreeable, while 5 mg/m^3 is severely objectionable. Normal exposure limit is 1 mg/m^3. Prolonged exposure to low concentrations may lead to a tolerance that can cause chronic lung damage.

Recommended Protective Equipment

Local exhaust ventilation should be applied wherever there is an incidence of point source emissions or dispersion of regulated contaminants in the work area. The most effective measures are the total enclosure of processes and the mechanization of handling procedures to prevent all personal contact with sulfuric acid. Electrical installations should be protected against the corrosive action of acid vapors. Smoking should be prohibited in areas in which sulfuric acid is stored or handled.

A NIOSH/MSHA-approved air-purifying respirator equipped with acid gas/fume, dust, or mist cartridges should be provided for exposure to concentrations up to 10 mg/m^3. An air-supplied respirator should be used if concentrations are higher or unknown.

Personal protective wear recommended is long-sleeved woolen, acrylic, or polyester clothing; PVC- or rubber-lined apron, gloves, and boots are required. Protective chemical safety goggles (or at least face-shields) and safety hats are required.

First Aid

In the case of physical contact with strong sulfuric acid, call a physician and follow these specific steps:[5]

- Inhalation: Remove to fresh air. If not breathing, give artificial respiration. If breathing is difficult, give oxygen.
- Ingestion: Do not induce vomiting. Give large quantities of water. Never give anything by mouth to an unconscious person.
- Skin contact: In case of contact, immediately flush skin with plenty of water for at least 15 minutes while removing contaminated clothing and shoes. Wash clothing before reuse. Excess acid on skin can be neutralized with a 2% solution of bicarbonate of soda. Do not apply salves or ointments.
- Eye contact: Immediately flush eyes with a gentle but large stream of water for at least 15 minutes, lifting lower and upper eyelids occasionally.

Disposal, Leak, and Spill Procedures

Early detection of acid spills is essential. The containment area may be blanketed with a silicone or fluorocarbon oil, or with acid-washed mineral oil, to minimize fumes.

Restrict access to the area until completion of cleanup. Ensure trained personnel conduct cleanup. Remove all ignition sources (no smoking, flares, sparks, or flames). All equipment should be grounded. Ventilate the area. Use appropriate personal protection equipment. Prevent liquid from entering sewers or waterways. Stop or reduce the leak if it is safe to do so.

In the case of small spills:
  Cover with *dry* earth, sand, or other incombustible material. Use clean, non-sparking tools to collect material and place it in loosely covered plastic containers for later disposal.

In the case of large spills:
  Prevent liquid from entering sewers or waterways. Dike with inert material (sand, earth, etc.). Collect in plastic containers for disposal. Consider *in situ* neutralization and disposal. Ensure adequate decontamination of tools and equipment following cleanup. Comply with federal, provincial/state, and local regulations on reporting releases.

Use of deactivating chemicals:
  Lime, limestone, sodium carbonate (soda ash), sodium bicarbonate, dilute sodium hydroxide, and dilute aqua ammonia are all deactivating chemicals. Additions of lime or soda ash minimize corrosion of metals and liberation of hydrogen gas. (Concrete can suffer "cement bacillus" attack by residual neutral sulfates.)

Waste disposal methods:
  Dispose of waste material at an approved waste treatment/disposal facility, in accordance with applicable regulations. Do not dispose of waste with normal garbage or to sewer systems.[6]

Add water only as a fog or foam and only outdoors (or in large open areas inside buildings) because of the violent reaction, which is still hazardous and fume-producing.

Fire Procedures

Although sulfuric acid is not flammable, it is highly reactive. It is a strong dehydrating agent, which may cause ignition of finely divided combustible materials on contact. It reacts violently with water with the evolution of heat and can react explosively with organic materials. It reacts with many metals to liberate hydrogen gas that can form explosive mixtures with air. Hydrogen, a highly flammable gas, can accumulate to explosive concentrations inside drums or any type of steel container or tank upon storage. Oxides of sulfur may be produced in fire.

For small fires, use carbon dioxide or dry chemical media; for large fires, use an all-purpose type AFFF foam. If only water is available, use it in the form of a fog. While fighting fires involving sulfuric acid, wear a NIOSH/MSHA-approved self-contained breathing apparatus if vapors or mists are present and wear full protective clothing. For fighting fires in close proximity to a spill or vapors, use acid-resistant personal protective equipment. Evacuate residents who are downwind of the fire. Prevent unauthorized entry to the fire area. Dike the area to contain runoff and prevent contamination of water sources. Neutralize the runoff with lime, soda ash, or other suitable neutralizing agents. Cool containers that are exposed to flames with streams of water until the fire is extinguished.[6]

Weak and Intermediate Strengths of Sulfuric Acid

Safety and health considerations in the range of 0–70% sulfuric acid arise primarily in two areas:

1. In the production of the weaker solutions by dilution of concentrated sulfuric acid, those problems associated with handling strong sulfuric acid and oleum, as previously described, are also present and should be prepared for.
2. The higher strengths of the dilute/intermediate solutions pose some of the problems associated with concentrated sulfuric acid, although the concentrations below about 5% are relatively harmless to personnel.

In general, the precautions and actions to be taken are the same as for concentrated sulfuric acid. The following comments summarize the situation regarding the weaker acids and, in particular, the three commercial grades.

Fire/Explosion Hazard

Flash-point flammability limits do not apply to any of these grades. However, some general precautions apply in the case of fire involving these acids.

10N H_2SO_4 (UN1760) Dry chemical extinguishing media are recommended. DO NOT USE WATER. Firefighters should wear proper protective equipment and a self-contained breathing apparatus, with full-face piece operated in positive pressure mode.

The product is a powerful, acidic oxidizer and contact with oxidizable materials, reducing agents, and organic and combustible materials is to be avoided.

5N H_2SO_4 (UN1830)[7] Dry chemical extinguishing media or carbon dioxide is recommended. DO NOT USE WATER. Firefighters should wear proper protective equipment and a self-contained breathing apparatus, with full-face piece operated in positive pressure mode.

The product is severe relative to health, skin contact, and reactivity.

0.1N H_2SO_4[8] Use extinguishing media appropriate for the surrounding fire. Firefighters should wear proper protective equipment and a self-contained breathing apparatus, with full-face piece operated in positive pressure mode.

This product is severe relative to skin contact, moderate relative to health, and only slightly reactive.

Health Hazard Data

10N H_2SO_4 (UN1760) This concentration is corrosive to eyes and skin and irritating to the respiratory tract. Its acute toxicity is described as moderate, with all body tissues subject to damage upon ingestion, inhalation, or contact. Chronic toxicity has not been determined but would arise from repeated contact with any and all body tissues. Overexposure would result in lung damage, severe burns, and/or deterioration of teeth.

5N H_2SO_4 (UN1830) On contact with skin or eyes, this concentration of acid may cause severe irritation or burns. The vapors may be irritating to eyes, nose, and throat. Call a physician.

0.1N H_2SO_4 This concentration acid, on contact with skin or eyes, may cause severe irritation or burns. The vapors may be irritating to eyes, nose, and throat.

Precautionary Measures

For all three grades, wash thoroughly after handling and avoid contact with eyes, skin, and clothing. Do not breathe the chemical. Suitable protective equipment consists of a hood, disposable gloves, side-shield safety glasses with shield, and a lab coat.

First Aid

The 10N grade:

- After eye or skin contact, immediately flush eyes or skin with water for 15 minutes. Remove contaminated clothing and call a physician.
- If ingested, do *not* induce vomiting. Give large quantities of water, as well as at least 1 ounce of milk of magnesia in an equal amount of water (or the whites of three eggs). Never give anything by mouth to an unconscious person. Call a physician.
- If inhaled, remove to fresh air.

The other two grades:

- After eye or skin contact, immediately flush eyes or skin with water for 15 minutes. Remove contaminated clothing and wash before reuse.
- If ingested, do *not* induce vomiting. Give water, milk, or milk of magnesia. Never give anything by mouth to an unconscious person.
- If inhaled, remove to fresh air.

Spill and Disposal Procedures

Early detection of acid spills is essential to minimizing damage to neighboring structures. Additions of lime or soda ash will minimize the corrosion of metals and liberation of hydrogen gas. Neutralization alone is *not* adequate to arrest attack on cement or concrete, which are subject to attack by high concentrations of sulfate ions even under alkaline conditions.[9] All acid spills of any grade must be disposed of in accordance with all federal, state, and local regulations.

10N H_2SO_4 (UN1760) Wear a self-contained breathing apparatus and full protective clothing. Cover contaminated surfaces with soda ash or sodium bicarbonate, mixing and adding water if necessary. Scoop up the slurry and wash the neutral waste down the drain with an excess of water. Wash the site with soda-ash solution.

5N H_2SO_4 (UN1830) Wear a self-contained breathing apparatus and full protective clothing. Cover contaminated surfaces with soda ash or lime. Scoop up the material and place in a clean, dry container. Wash the spill area with water.

0.1N H_2SO_4 Wear a self-contained breathing apparatus and full protective clothing. Stop the leak if this can be accomplished without risk. Ventilate the area and neutralize with soda ash or lime. Scoop up the material and place in a clean, dry container. Wash the spill area with water.

General Considerations

Concentrations of about 10% or above will cause minor burns to human flesh. Eye contact with any concentration will cause damage. There are no fumes associated with this range of concentrations.

On exposure, remove contaminated clothing and shoes and flush with copious quantities of water for 5 to 10 minutes. Do not apply salves or ointments. Seek immediate medical assistance for acid burns.

Protective Clothing

The basic protective clothing recommended for strong acid should be used when handling any strength of acid: long-sleeved wool, acrylic, or polyester clothing, plus PVC- or rubber-lined apron, gloves, and boots. Protective chemical safety goggles, or at least face-shields, and safety hats are required.

References

1. Anon, "Recommended Safe Practices and Emergency Procedures for Sulfur Trioxide, Oleum and Chlorosulphonic Acid," The Soap and Detergent Industry Association, April (1979): 39 pp.
2. Anon, "Sulfuric Acid Chemical Backgrounder," NSC, National Safety Council (2003), http://www.nsc.org/library/chemical/sulfuric.htm.
3. Anon, "Material Safety Data Sheet: Sulfuric Acid, 77–100%," (Wilmington, DE: DuPont Company, January 1988).
4. Anon, "Material Safety Data Sheet: Oleum" (Wilmington, DE: DuPont Company, January 1988).

5. Anon, "Sulfuric Acid52—100%," Material Safety Data Sheet (Phillipsburg, NJ: J.T. Baker Co., 2002): 8 pp.
6. Anon, "Sulfuric acid," MSDS\70–100% H2SO4 02_02 (North York, ON, Canada: Marsulex Inc, 2002): 11 pp.
7. Anon, "Material Safety Data Sheet (5N)," (Phillipsburg, NJ: J.T. Baker Co., 2002).
8. Anon, "Material Safety Data Sheet (0.1N)" (Phillipsburg, NJ: J.T. Baker Co., 2002).
9. C. P. Dillon, "Corrosion Control in the Chemical Process Industries," 2nd Edition, publication No. 45. (St Louis, MO: MTI Inc., 1994): 420 pp.

5

Production of Sulfuric Acid and Oleum

Sulfuric acid is made by converting sulfur oxides, derived from various sources, using one of a number of common processes. Some of these processes depend on the source of sulfur species; others depend on the strength and quality of acid that is required. Typical sulfur oxide sources include the following:

- burning of native sulfur
- roasting of pyrites ores
- roasting and smelting of metal sulfides
- regeneration of used sulfuric acid
- roasting of metal sulfates
- combustion of H_2S or other sulfur-containing gases
- other processes

The possible methods used to produce sulfuric acid can be divided into the following two groups based on the level of SO_2 in the process gas.[1]

Poor gas processes with >3 vol. % SO_2:

- Single contact process
- Double contact process
- Wet Contact Process (WCP)

Tail gas processes with <3 vol. % SO_2:

- Modified Lead Chamber Process (MLCP)
- H_2O_2 process
- Activated carbon
- Other processes

Of these production processes, by far the most common is the double contact process, which is used to make the vast majority of all concentrated sulfuric acid produced. This is the process that will be discussed throughout this book. The overall method of making sulfuric acid involves producing sulfur dioxide, converting it to sulfur trioxide, and absorbing that in sulfuric acid.[2] Figure 5.1 shows the three common front-end processes for producing the sulfur dioxide, viz., sulfur burning, metallurgical, and spent acid regeneration, followed by the double contact conversion for producing the strong acid.

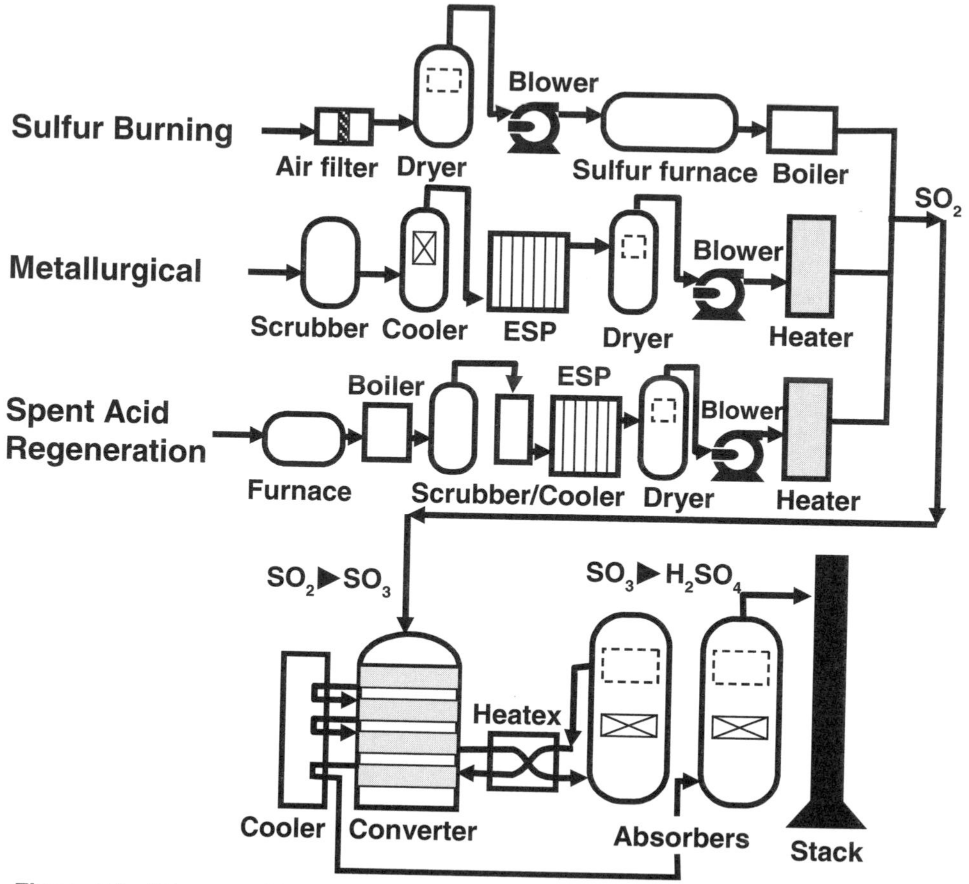

Figure 5.1 Diagram of Sulfuric Acid Production Showing Three Possible Feedstocks

The metallurgical and spent acid regeneration processes are collectively known as wet-gas processes. They are distinguished from sulfur burning by the amount of gas cleaning that must be undertaken before the dry, clean gas can be fed to the contact process in a manner similar to that of sulfur burning. The gas cleaning processes used vary depending on the source of the gas and the operating conditions but typically include scrubbing/quenching, gas cooling, mist precipitation, and drying stages. Details of some of the equipment used in gas drying are given in Chapter 12.

Sulfur Burning

Filtered ambient air is drawn through a drying tower by the main compressor. The drying tower removes moisture by contact with concentrated sulfuric acid. The compressed dry air enters a refractory-lined furnace where molten sulfur is burned to produce relatively high-strength sulfur dioxide (SO_2). The hot SO_2 combustion gas is then cooled in a steam boiler to the proper temperature to promote catalytic conversion to SO_3 in the converter, Figure 5.2.[3]

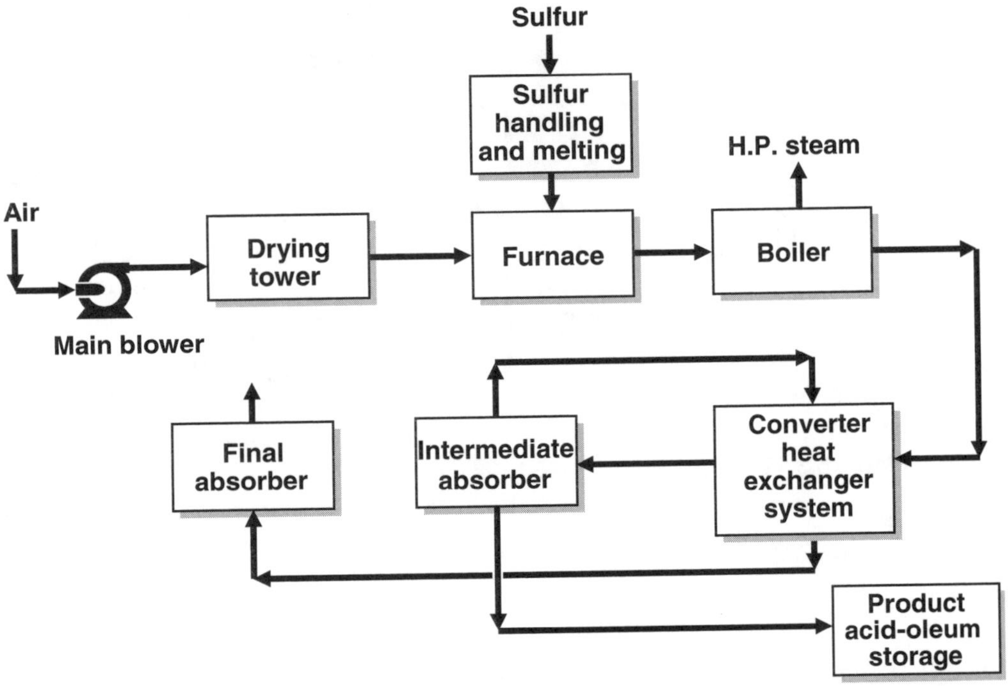

Figure 5.2 Flow Diagram of Sulfur-Burning Process of Sulfuric Acid Production

Spent Acid Regeneration

Spent acid and/or hydrogen sulfide are decomposed at elevated temperatures in a fuel-fired furnace. H_2S decomposition does not normally require supplemental fuel firing. The hot, "wet" SO_2 is then cooled in a steam boiler. The cooled gas enters a quench scrubber system where the gases are further cooled, removing excess water vapor, particulate, and potential fume contamination. The cooled and cleaned gas is then pulled through a drying tower by the main compressor to remove any remaining water. The dry gas is then heated in a train of gas-to-gas heat exchangers to promote conversion to SO_3 in the converter,[4] Figure 5.3.[5]

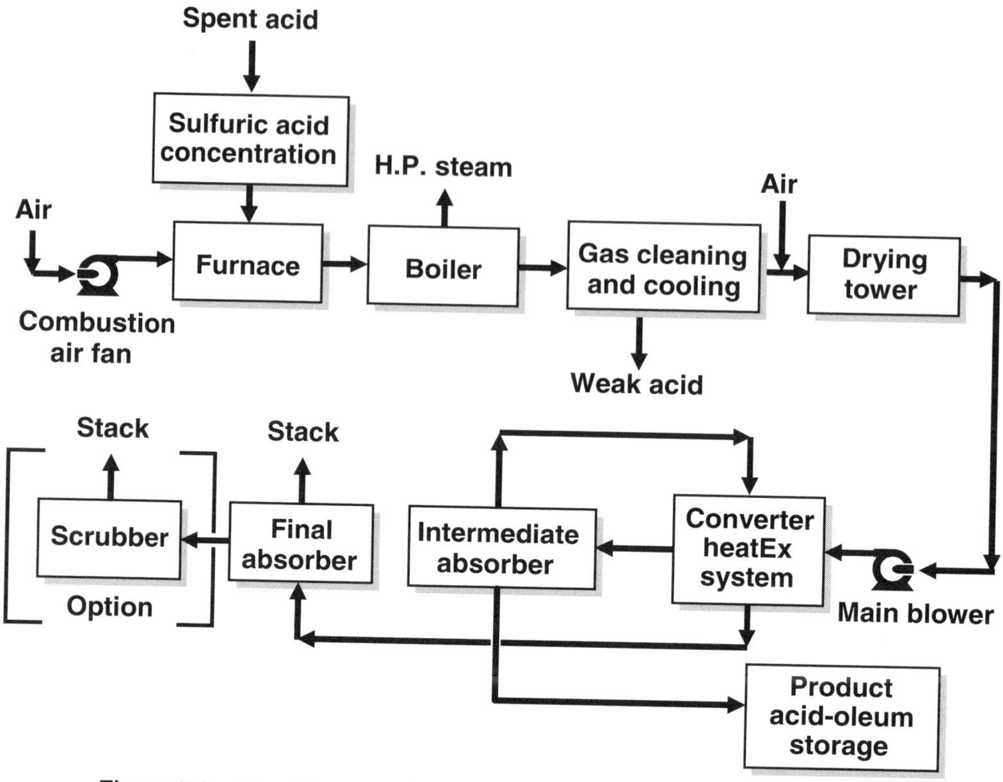

Figure 5.3 Flow Diagram of Acid-Regeneration Sulfuric Acid Production

Metallurgical

Hot, dirty, and "wet" off-gas from the smelter or roaster is pulled through a quench scrubber system where the gas is cooled to its adiabatic saturation temperature, removing excess water vapor, particulate, and potential fume contamination. Acid mist and additional contaminants are removed in an electrostatic mist precipitator. The cooled and cleaned gas is then pulled through a drying tower by the main compressor to remove any remaining water. The dry gas is then heated in a gas-to-gas heat exchanger to promote conversion to SO_3 in the converter, Figure 5.4.[6]

Conversion

The heart of the conversion process for sulfuric acid manufacture is the catalytic oxidation of sulfur dioxide (SO_2) to sulfur trioxide (SO_3), according to the equation:

$$SO_2 + 1/2\ O_2 = SO_3 + \text{Heat}$$

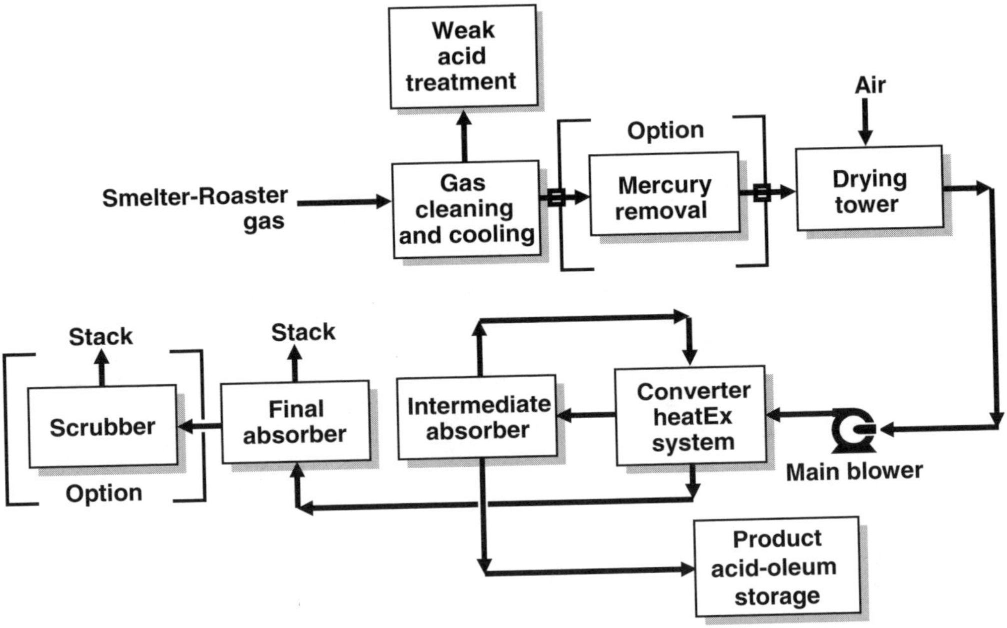

Figure 5.4 Flow Diagram of Metallurgical Sulfuric Acid Production

The reaction is highly exothermic and reversible so the conversion reaction needs to be carried out in stages (converter passes) with heat removed between the passes. This is necessary to obtain the high degree of conversion required to meet current environmental emissions standards. To achieve the final increment of required conversion, SO_3 must be removed from the process gas stream prior to passing through the final conversion pass.

Absorbing

After passing through the first three catalyst passes, SO_3-rich gas is cooled and absorbed with 98% sulfuric acid in the interpass absorbing tower. The lean SO_2 gas flows through a mist eliminator to remove fine mist particles, thus protecting downstream equipment from corrosion. The clean and lean gas is then reheated and enters the last catalyst pass to complete the conversion of the remaining SO_2 to SO_3. The SO_3 is then absorbed in 98% acid in the final tower. Mist eliminators are installed prior to the gas exiting to the stack to remove acid mist and ensure compliance with environmental regulations.

Oleum of 35% concentration (107.87% acid) is made by incorporating an additional oleum tower in the contact process. If 40% oleum (109% acid) is to be made, two towers in series are employed. The upper concentration limits are governed by the partial pressure of sulfur trioxide in the gas feed and the temperature of the oleum in the oleum tower. Higher strengths of oleum are made by absorbing SO_3 in 20–36% product.[7]

Heat Recovery System

There is a patented heat recovery system (HRS) that provides high thermal efficiency by recovering the heat produced in the absorption stage. This heat is normally lost through the coolers from the acid circulating around the system. This system has been installed in a number of acid plants including sulfur burners and metallurgical plants. It basically comprises an absorber that operates at 400°F (204°C) and uses a boiler to recover the absorption heat as steam (at up to 150 psig). The hot gases leaving the first stage are then cooled in the second stage and the remainder of the SO_3 is absorbed. Gases leaving the absorption tower are essentially the same as gas leaving a conventional interpass tower. The basis of the system is to keep the acid concentration in a narrow range very close to 100% where the corrosion rate of type 310 (S31000) stainless steel in the hot acid is very low (see Figure 5.5[8]), while at the same time maintaining absorption efficiency.[9] In practice, the concentration band turns out to be narrower than was originally expected. The system has had a number of problems that are said to have been resolved, and the HRS is now reported to be working well.[10]

Production of 93% Sulfuric Acid

Three common methods of manufacture for 93% acid use a vacuum concentrator, waste acid concentrator, and drum concentrator. It is also made by dilution of strong acid with water or taken as product acid from the drying tower circuit.

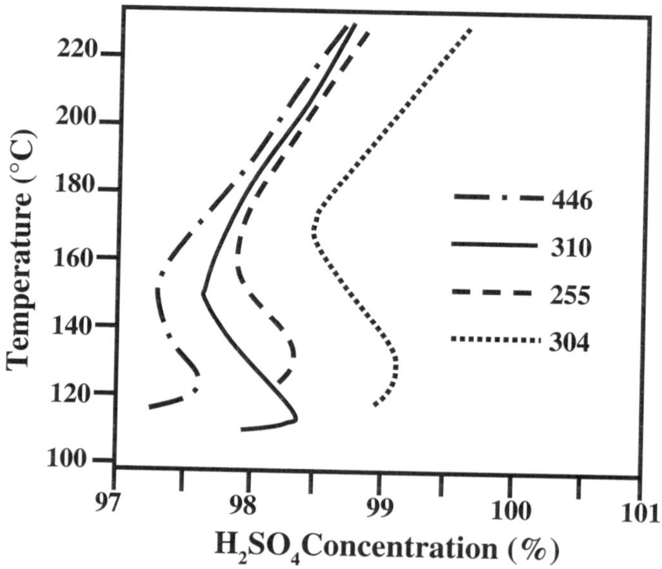

Figure 5.5 Isocorrosion Curve at 5 mpy (0.13 mm/y) for Various Stainless Steels in Hot, Concentrated Sulfuric Acid

Acid Concentrators

Much of the sulfuric acid used in chemical or other processes, such as nitration or alkylation processes, ends up as weak acid, often contaminated with other chemicals. Much of this weak, contaminated acid used to be dumped, with or without prior neutralization, but this is rarely practiced now. Current practice is to use some weak acid in other chemical production, such as in the manufacture of phosphoric acid, where some is burnt to produce fresh acid in the contact process, but much of it needs to be concentrated and cleaned for subsequent use. The obvious way to do this is by evaporation using heat.

The original method of acid concentration used cast-iron pots, set in refractory frames and heated from beneath, usually by open gas or oil flame. Acid is fed to this type of pot concentrator (also known as a Plinke or Pauling concentrator) through a packed column counter-current to the vapors leaving the pot. It operates at the atmospheric boiling point of around 300°C (572°F) to produce 95–96% acid in which much of the organic material has been thermally decomposed. Pot life can be very short, a year or less, and occasionally pot failure can be sudden, producing vast quantities of acid vapor. Pot concentrators are still being used, particularly for small-batch operations, but are becoming less common.

An alternative is to use a drum concentrator, commonly known as a Chemico drum concentrator. In this process, waste acid, typically around 70%, flows through three horizontal drums counter-current to furnace gases that are blown over the acid surface. The acid gradually becomes hotter and stronger and leaves the final- or high-stage drum at around 93% acid at 220°C (428°F). The vapors leaving the concentrator are cleaned in a venturi scrubber. The drums are carbon steel lined with homogeneously bonded lead and acid-resistant brick. This type of concentrator requires regular maintenance and, if treating nitration waste acid, can produce a colored stack due to the presence of NO_x species.

In order to reduce the boiling point of the strong acid being produced, vacuum can be applied to the heated acid. This is the basis of the Simonson-Mantius process that is able to produce up to 93% acid from a wide range of feed acids. These concentrators have been fabricated from various combinations of lead, silicon iron, graphite, Hastelloy® D, and acid-resistant brick. Glass-lined steel was later introduced for the concentrator bodies.

Modern acid concentrators use glass or glass-lined vessels heated with tantalum bayonet heaters. These systems also make use of zirconium- and PTFE-lined steel within their temperature and acid strength limits. The process uses multiple stages, with the acid becoming more concentrated at each stage, Figure 5.6.[11] Acid flow between stages is achieved through gravity and pressure differences, so no pumps are needed. The use of vacuum to lower the boiling point means that clean acid up to 96% can be produced at temperatures within the temperature limits (approximately 200°C [392°F]) of glass-lined vessels.[12,13] The starting acid strength can be as low as 7% H_2SO_4 but is more usually around 70%.

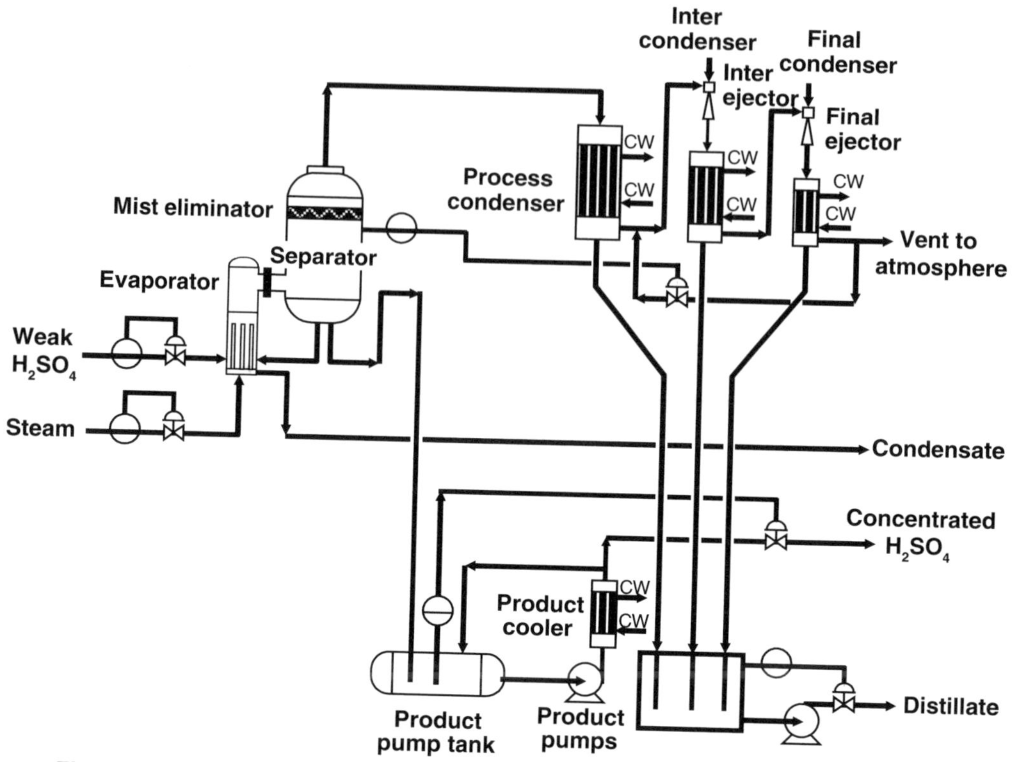

Figure 5.6 Flow Diagram of a Single Stage of a Sulfuric Acid Concentration Plant

Production of Weak Acid

Most weak sulfuric acid is produced by diluting concentrated industrial acid (usually 93%) with demineralized or other suitable water. This diluted acid (typically 2–10%) is used, for example, for pH control of feedwater to a cooling tower, or to feed a reverse osmosis or electrodialysis unit for the production of demineralized water.

Water must never be added to strong acid. The acid must be added to water, slowly and carefully, because the reaction is violent. Even with proper procedures, temperatures up to the boiling point can develop, especially if done without supplemental cooling. Acid dilution must be carried out in such a way that the heat produced is limited and the concentrated acid and water cannot suddenly come into contact with each other, Figure 5.7.[14]

Concentrated-acid entry piping should approach the dilution tee from below, the effect of gravity further ensuring against ingress of the more dilute and less dense acid solution into the bulk concentrated acid.

Check valves should be provided in both the acid and water lines to prevent backflow and unwanted mixing, which could occur if there were a flow interruption of either stream. To add a margin of safety, it is recommended that PTFE-lined pipe be utilized in the immediate mixing area, upstream of both check valves and approximately 20′ (6 m) downstream of the mixing area. Piping should be sized to ensure

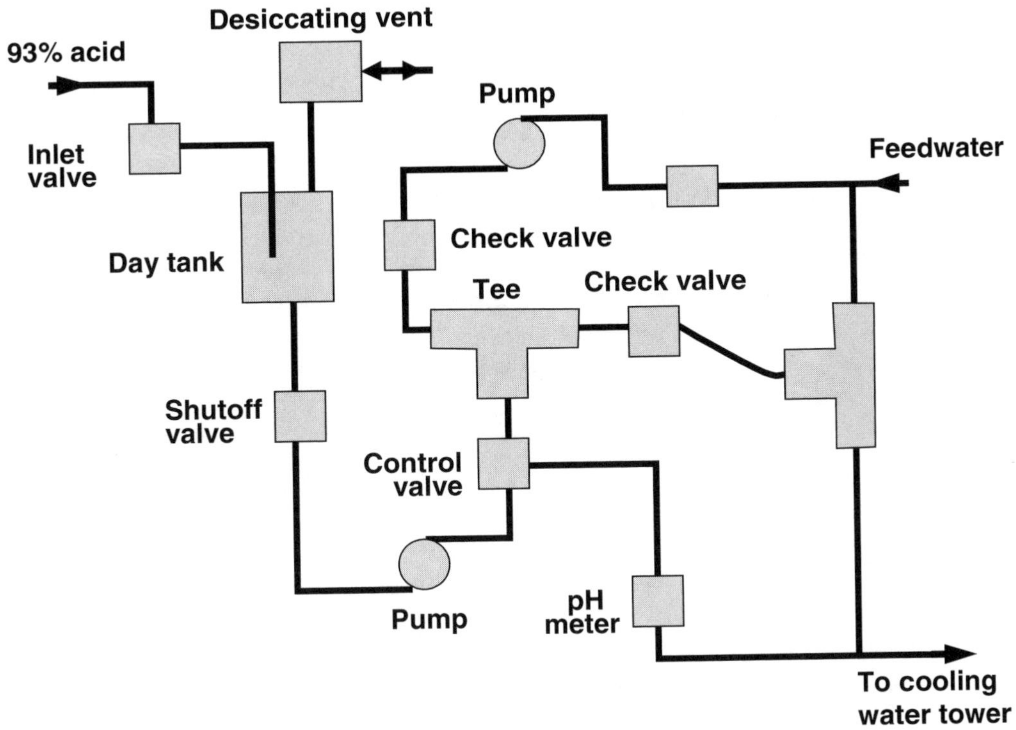

Figure 5.7 Acid Dilution System

turbulent flow for good mixing and a minimum of hot spots. Acid dilution should take place as close as possible to the point at which the diluted acid will be used or stored. This will minimize the amount of piping required to handle the hot acid.[15]

For smaller quantities of acid or when acid contamination must be minimized, acid dilution systems made from borosilicate glass are available. These systems can handle up to 3,800 gph of acid.[16]

Sulfuric acid in the 5–25% range is either produced by dilution of stronger acid or as a byproduct of some other process. The manufacture of titanium dioxide, for example, produces spent acids in the 20–23% concentration range, with additional metallic sulfates at 7–15% by weight.

Sulfuric acid in the 25–70% range also usually occurs as a byproduct. Contaminated 70% acid is recovered from nitration processes, nitric acid concentration, and chlorine-drying operations.

References

1. Anon, "Production of Sulphuric Acid," Vol. 3 (Brussels, Belgium: EFMA, European Fertilizer Manufacturers' Association, 2000): 68 pp.
2. C. M. Schillmoller, "Selection and Performance of Stainless Steels and Other Nickel-Bearing Alloys in Sulphuric Acid," NiDI Technical Series no. 10 057 (Toronto, ON, Canada: NiDI, 1990): 10 pp.

3. J. Thomson, "Sulphur Burning Sulphuric Acid Plant" (Vancouver, BC, Canada: Chemetics, A division of Aker Kvaerner Canada Inc., 2004).
4. Anon, "Spent Acid Regeneration (SAR)" (St Louis, MO: EnviroChem Systems, Inc., 2003), http://www.enviro-chem.com/plant-tech/3rdtier/acidprocesstop.html.
5. J. Thomson, "Acid Regeneration Sulphuric Acid Plant" (Vancouver, BC, Canada: Chemetics, A division of Aker Kvaerner Canada Inc., 2004).
6. J. Thomson, "Metallurgical Sulphuric Acid Plant" (Vancouver, BC, Canada: Chemetics, A division of Aker Kvaerner Canada Inc., 2004).
7. Anon, "The Acid Process—Sulfuric Acid" (St Louis, MO: EnviroChem Systems, Inc., 2003), http://www.enviro-chem.com/plant-tech/3rdtier/acidprocesstop.html.
8. J. E. Niesse, D. R. McAlister, "Stainless Steels for Heat Recovery from High Temperature Sulfuric Acid," Corrosion '87, paper no. 22 (Houston, TX: NACE International, 1987): 11 pp.
9. D. R. McAlister, A. G. Corey, L. J. Ewing, S. A. Ziebold, "Economically Recovering Sulfuric Acid Heat," Chemical Engineering Progress (July 1986): pp. 34–38.
10. S. M. Puricelli, J. R. Shafer, D. L. Randolph, "Sulfuric Acid Heat Recovery—A Technology Update" (St Louis, MO: EnviroChem Systems, Inc., 2003), http://www.enviro-chem.com/plant-tech/3rdtier/hrstop.html.
11. J. Thomson, "Schematic Flowsheet of Typical Stage Sulphuric Acid Concentrator" (Vancouver, BC, Canada: Chemetics, A division of Aker Kvaerner Canada Inc., 2004).
12. I. Rodger, "Developments in the Concentration of Sulfuric Acid," AIChE (1982): pp. 151–155.
13. G. M. Smith, E. Mantius, "The Concentration of Sulfuric Acid," CEP (September 1978): pp. 78–83.
14. Anon in C. P. Dillon, ed., "Concentrated Sulfuric Acid and Oleum," vol. MS-1, Materials Selector for Hazardous Chemicals (St. Louis, MO: MTI Inc., 1997): 212 pp.
15. Anon, "Handling Sulfuric Acid" (Marion, NC: Crane Resistoflex, 2000), http://www.resistoflex.com/sulfuric.htm.
16. Anon, "Sulfuric Acid Dilution," Process Profile 13-2 (West Union, NJ: De Dietrich Process Systems, Inc., 1995): 2 pp.

6

Uses of Sulfuric Acid

Most of the uses of sulfuric acid are indirect because the acid is used as a reagent, not an ingredient. The largest single sulfuric acid consumer by far is the fertilizer industry. Sulfuric acid is used to digest phosphate rock in the manufacture of phosphate fertilizers and smaller amounts are used in the manufacture of ammonium and potassium sulfate. Large quantities are used as an acidic dehydrating agent in organic chemical and petrochemical processes, as well as in oil refining. In the metal processing industry, sulfuric acid is used for pickling and descaling steel; for the extraction of copper, uranium, and vanadium from ores; and in nonferrous metal purification and plating. In the inorganic chemical industry, it is used most notably in the production of titanium dioxide.

Certain wood pulping processes for paper require sulfuric acid, as do some textile and fiber processes (such as rayon and cellulose manufacture) and leather tanning. Other end uses for sulfuric acid include effluent/water treatment, plasticisers, dyestuffs, explosives, silicate for toothpaste, adhesives, rubbers, edible oils, lubricants, and the manufacture of food acids such as citric acid and lactic acid. Probably the largest use of sulfuric acid in which this chemical becomes incorporated into the final product is in organic sulfonation processes, particularly for the production of detergents. Many pharmaceuticals are also made by sulfonation processes.[1]

Between 60% and 70% of the sulfuric acid used in the United States is used by the fertilizer industry to convert phosphate rock to phosphoric acid. All other uses account for <1% to <10% of total consumption. Sulfuric acid use is declining in some industries. There is a trend in the steel industry to use hydrochloric acid instead of sulfuric acid in pickling, and hydrofluoric acid has replaced sulfuric acid for some uses in the petroleum industry. The primary consumer product that contains sulfuric acid is the lead-acid battery; however, this accounts for a small fraction of the overall use. It is also used as a general-purpose food additive.[2]

There are many applications that use concentrated sulfuric acid, ranging from simple dehydrating (drying) processes to reactions with inorganic or organic species. Strong acid may be recovered from process reactions, e.g., by evaporation or by enhancement with sulfur trioxide. There exists a constant boiling mixture (CBM) at 98% acid, which limits further concentration by simple evaporation. The following sections describe typical applications of sulfuric acid and oleum and give basic guidance as to appropriate materials of construction to be used.

Concentrated Acid

In the utilization of concentrated sulfuric acid as a drying agent or as a reactant for either inorganic or organic processes, the corrosion characteristics may be profoundly altered by dilution, the resulting exotherm, contamination, or by a combination of these factors. Iron contamination problems may limit the use of steel or cast iron where otherwise suitable, while the number and kind of oxidizing and/or reducing species will determine which groups of materials are suitable. Often, only experimentation or past experience can determine which materials of construction are safe, reliable, and cost-effective.

Drying Operations

Strong acid, usually 98–99% concentration, may be used to dry air or sulfur dioxide, as previously described for the contact process, as well as in drying chlorine, methyl chloride, and other chemicals.

Chlorine Chlorine is typically dried by passing it through acid in two or three packed towers in series. The first tower, which sees a somewhat diluted acid, is lined with acid brick set in potassium silicate cement. The covers are flat and lined with PVDF, designed to relieve an explosion in the event chlorine becomes contaminated with hydrogen. The last tower, where 98% acid is fed countercurrent to the wet chlorine gas stream, is made of carbon steel. If chlorine drying with 98% sulfuric acid is done in a carbon brick-lined tower, control of the dilution is critical, as hypochlorites formed in acid below 90–93% acid will attack carbon, while saturation of stronger concentrations with chlorine does not cause deterioration. Alloy C-276 (N10276) columns and piping are also used, with cast CW-12MW (N30002) valves and pumps.[3]

Methyl Chloride Drying methyl chloride with 93% acid results in some hydrolysis, forming traces of HCl and causing SCC of type 316L (S31603) equipment. Replacement with alloy 20Cb-3 (N08020) resolves this problem.

Methyl-chloride gas produced in and for chloromethane plants is usually dried with concentrated sulfuric acid in acid brick-lined towers.

Organic Syntheses

Many organic syntheses utilize concentrated acid in the processes of sulfation, where an OSO_2OH group is attached to a carbon atom, and of sulfonation, where a sulfonic acid group, SO_2OH, or its corresponding salt or sulfonyl halide is attached. Such syntheses produce surface-active materials like detergents, emulsifiers, wetting agents, and penetrants from fatty alcohols, aromatics, and other hydrocarbons.[4]

Bismethylphenethane This product is formed by reaction of toluene and acetylene in 90 to 95% acid at 8 to 20°C (45 to 70°F) in the presence of a suspended mercuric sulfate catalyst. Type 304L (S30403) vessels and piping should be suitable.

Wetting Agents and Penetrants Fatty alcohols (e.g., lauryl, oleyl) are sulfated in 93% sulfuric acid at temperatures between 25 and 60°C (77 and 140°F), the products being salted out, washed, and neutralized with caustic soda in an alloy 400 (N04400) vessel.[5] In sulfations in which the acid concentration is kept above 80%, steel and cast-iron equipment may be used, provided iron contamination is not objectionable. Otherwise, alloys 400 (N04400) or 20Cb-3 (N08020) find application. Linear alkylbenzene (e.g., dodecylbenzene) may be sulfonated with oleum to produce biodegradable surfactants. Equipment is usually types 304L (S30403) and 316L (S31603) with higher-alloyed materials used where velocity or dilution poses potential problems. Anodic protection (AP) is sometimes used to extend the reliability of equipment.

Alkylation Alkylation combines low-molecular-weight olefins with isobutene in the presence of a catalyst, either sulfuric acid or hydrofluoric acid. The product is called alkylate and is a mixture of high-octane, branched-chain paraffinic hydrocarbons. Alkylate is a premium blending stock because it has exceptional anti-knock properties and is clean burning. The octane number of the alkylate depends mainly upon the kind of olefins used and upon operating conditions.

In cascade-type sulfuric acid alkylation units, the feedstock (propylene, butylene, amylene, and fresh isobutane) enters the reactor and contacts the concentrated sulfuric acid catalyst (in concentrations of 85 to 95% for good operation and to minimize corrosion). The reactor is divided into zones, with olefins fed through distributors to each zone and the sulfuric acid and isobutanes flowing over baffles from zone to zone.

The reactor effluent is separated into hydrocarbon and acid phases in a settler, and the acid is returned to the reactor. The hydrocarbon phase is hot water washed with caustic for pH control before being successively depropanized, deisobutanized, and debutanized. The alkylate obtained from the deisobutanizer can then go directly to motor-fuel blending or be rerun to produce aviation-grade blending stock. The isobutane is recycled to the feed.[6] Some alkylation waste acid (AWA) is also produced.

Carbon steel equipment is generally satisfactory. However, austenitic stainless steels (e.g., types 304L [S30403] and 316L [S31603] and analogous castings) are used for high-velocity or turbulent areas, such as pumps, valves, and piping return bends.

Sulfonation Oleum is commonly used to sulfonate alkylates in the manufacture of soaps, detergents, greases, etc. The alkylates are digested with oleum and then water is added to separate out the sulfonic acid. Waste sulfuric acid, typically at around 70 to 80%, is also produced. At moderate temperatures, cast-iron sulfonators have been used, with alloy 20-type pumps, valves, and other accessory parts. Glass-lined sulfonators are used for sulfonations carried out with strong oleum or weaker sulfuric acid solutions at high temperatures, such as 150°C (302°F). Alloy 400 (N04400) can be used for sulfonations at moderate temperatures, when acid concentrations fall below 80%. A complete loop comprising tower, pumps, pump tanks, and piping was built using SX® alloy (S32615) at a sulfonation plant in Germany in 1994.[7]

Acid-Washed Oils Lubricating oils and other refinery distillates are acid-treated at temperatures varying from 66 to 104°C (151 to 219°F) with 83% acid. Subsequently, the emulsion is diluted with water to separate oil from sludge. Alloy 400 (N04400) is commonly used for vessels, piping, and centrifuges.

Sulfated Fatty Acids The reaction of, for example, castor oil with cold 96% sulfuric acid in about a 70/30 acid : oil mixture produces a sulfated product, which is salted out, washed, and neutralized in a batch process, using alloy 400 (N04400) or molybdenum-bearing stainless-steel equipment. Other animal or vegetable oils may be similarly treated. Alloy C-276 (N10276) may be used when nickel or copper contamination is objectionable.

Organic Esterifications Small amounts (0.1–2%) of 93–98% sulfuric acid are used as catalyst for esterification reactions between alcohols and organic acids. Type 316L (S31603) is often used, but corrosivity will vary with acid concentration, amount of water present in the process stream, and temperature. The corrosion rate of type 316 (S31600) in a boiling solution of 25% acetic acid, 59% butyl acetate, 10% water, and 6% butanol varied from nil without sulfuric acid to 0.48 mm/y (19 mpy) with 0.1% sulfuric acid (added as concentrated acid). The corrosion rates were 5.92 mm/y (233 mpy) with 0.5% acid present, and 17.53 mm/y (690 mpy) with 1% sulfuric acid present.[4] High-strength copper alloys, such as a proprietary inhibited aluminum bronze, are also used for esterification kettles.

Nitration

93% sulfuric acid is used with nitric acid in the manufacture of trinitrotoluene and in the manufacture of nitroglycerine from glycerol. Type 304L (S30403) equipment is normally suitable.

Inorganic Preparations

Ammonium Sulfate Although many ammonium-sulfate plants absorb ammonia in acidified ammonium-sulfate solutions (i.e., dilute sulfuric acid), the direct reaction process uses ammonia absorbed directly into 93–98% acid. Type 316L (S31603) has been used but is a borderline material in the 110°C (230°F) process, occasionally going active and corroding at high rates. Higher alloys (e.g., alloys 20Cb-3 [N08020], 825 [N08825], and G-30 [N06030]) are preferred.[4]

Phosphoric Acid Concentrated sulfuric acid is employed in the hemihydrate variant of wet-process phosphoric acid, in which phosphate rock is reacted with sulfuric acid to release phosphoric acid and precipitate calcium sulfate hemihydrate. The temperature is about 90–95°C (194–203°F).

While the corrosion characteristics vary with the chemistry of the phosphate rock, which introduces a number of contaminants, nickel-chromium-molybdenum alloys (e.g., alloys C-276 [N10276] and 625 [N06625]), high-molybdenum austenitic alloys (e.g., alloy G-30 [N06030]), and 6% molybdenum superaustenitic stainless steels are usually suitable.

Hydrofluoric Acid Hydrogen fluoride (HF) is produced by the action of concentrated sulfuric acid on ground fluorspar. Concentrated sulfuric acid is also used to dehydrate the HF produced. Nickel-based alloys such as alloys C-276 (N10276) and B-2 (N10665) are widely used in this process. HF is also produced by the reaction of concentrated sulfuric acid with fluorosilicic acid, a byproduct of wet-process phosphoric acid manufacture.

Traces of HF act as an inhibitor in concentrated sulfuric acid for chromium- and copper-bearing nickel alloys, e.g., alloy G-30 (N06030), apparently by incorporation of fluorine in a defective oxide film.[8] In the copper-free grades, such as alloys 625 (N06625) and C-276 (N10276), traces of HF accelerate attack below about 80% sulfuric acid concentration, above which (the "switching concentration") the fluorides become inhibitive. An abrupt increase in corrosion at a critical velocity indicates a shearing effect on the protective film rather than a change in acidity or redox potential. Alloys 400 (N04400) and B-2 (N10665) suffer some acceleration of attack under the same conditions.

Alloy 825 (N08825) has been recommended for HF production plants. However, knife-line attack has been observed in weld heat-affected zones (HAZ) and seamless or full-finished welded tubes are required.[9]

0–5% Acid

Nominally a simple reducing acid, solutions of sulfuric acid in the 0–5% range vary from simple acidified water solutions to commercially important reactants for hydrometallurgical applications. In every practical instance, corrosion characteristics are determined by the number and kind of contaminants, rather than the sulfuric acid itself. Selection of metals and alloys for these services is primarily contingent upon an understanding of the chemistry of the application. Utilization of plastics and elastomers is primarily constrained by considerations of temperature and pressure, but organic contaminants can also play an important role.

Hydrometallurgy

Sulfuric acid solutions used in hydrometallurgical processes for leaching ores vary in concentration from <1 to 98% and in temperature from ambient to 270°C (518°F).[10] Stainless steels are used extensively in dilute leaching operations limited by chloride contamination and extended by the presence of oxidizing agents.

Hydrometallurgical autoclaves are exposed to sulfuric acid concentrations of <1% at temperatures well in excess of 100°C (212°F). In this environment, nickel-based alloys can be severely corroded, and duplex stainless steels with at least 25% chromium are preferred.[11]

Copper Leaching Heap and dump leaching of oxide-containing copper ores and pyritic copper sulfide ores is performed in type 316L (S31603) equipment, using acid of less than 1 to 5%. Concentrations in the solvent extraction and stripping units are substantially higher. Copper oxides dissolve readily, while exothermic dissolution of sulfidic ores is effected through oxidation with the help of bacteria, oxygen, and ferric sulfate at about 35°C (95°F). Even type 304L (S30403) has been successfully used under some conditions. However, type 316L (S31603) showed high rates of crevice corrosion in a 3–6% acid / 4–6% cupric sulfate solution from sulfidic ores, apparently due to unreported chloride contamination.

Type 316L (S31603) is also successfully used in a liquid ion exchange (solvent extraction) process at pH 2.5 and 20–35°C (68–95°F). The resistance of stainless steels in this application is due to their passivation by cupric and ferric ions.

Uranium Leaching This strength of acid is also used for leaching uranium ores in a temperature range of about 35 to 70°C (95–160°F), ferric cations and chlorate anions promoting passivity of the conventional austenitic stainless steels. However, chloride and fluoride contamination cause pitting and crevice corrosion. The high-performance nickel-rich alloys (e.g., N08020, N08825, N08904) seem to provide satisfactory resistance in both laboratory and field tests.[10]

Other Metals Both zinc and manganese ores may be leached with about 3% sulfuric acid at slightly elevated temperatures. Oxidizing species such as ferric and cupric ions provide inhibition for type 316L (S31603) coolers, while pumps are of CN7M (N08007) or similar alloys to resist erosion-corrosion.

Inorganic Sulfates

Aluminum Sulfate In the production of aluminum sulfate from bauxite ores, the boiling dilute sulfuric acid is corrosive to most common alloys other than lead if the ore contains substantial amounts of iron. Alloy 400 (N04400) will resist low-iron ores during the digestion process.

When alum solutions (e.g., 15–25%) of pH 2–3 are prepared, type 316L (S31603) is preferred for dissolving tanks, because of the presence of ferric ions, which render alloy 400 (N04400) and related alloys susceptible to attack. Higher nickel-rich, chromium-bearing alloys (e.g., CN7M, N08007) are used for pumps and valves. The wrought materials, alloy 20Cb-3 (N08020) or alloy 825 (N08825), are also used where type 316L (S31603) is borderline. In some cases, the high-chromium alloy 28 (N08028) has proven superior to type 316L (S31603) when oxidizing contaminants were present in smaller concentrations than usual.

Non-Oxidizing Acid Sulfates Slightly acidic solutions of ferrous, nickel, stannous, and zinc sulfates are handled successfully in alloy 400 (N04400) or alloy 200 (N02200) under a variety of conditions.

Ammonium sulfate solutions in the range of concentrations considered occur as byproducts of caprolactam production. A pH of 3 for type 316 (S31603) and 4 for type 304 (S30403) is considered a practical limit of compatible acidity by many operators.

Oxidizing Acid Sulfates In dilute acidic solutions of oxidizing salts, the chromium-bearing alloys (e.g., alloy 600 [N06600]), the nickel-chromium-molybdenum-iron alloys (e.g., N08825), and the nickel-chromium-molybdenum alloys (e.g., N06625, N10276, N06022) are needed where stainless steels are unacceptable.

Pulp Digestion Applications

Zircadyne® 705 (R60705) has been used in the production of ethanol from wood pulp. The hardwood pulp is dissolved in 5% sulfuric acid at a temperature of 220°C (428°F) and pressures of up to 400 psi. This alloy is susceptible to delayed hydride cracking in this environment, so the pressure vessels need to be stress-relieved at 550°C (1,022°F) for 4–6 hours at temperature.[12]

5–25% Acid

Sulfuric acid in the 5–25% range is produced only as a byproduct. The manufacture of titanium dioxide, for example, produces spent acids in the 20–23% concentration range, with additional metallic sulfates at 7–15% by weight.

Acid Pickling

Pickling acids typically contain about 10% sulfuric and about 8% iron. The iron should be in the ferrous state from hydrogen evolution during pickling but may oxidize to Fe^{+++} during handling, storage, or shipment; typically, the net result is an acid solution of oxidizing characteristics. Titanium Dioxide Strong Spent Acid is typically 20% sulfuric plus 1% iron and 2% other metals. Such acids may be recycled for concentration to strong acid, neutralized to gypsum (calcium sulfate), or neutralized for disposal. However, they are expensive to transport because of high water content and attendant corrosivity.

Pickling in 5–15% sulfuric acid at 60–95°C (140–200°F) is used to remove oxide scales from hot-rolled, forged, or heat-treated steel parts.[4] The corrosion during pickling releases gaseous hydrogen, which tends to consume dissolved oxygen (DO) and keep metallic cations in the reduced state. Alloy 400 (N04400) is routinely used for hardware (e.g., racks, baskets, hooks, chains). Presumably, there is some cathodic protection afforded by contact with the steel components. Brass and copper may be pickled (e.g., prior to enameling) and alloy 400 (N04400) is used in such applications.

Alloy 825 (N08825) is also used in pickling operations. In one case, this alloy replaced an alloy 400 (N04400) hook that went first into 12% sulfuric acid containing sodium dichromate at 70°C (158°F), then into 60% H_2SO_4, 25% HNO_3, and 0.2% HCl. Alloy 825 (N08825) performed well in this application and also in the heating coils in the pickling baths.[5]

Steel strip is sometimes pickled using 20% sulfuric acid solutions at 100°C (212°F). Hooks (80 mm diameter) made from aluminum bronze are used to lift strip from one tank to the next and last a maximum of 18 months. Replacement hooks in alloy 31 (N08031) showed no visible corrosion after 12 months' service.[13]

Pickling tanks in a rubber factory needed heat-transfer coils to maintain 18% H_2SO_4 at 180°F (82°C). Thermoplastic resin tubing was not attacked but was abraded. Black iron pipes encased in lead failed by the formation of cracks in the lead, permitting acid attack of the iron. Zirconium heating coils were found to be the answer, and fewer coils are required because their heat transfer is better than previous heaters and higher-pressure steam can be used in them, unlike in the previous thermoplastic coils.[14]

Hydrometallurgy

In electrowinning operations, acid concentrations of 13–15% at 50–65°C (120–150°F) are encountered.[4,5] Type 316L (S31603) is used for tanks, piping, and so on in a copper refinery, being inhibited by the attendant cupric ion contamination. Heating coils, however, require alloy 20Cb-3 (N08020) or alloy 904L (N08904) because of hot-wall effects. In zinc refineries, type 304 (S30400) is used for cathode starting sheets and type 316L (S31603) for evaporators and crystallizers (again inhibited by cupric sulfate contaminants).

Uranium is extracted with, for example, 5–6% acid at 45–48°C (113–118°F), which becomes contaminated with chlorates, iron, and so on. The molybdenum-bearing stainless steels (S31603, S31703) resist general corrosion but are subject to severe crevice corrosion, so higher alloys (e.g., N08825, N06030, N10276) are generally more suitable.

Ammonium Sulfate

An ammonium sulfate liquor containing 4–10% free sulfuric acid at 50°C (120°F) is used to absorb ammonia. The scrubber tails stream is crystallized and the product centrifuged or filtered and dried; the recovered mother liquor is fortified with sulfuric acid and recirculated.[5]

Both alloy 400 (N04400) and type 316L (S31603) have been used in this application. Higher alloys (e.g., CN7M, N08007) are used to combat erosion-corrosion and other conditions too severe for 316L. The nickel-chromium-molybdenum alloys (e.g., N10276, N06022) have been used in the more severe applications, as have high-silicon irons. Impervious graphite has replaced alloy 400 (N04400) in some heat exchangers.

Phosphate Fertilizers

Phosphate rock is reacted with a mixture of concentrated sulfuric acid and recycled partially spent acid, resulting in a 28% phosphoric, –21% sulfuric acid mixture contaminated with fluorides (mostly as hydrofluorosilicic acid), chlorides, and metal cations (e.g. Fe^3). The reactor itself is usually brick-lined, but the internals and the predilution and other vessels and piping are of high-alloy construction (e.g., N08028, N06030, N06007). When contamination is at its worst (which depends upon the source of the phosphate rock), nickel-chromium-molybdenum alloys (e.g., N06625, N10276) may be required.[15]

Starch

Copper equipment has been used in the saccharification of starch with dilute sulfuric acid at 120°C (248°F) and 0.4 MPa.[16]

Furfural

Copper-lined autoclaves have been used for the production of furfural (2-furaldehyde). The reaction involves 20% sulfuric acid and 1–5% acetic acid and oat husks at 120°C (248°F) at 4–5 atm.

Viscose

In vacuum evaporation of viscose spinning bath solutions (6–20% acid at 40–50°C [104–122°F]), alloy 200 (N02200) is successfully used, apparently because of the absence of dissolved oxygen and/or the influence of inorganic sulfates (sodium and zinc).

25–70% Acid

Sulfuric acid in the 25–70% range usually occurs only as a byproduct. Contaminated 70% acid is recovered from nitration processes, nitric acid concentration, and chlorine-drying operations.

Metal Cleaning

Contrary to the data for commercially pure acid, a 30% acid at 54°C (130°F) from a cleaning operation in a metal industry plant showed corrosion rates of <1 mpy (<0.025 mm/y) for types 304 (S30400), 316 (S31600), 317 (S31700), and alloy 20. The very low rates are undoubtedly due to inhibition by oxidizing cations (e.g., Fe^{+++}, Cu^{++}).

Tall Oil

Tall oil from pulp and paper mills is split by reaction with 98% sulfuric acid before distillation, but the spent acid in the chlorate plant (which feeds the rosin boiling plant) is about 30% concentration at 45°C (113°F). Contaminated with both chlorates and chlorides, this acid attacks type 316L (S31603) and pits alloy 904L (N08904) within about two years. Alloy 28 (N08028) has outperformed the 4% molybdenum grade, giving more than four years' service.[17]

Nitration Acid

Nitration acid typically contains up to 3,000 ppm nitrated organics and 500 ppm nitric acid. This acid may be used in the manufacture of alum or phosphoric acid, or burned to produce SO_2 for strong acid manufacture. Nitrated organic contaminants are sometimes difficult to oxidize. Field corrosion tests or prior experience should be the basis for alloy selection.

Nitric Acid

Sulfuric acid recovered from nitric acid concentration operations typically contains about 500 ppm nitric acid. The resulting evolution of NO_X may be a problem in some reuse applications. Materials selection should be based on field tests and/or prior experience, but the presence of the oxidizing nitric acid may reduce corrosion in some alloys.

Chlorine Drying

Acid from chlorine-drying operations typically contains about 30 ppm organic chlorides. It has found use in the manufacture of chlorine dioxide, phosphoric acid, and alum, as well as for coke oven gas scrubbing. However, the chlorinated organics are toxic and may pose a problem in some reuse applications, while chloride ions cause their usual problems of pitting, crevice corrosion, and SCC. Plate-type heat exchangers of alloy C276 (N10276) have given satisfactory service in this service for many years. More recently, alloy C22 (N06022) is being chosen.

Fuel-Grade Ethanol

In a strong-acid process for production of fuel-grade ethanol, cellulosic wastes are treated with 30–70% sulfuric acid at 60–80°C (140–175°F) at atmospheric pressure. The sugar solution produced is separated from the acid in an ion-exchange chromatograph. The acid runs quickly through and is recycled, while the sugar is slowed by the packed resins, removed by backflushing with water, and sent to the fermentation step. This process avoids the production of unwanted byproducts (e.g., furfurals and aldehydes) incurred in the weak-acid (3%) process operating at 250°C (480°F) and 300 psig, as well as the acid losses incurred by neutralization of the weak acid.

Methyl Methacrylate and Butyl Alcohol

In the production of methyl methacrylate, the nominal sulfuric acid concentration is in the range of 25–35% and temperatures exceeding 150°C (302°F). The effective concentration is actually in the 45–55% range due to the presence of other constituents. In a secondary butyl alcohol application the sulfuric acid concentration is in the 50–70% range at temperatures from 90 to 160°C (194 to 320°F). Zirconium 702 has been found to be satisfactory in this application, although there is some attack of the welds that can be avoided by stress relief.[18]

References

1. Anon, "Production of Sulphuric Acid," Vol. 3 (Brussels, Belgium: EFMA, European Fertilizer Manufacturers' Association, 2000): 68 pp.
2. Anon, "Strong Inorganic Acid Mists Containing Sulfuric Acid," from 10th report on carcinogens, U.S. Department of Health and Human Services (2003), http://ntp-server.niehs.nih.gov/NewHomeRoc/AboutRoC.html.
3. C. P. Dillon, ed., "Concentrated Sulfuric Acid and Oleum," vol. MS-1, Materials Selector for Hazardous Chemicals (St. Louis, MO: MTI Inc., 1997): 212 pp.
4. Anon, "The Corrosion Resistance of Nickel-Containing Alloys in Sulfuric Acid and Related Compounds," CEB-1 (Suffern, NY: The International Nickel Co. Inc., 1983): 90 pp.
5. C. M. Schillmoller, "Selection and Performance of Stainless Steels and Other Nickel-Bearing Alloys in Sulphuric Acid," NiDI Technical Series no. 10 057 (Toronto, ON, Canada: NiDI, 1990): 10 pp.
6. Anon, "Alkylation," Set Laboratories (2003), http://www.setlaboratories.com/alkylation.htm.
7. F. Kodeda, G. Berglund, "Edmeston SX System & Sandvik SX® Stainless Steel—Review of 15 Years' Experience in the Sulfuric Acid Business," Sulphur 2000 Conference, San Francisco, CA (2000): 8 pp.
8. N. Sridhar, S. M. Corey, "Prediction of Corrosion Behavior in Acid Mixtures," Proceedings of First NACE International Symposium (Houston, TX: NACE International, 1990).
9. T. F. Degnan, private communication, in C. P. Dillon, ed., "Concentrated Sulfuric Acid and Oleum," vol. MS-1, Materials Selector for Hazardous Chemicals (St. Louis, MO: MTI Inc., 1997): 212 pp.

10. Anon, "Corrosion Handbook," Vol. 8, Table 91 (Frankfurt, Germany: Dechema eV, 1991): p. 100.
11. G. Coates, "Duplex Stainless Steel," in CASTI Handbook of Stainless Steels and Nickel Alloys, S. Lamb, ed. (Edmonton, AB, Canada: CASTI Publishing Inc., 2000), p. 233.
12. Anon, "Zircadyne 705 Chosen for Use in TVA's Ethanol-from-Wood Process," Outlook 7, 3 (1996): pp. 1–2.
13. E. Altpeter, R. Kirchheiner, F. E. White, "Sulphuric Acid Corrosion of Some Special Stainless Steels and Nickel Alloys: Laboratory Tests and Plant Experience," NACE Conference, Grado, Italy (1995): 9 pp.
14. B. Adams, "Heat Transfer Coils of Zirconium Withstand H_2SO_4 Pickling Solution," Chemical Processing (August 1981): pp. 26–27.
15. C. P. Dillon, "Corrosion Control in the Chemical Process Industries," 2nd Edition, publication No. 45 (St Louis, MO: MTI Inc., 1994): 420 pp.
16. Anon, "Corrosion Handbook," Vol. 8, Sulfuric Acid (Frankfurt, Germany: Dechema eV, 1991).
17. J. R. Davis, ed., "Corrosion," ASM Metals Handbook, 9th ed., vol. 13 (Metals Park, OH: ASM International, 1987): p. 795.
18. R. C. Sutherlin, "Zirconium and Zirconium Alloys for Use in Sulfuric Acid Applications," ACHEMA 2003, Frankfurt am Main, Germany (2003): 112 pp.

7

Corrosion by Sulfuric Acid

Introduction

Corrosion is the deterioration of a material by reaction with its environment. The material itself may deteriorate, or it may suffer some impairment of physical or mechanical properties as a consequence of exposure to the corrosive environment. With metals and alloys in aqueous media in particular, corrosion is an electrochemical process involving anodic and cathodic reactions. In nonmetallic materials (e.g., plastics, elastomers, ceramics), the process is chemical in nature and the materials deteriorate due to thermal degradation, oxidation, or attack by contaminants in the acid.

The corrosion resistance of an alloy in sulfuric acid depends on the concentration of the acid, the temperature, the presence of impurities, and the velocity. The effect of concentration can be divided into three ranges. At low concentration (<20%), nickel-based alloys are more resistant than iron-based alloys of the same chromium and molybdenum content. Copper does not have any significant effect. In the intermediate range (20–60%), molybdenum is beneficial, as is copper in chromium and molybdenum-bearing alloys. In high concentrations, especially >90%, chromium and silicon above a critical level are most beneficial.[1]

The corrosion resistance of individual metals and alloys in the various ranges of acid strength will be described in detail in following chapters. General materials selections can be made on the basis of acid concentration and temperature, Table 7.1.[2]

Table 7.1 Materials Choices for Various Concentrations and Temperatures

Concentration Range % H_2SO_4	Temperature °C (°F)	Materials Selection
<10	<40 (<104)	Standard SS, e.g., 304L, 316L
<20	<80 (<176)	High-alloy SS, e.g., 2205, 2507, 904L, alloy 28
20–70	<50 (<122)	SS with copper, e.g., alloy 28, 904L
20–70	>50 (>122)	Ni-based alloy, e.g., alloy B (N10665), zirconium
70–96	<50 (<122)	High-alloy SS, e.g., 2205, 2507, alloy 28
70–96	>50 (>122)	Tantalum, Si-iron
98	>60 (>140)	SX®

The areas of use of various materials have been summarized for the complete range of sulfuric acid concentrations, Figure 7.1.[3,4] The various areas represent corrosion rates of <0.5 mm/y (19.7 mpy). These data were generated some time ago and came from tests in alloys that are now somewhat changed and using acid that probably contained more iron than does current acid. The information presented in this figure should be viewed with these provisos in mind and used only as a preliminary guide to materials selection.

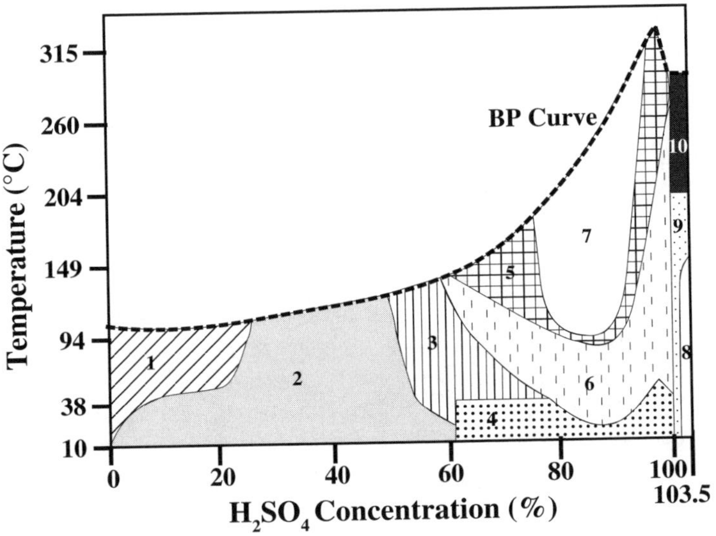

Legend

Zone 1

Impervious graphite
Tantalum
Gold
Platinum
Silver
Zirconium
Ni-o-nel ~ alloy 825
Tungsten
Molybdenum
Type 316 up to 10% aerated
10% Al Bronze, copper, alloy 400 if air-free
Illium® G
Glass
Hastelloy® B and D
Durimet 20
Worthite®
Lead
Haveg 43
Rubber up to 77°C (170°F)
Chlorimet 2 up to 70°C (158°F)
Cast alloy C

Zone 2

Ni-Resist® up to 20% at 24°C (75°F)
Impervious graphite
Tantalum
Gold
Platinum
Silver
Zirconium
Ni-o-nel ~ alloy 825
Tungsten
Molybdenum
Type 316 up to 25% at 24°C (75°F) aerated
10% Al Bronze, copper, Monel if air-free
Glass
High-silicon cast iron (14.5% Si)
Hastelloy® B and D
Durimet 20 up to 66°C (150°F)
Worthite® up to 66°C (150°F)
Lead
Haveg 43
Rubber up to 77°C (170°F)
Chlorimet

Zone 3

Impervious graphite
Tantalum
Gold
Platinum
Zirconium
Molybdenum
Monel if air-free
Glass
High-silicon cast iron (14.5% Si)
Hastelloy® B, B-2, and D, not up to BP
Cast alloy C
Durimet 20, Worthite® up to 66°C (150°F)
Lead
Chlorimet

Zone 4

Impervious graphite up to 96%
Tantalum
Gold
Platinum
Zirconium
Ni-Resist®
Carbon steel
Type 316 (>80%)
Glass
High-silicon cast iron (14.5% Si)
Hastelloy® B, B-2, and D
Durimet 20
Worthite®
Lead up to 96%

Zone 5

Tantalum
Gold
Platinum
Glass
High-silicon cast iron (14.5% Si)
Hastelloy® B, B-2, and D, not up to BP

Zone 6

Tantalum
Gold
Platinum
Impervious graphite and lead up to 80°C (176°F) and 96% sulfuric
Glass
High-silicon cast iron (14.5% Si)
Hastelloy® B, B-2, and D, not up to BP
Duriment 20 and Worthite® up to 66°C (150°F)

Zone 7

Gold
Platinum
Glass
High-silicon cast iron (14.5% Si)
Tantalum

Zone 8

Alloy C-276
Illium® G
Gold
Platinum
Glass
304
Durimet 20, alloy 20Cb-3
Carbon steel

Zone 9

Worthite®
Gold
Platinum
304
Glass
Durimet 20, alloy 20Cb-3

Zone 10

Platinum
Glass
Gold

Figure 7.1 Areas of Use of Materials in 0–103.5% Sulfuric Acid

Oxidizing and Reducing Conditions

Corrosive media are often described as oxidizing or reducing in an attempt to simplify materials selection. The concept is valid insofar as some environments are wholly oxidizing in nature (e.g., nitric acid) and some are substantially reducing in nature (e.g., dilute hydrochloric acid).

Typically, a reducing acid is one in which the cathodic reaction is the reduction of protons (hydrogen ions, H^+) and the evolution of hydrogen or, in very dilute solutions, the reduction of dissolved oxygen to form water.[5] An oxidizing acid typically suffers reduction of an oxidant (e.g., an anion such as nitrate to oxides of nitrogen, of sulfate to sulfite or sulfide; a cation, such as ferric or cupric ions) by reaction with protons and electrons.

Because metallic corrosion involves both oxidizing reactions (i.e., of the metal to a positively charged cation at the anode) and reducing reactions (e.g., of protons or other species at the cathode), the net result depends not only on the corrosive media but also upon the particular metal or alloy system. The net result of all possible oxidizing and reducing reactions in a dilute acid mixture gives the redox potential of the solution. In a general way, it can be observed that the result is a more oxidizing or more reducing environment. Unfortunately, only rarely can the redox value be correlated with particular corrosion effects relative to specific materials.

It has been found that as acid strength increases, there is a sudden change in redox potentials on platinum at around 68% sulfuric acid, Figure 7.2.[1] In the lower strength range the redox potential has a slope of $-0.059 \times \text{pH}$, which is the slope at room temperature representing the hydrogen evolution reaction. At higher concentrations the potentials rise more steeply with pH and H° (acidity function). This steeper slope ($-0.9 \times \text{pH}$) can be explained in terms of the presence of a mixture of redox reactions. The reactions occurring in strong sulfuric are strongly influenced by the absence of water and by the formation of various hydrated and anhydrous ferrous sulfate species on iron alloys.[6]

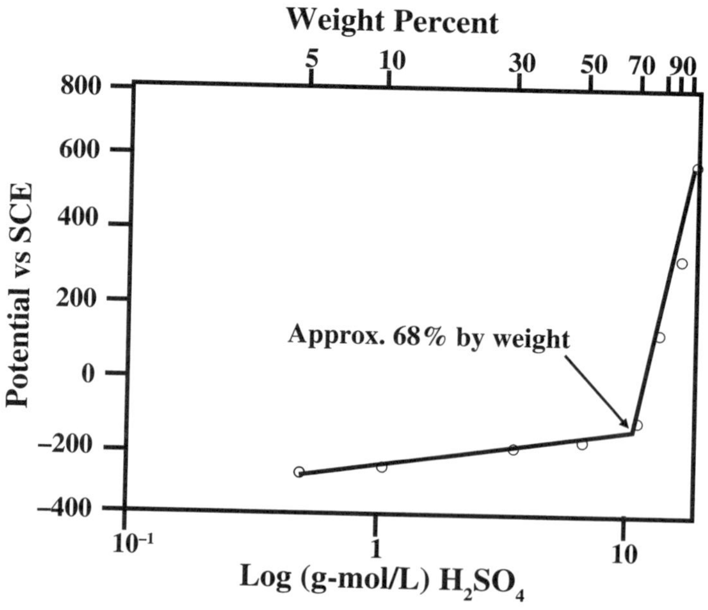

Figure 7.2 Effect of Sulfuric Acid Concentration on Redox Potential on Platinized Platinum

The effect on corrosiveness of the oxidizing and reducing nature of sulfuric acid can be explained in terms of hydrogen ion concentration and oxygen solubility. The hydrogen ion concentration at 25°C (77°F) increases with acid concentration until it reaches a maximum at about 30% acid and then decreases. The oxygen solubility, however, is highest at low acid concentration and reaches a minimum at about 75% sulfuric acid. Oxygen solubility increases with temperature up to a maximum at about 80°C (176°F), then reduces to effectively zero at the boiling point. The result of these factors is that boiling dilute and intermediate-strength acid is highly reducing and corrosive because it has high hydrogen concentration and low dissolved oxygen (DO). Concentrated acid is strongly oxidizing because it has high DO and low hydrogen ion concentration.[7]

Dilute solutions of sulfuric acid act as a typical reducing acid, attacking many metals (e.g., zinc, iron) with the evolution of hydrogen. At about 25% concentration, sulfuric acid starts to become oxidizing in nature, undergoing reduction of the sulfate ion at room temperature by finely divided nickel and at the boiling point by alloy 400 (67% Ni, 30% Cu; N04400). Warm 25% acid is still reducing toward steel or stainless steels. (The hydrogen ion concentration in dilute sulfuric acid reaches a maximum of about 2 g/L at 25–30% acid and diminishes to about 0.8 g/L at 70%.) In the 53–57% range, boiling acid suffers anion reduction in contact with an 85% Ni, 10% Si, 3% Cu, 2% Al cast alloy (now obsolete), with attendant high rates of attack, while a siliceous film otherwise protects this composition at concentrations on each side of this narrow range.

In the broad range of 0–70% sulfuric acid, the oxidizing or reducing nature of the acid is largely determined by specific contaminants, as described further in following chapters.

Strong sulfuric acid can act as either a reducing acid or an oxidizing acid, depending on the specific material exposed and the contaminants present. In contact with steel or cast iron, the cathodic reaction is hydrogen evolution typical of a reducing acid, and corrosion is stifled only by the accumulation of insoluble ferrous sulfate (i.e., by anodic polarization). This permits the use of nickel 28% molybdenum alloys (e.g., alloys B-2 [N10665], B-3 [N10675] and B-4 [N10629]) at elevated temperatures *in the absence of oxidizing contaminants.* On the other hand, although reduction of the sulfate ion is the cathodic reaction only in special cases, the oxidizing capacity is sufficient to maintain passivity of austenitic stainless steels.

Aluminum is resistant at ambient temperatures above 95%, although rates are excessive below this concentration. Copper, nickel, and nickel-copper alloy 400 (N04400) are directly attacked by strong acids, presumably by anion reduction as these materials cannot displace hydrogen. Concentrated acid also carbonizes some plastic compounds (e.g., polyvinylidene chloride [PVDC]) and chars wood, although the charring character is more dehydration than oxidation.

The Electrochemistry of Corrosion

Corrosion is essentially an electrochemical reaction; there must be an anode and a cathode, an internal circuit (electrolyte), and an external circuit, usually a metallic junction.

These conditions are fulfilled when a metal is immersed in dilute acid. Discrete areas on the metal act as anodes and cathodes, the metallic structure itself comprises the external circuit, and the acid is the electrolyte.

Corrosion of metals entails conversion from the electrically neutral atomic state to positive metal ions, as the result of oxidation (i.e., loss of electrons). This *anodic* reaction

$$M^{\circ} \longrightarrow M^{+} + ne$$

produces a positively charged metal ion and one or more free electrons, depending on the valence state of the metallic ion.

Iron and steel, for example, typically form ferrous ions and release two electrons

$$Fe^{\circ} \longrightarrow Fe^{++} + 2\,e$$

under reducing conditions.

In order for corrosion to proceed, there must be a corresponding and equivalent *cathodic* reaction. This condition may be fulfilled either by (a) the evolution of hydrogen by reduction of protons (i.e., hydrogen ions; H^+), or (b) the reduction of oxygen to form water.

$$2\,H^{+} + 2\,e \longrightarrow 2\,H^{\circ} \longrightarrow H_2 \text{ (a)}$$

$$O_2 + 4\,H^{+} + 4\,e \longrightarrow 2\,H_2O \text{ (b)}$$

As previously noted and further discussed below, other anodic and cathodic reactions are possible when the system involves species other than simply the metal and a pure aqueous solution of sulfuric acid.

The change in behavior of sulfuric acid as concentration increases occurs because other cathodic reactions such as sulfate reduction begin to dominate.

Passivity

The corrosion resistance of the oxide film-forming alloys is due to the phenomenon known as passivity. Stainless steels and some other iron-chromium and nickel-chromium alloys, as well as titanium, zirconium, and aluminum, achieve surface passivity by developing a tenacious surface oxide layer that is relatively inert. This can occur under conditions in which the metal would be highly active in electrochemical terms in the absence of this protective film. Interruptions to this film by inclusions, embedded metal particles, or mechanical breaks provide unprotected sites for corrosion to occur.

In the case of austenitic stainless steels, inclusions and other foreign materials can be removed from the surface by immersion in 20% nitric acid at about 75°C (167°F). The purpose of this treatment is to clean the metal surface and remove embedded particles before allowing the oxide film to re-form in a more continuous, thicker, and more tenacious form than might occur naturally.[8] A similar process involving the use of caustic solutions is applied to aluminum alloys. The protective surface oxide re-forms spontaneously upon reexposure to air.

The terms "active" and "passive" are used to describe the corrosion behavior of stainless steels. When active, the protective oxide film is breached or removed and corrosion proceeds on the base material; e.g., in warm, dilute sulfuric acid. When passive, the protective oxide film remains intact and minimizes corrosive activity, as in strong sulfuric acid or dilute sulfuric acid containing oxidizing agents (e.g., Fe^{+++}, Cu^{++}). Under some conditions of temperature, concentration, and contaminants, the potential of a metal such as stainless steel can oscillate between the active and passive condition at regular intervals. In the active state, corrosion rates are very high; when passive, corrosion rates are very low.

Temperature has a strong influence on whether stainless steel remains active or passive. For example, type 316 (S31600) stainless steel in 93% sulfuric acid, containing sulfur dioxide and other impurities, was found to be in the stable, passive state at 25°C (77°F) with a corrosion rate of <5 mpy (<0.13). As the temperature increased to 75°C (167°F) it was occasionally active, corroding at about 50 mpy (1.27 mm/y), and at 95°C (203°F) the sample become active regularly every minute or so and corroded at >200 mpy (>5.1 mm/y).[9]

Changes in environment, such as the presence of strong oxidizers, can cause metal potentials to increase out of the passive range and the passive film starts to break down. In this transpassive condition, corrosion rates can also be very high.

It is possible to maintain the metal surface in the passive state by providing an external source of power. This is the principle of anodic protection that is widely applied to steel and stainless steel in strong sulfuric acid.

Anodic Protection (AP)

Principle of Anodic Protection

A freely corroding metal develops a corrosion potential between itself and its ions in a corrosive medium. Such a potential is measured against a standard reference electrode. The potential can be changed by the application of an external direct current (DC) polarizing the metal and changing the extent of corrosion. When the potential is shifted in a more negative direction, the reduction in corrosion is due to cathodic protection (CP), but this is not applicable to sulfuric acid.

For specific combinations of metals and environments, the potential may be shifted in the more noble (positive) direction to enhance a protective surface film. The resultant passivity is said to have been achieved by anodic protection, or AP. Nonpassivating metals suffer accelerated attack when made the anode in such a system.

Steels, stainless steels, and similar alloys form stable, insoluble corrosion products in concentrated sulfuric acid, and these films can be enhanced by AP. At some potential, an insoluble protective film begins to form with an attendant diminution in corrosion rate. The upper service temperature limit of the alloy is also raised by the electrochemically induced enhanced film, and this also increases the alloy's resistance to velocity. A critical current density is required to initiate passivity and, subsequently, sufficient maintenance current must be applied to keep the complete metal surface within the passive range of potentials (see Figure 7.3). There cannot be excessive current, either, as too positive a potential can cause corrosion in the transpassive region.

In the transpassive region, corrosion rates of steel and stainless steels can be excessive, Table 7.2.[10] A malfunctioning AP system caused severe transpassive corrosion of stainless steel valves and other fittings in a large storage tank in the early days of the application of this technology.[11]

AP systems must be properly designed, installed, and maintained for the particular combination and configuration of metal and environment.

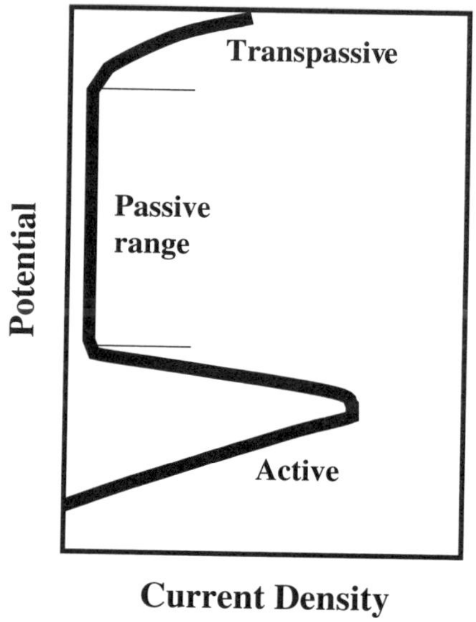

Figure 7.3 Schematic Diagram of Polarization Curve Showing Active and Passive Regions

Table 7.2 Corrosion Rate mm/y (mpy) of Steel and Stainless Steels at High Anodic Potentials

Alloy (UNS No.)	93.5% H_2SO_4, 25°C (77°F) at 1.5 V vs Pt	20% Free SO_3, 40°C (104°F) at 1.4 V vs Pt
Mild Steel	0.11 (4.3)	0.15 (5.8)
316L (S31603)	2.54 (100)	13.2 (520)
304L (S30403)	5 (196)	10 (395)
29/4/2 (S44800)	1.6 (63)	5.46 (215)

Application of AP

AP is a highly effective method of controlling corrosion of either carbon steel or stainless steel in strong sulfuric acid service. Essentially, the equipment to be protected comprises the anode in an impressed-current electrolytic cell, its potential being continuously monitored by means of strategically located reference half-cell electrodes. External cathodes are attached to the equipment, via external connection, to a suitable D.C. source (e.g., a rectifier).

Such systems are now routinely applied to tanks, piping systems, and acid coolers, either to diminish the extent of iron contamination in the product or to reduce corrosion rates effectively and prolong the life of equipment. Practically indefinite life can be achieved, even in hot acid, for strongly passivating alloys (e.g., types 304 and 316 [S30400 and S31600], 904L and 254SMO [N80904 and S31254], and alloy 20Cb-3 [N08020]).

Tanks Carbon steel storage tanks are standard for storage of concentrated sulfuric acid, with corrosion rates in the order of 10 to 20 mpy (0.25 to 0.51 mm/y). Anodic protection is typically applied by a working electrode inserted from the top of the tank. Corrosion rates can be reduced to about 2–5 mpy (0.05–0.13 mm/y) with an attendant diminution of iron contamination and elimination of localized corrosion effects caused by hydrogen released by corrosion (e.g., blistering and grooving).

While the platinum reference electrode normally used is stable in commercially pure acid, its potential is markedly affected by oxidizing species present in the acid.[10] For example, hydrogen peroxide added to bleach acid, or nitric acid present in nitration acid, can change the reference potential value by around 0.5V. While NO_x present in smelter acid can have a strong effect on corrosion, it was found to have little influence on the reference potential of platinum.

Piping Traditional cast-iron piping, with velocity limits of about 3 to 5 ft/s (1 to 1.5 m/s) had a limited service life. Its flanged joints are a common source of leaks and attendant external corrosion. With AP, piping may be installed in austenitic stainless steel (e.g., type 304L; S30403), using thinner wall and welded construction. Temperatures well above the 60°C (140°F) limitation for unprotected type 304L (S30403) piping are permissible. Velocities in the 6 to 10 ft/s (2 to 3 m/s) range are feasible, with an attendant reduction in piping size and weight. While this is technically feasible and is occasionally used, it is more common these days to use welded piping systems of appropriate stainless steels, such as those containing silicon levels of at least 5%.

Acid Coolers At one time acid coolers were typically made from cast iron with bolted joints. Since AP became a viable, commercial option, cast iron has largely been replaced by protected austenitic stainless-steel shell and tube exchangers. High-velocity cooling water is in the tubes and AP is applied to the shell side containing the hot acid. The power to anodically protect a cooler is provided via cables and inert cathodes with a control system receiving signals from installed reference electrodes, Figure 7.4.

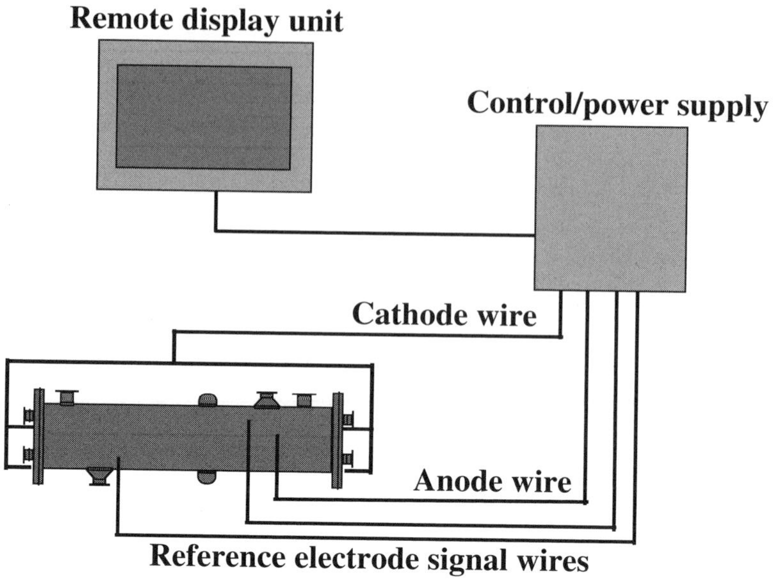

Figure 7.4 Typical Layout of Anodic Protection for an Acid Cooler (Courtesy of Chemetics, a Division of Aker Kvaerner)

A rectifier operating at about 5 volts with 60 amp capacity usually supplies power. The exact current and voltage utilized depends on the specific design of individual coolers. PTFE-sheathed cathodes are installed through the water boxes into the tube bundle and also in the nozzles, Figure 7.5. Reference electrodes are strategically placed in the tube-bundle/shell and nozzles to produce the completed AP exchanger, Figure 7.6.

Electronic instrumentation measures and maintains the necessary voltage and current. If acid coolers are designed for AP, the water velocity must be above the minimum limit to prevent fouling that can occur at low flow rates (e.g., <3 ft/s [<1 m/s]). The acid velocity on the shell side must not be so high as to erode passive films, and the tube-wall temperature in the acid must be below some maximum value, which depends on the acid concentration and purity and the alloy being protected.

Usually, type 304L (S30403) is used for 98% acid and type 316L (S31603) for 93–95% acid. 6% molybdenum alloy (e.g., alloys 254 SMO [S31254], AL-6XN [N08367], or 926 [N08926]) tubes may be used to resist chloride pitting or stress corrosion cracking by high-chloride waters. The superferritic stainless steel, Cronifer® 2803 (S44660), has also been used in acid coolers and is said to be well suited for applications exposed to contaminated cooling waters.[12]

Alloy G (N06007) and its variants are even more amenable to AP than is type 316 (S31603), Table 7.3.[1]

Figure 7.5 Water Box of an Anodically Protected Cooler Showing Cathodes, Tubesheet, and Tube Ends (Courtesy of Chemetics, a Division of Aker Kvaerner)

Figure 7.6 Installed Anodically Protected Cooler at a Metallurgical Acid Plant (Courtesy of Chemetics, a Division of Aker Kvaerner)

Table 7.3 Effect of Anodic Polarization on Corrosion Rates of Stainless Steels and Nickel-Rich Alloys in 99% Sulfuric Acid at 160°C (320°F)

Alloy	No AP	With AP
Type 316L	110 (2.8)	8.5 (0.2)
Alloy 255	34 (0.9)	1.9 (0.05)
Alloy G-3	58 (1.5)	0.9 (0.02)
Alloy G-30	42 (1.1)	0.2 (0.005)
Alloy C-22	68 (1.7)	355 (8.9)

Forms of Corrosion

The forms of corrosion most often encountered in sulfuric acid are briefly described in this section.

General Corrosion

General or uniform corrosion is the common form of metal loss in most corrodents in the absence of passivating films. In this form of corrosion the metal is removed uniformly over the entire exposed surface. This is the predominant mode of attack on base metals such as iron alloys and lead, although localized corrosion can also occur.

Localized Corrosion

Localized corrosion is a term used to describe several related forms of attack, including pitting, crevice corrosion, and intergranular attack. Localized corrosion is often encountered on metals whose normally protective surface film has been damaged or whose passive film has been penetrated by aggressive ions. This is often the form of attack on stainless steels and other film-forming metals.

Pitting Pitting is a localized type of corrosion usually taking the form of small but deep cavities. It is uncommon in sulfuric acid except in the presence of chlorides or when dilute acid contacts steel.

Crevice Corrosion Crevice corrosion is induced by a similar mechanism to pitting: breakdown of the passive film leading to active localized attack. Crevice corrosion occurs on closely abutting surfaces, such as flange faces and under gaskets where the bulk liquid cannot penetrate. Chemical changes within the crevice, due to depletion or accretion of aggressive species such as chlorides, cause pitting and attack within the crevice.

Intergranular Attack Intergranular attack (IGA) is a form of attack that occurs when constituents of an alloy in or near the grain boundaries are susceptible to preferential attack by the environment. It occurs in austenitic stainless steels because of differences in chromium content between the matrix and the grain boundaries. These chemical differences within the alloy result from thermal processing in a temperature range, usually 425–815°C (800–1,500°F), favoring chromium carbide precipitation at the grain boundaries. Chromium depletion at the grain boundaries results in diminished corrosion resistance, compared with the matrix material. The stainless steel is then described as "sensitized" (i.e., susceptible to IGA).

Intergranular attack of austenitic stainless steels may occur in the lower range of sulfuric acid concentration unless low-carbon or stabilized alloys are used. This type of attack is less prevalent above about 90% acid concentration.

Galvanic Corrosion

Galvanic corrosion can occur when two dissimilar metals are in contact in the presence of an electrolyte. Galvanic corrosion is rare in concentrated acid, but it can occur in the lower concentration range (e.g., alloys in contact with carbon products or sulfated lead surfaces) or if conductive films (e.g., mill-scale or sulfide products) are present on carbon steel. Either anodic corrosion or cathodic phenomena due to hydrogen effects may be encountered.

In one case the top section of a lead-lined tower handling 70% sulfuric acid was replaced with alloy 20 (N08020), used to avoid making a complicated lead-lined head with many small branches. Initially the lead was anodic to the alloy, corroding slowly and protecting the alloy. Once a substantial layer of lead sulfate was formed, however, it was cathodic to the alloy and the alloy head suffered severe local corrosion.

Welds in, for example, storage tanks can be subject to galvanic corrosion caused by differences in composition between the weld and adjacent plate. The weld may be cathodic to the plate, in which case the plate corrodes locally and leaves the weld standing proud, or it may be anodic to the plate, in which case it is preferentially corroded away, Figure 7.7.[13]

Figure 7.7 Section Through a Weld in a Tank Floor Plate Showing Galvanic Corrosion of the Weld

Velocity-Related Corrosion

This term covers a number of types of corrosion including erosion, cavitation, and impingement. They all result from the loss of protective surface oxide due to excessive velocity or turbulence of contacting fluids or abrasion by moving particles. Metals that rely on a film of insoluble corrosion products for resistance, such as carbon steel, cast iron, and lead, are particularly prone to this type of attack. Centrifugal pumps often exhibit this type of corrosion. This problem is controlled by alloy selection and the specification and maintenance of sulfuric acid velocities that are appropriate for the alloy and the environment, acid temperature, strength, and so on.

Environmental Assisted Cracking (EAC)

EAC occurs in susceptible materials if specific contaminants and environmental conditions are present. Either anodic stress corrosion cracking (SCC) or cathodic hydrogen-assisted corrosion (HAC) may occur. Unless properly cleaned, high-nickel alloys may suffer liquid metal corrosion (LMC) during welding or joining, from sulfur contamination by acid or acid salt residuals.

SCC is an environmentally assisted form of attack that results from interaction between the environment and a specific alloy system under tensile stress. It is a temperature-sensitive phenomenon. Classic examples are SCC of copper alloys in ammonia, steel in caustic soda, and stainless steels in chloride environments. This type of chloride SCC can occur in sulfuric acid solutions.

Hydrogen Effects

Hydrogen formed by the corrosion reaction may cause hydrogen grooving of carbon steels. Hydrogen bubbles are formed and remove the protective iron sulfate film that produces deep grooves in walls of tanks or piping. Atomic hydrogen can also enter the steel plates and coalesce at discontinuities in the structure, causing sufficient internal pressure to form blisters. Both of these phenomena will be described in detail later.

Microbiologically Influenced Corrosion (MIC)

MIC is attack initiated by microorganisms or their byproducts, usually in a biofilm, inducing localized pitting or crevice corrosion under the film. This type of corrosion does not occur in strong acids but can occur in process equipment prior to use and during extended lay-ups. Water used to flush or hydrotest equipment will leave a heel if not completely drained. The water residue may sit for months before startup, and it is during this period that MIC may occur. Usually the first evidence of MIC is leakage during startup. Stainless-steel equipment rinsed or hydrotested with water is very susceptible to MIC. The condition has also been seen in aluminum alloys. Prevention requires removal of *all* residual water and proper drying of the equipment and piping internals prior to shipping or storage.

Vapor-Phase Attack

Most corrosion data are derived from laboratory tests made in the liquid phase. However, in actual plant equipment, there is often a vapor space above the liquid acid where acid vapor composition may differ from the bulk liquid. This acid vapor may condense on the equipment's inner surface and be either more or less corrosive than the bulk liquid. In sulfuric acid, a form of vapor-phase attack occurs when moisture enters the vapor space—in storage tanks, for example—and dilutes the surface acid to more aggressive concentrations. Great care needs to be taken in storage tanks to prevent ingress of moisture causing attack in the vapor phase.

The boiling point-composition curve for sulfuric acid in the 0–70% range is such that there is only water (*no* acid) in the condensing vapors. About 7% acid will condense from boiling 85% sulfuric at 223°C (433°F), while the vapor and liquid are the same composition at the azeotrope (constant boiling mixture [CBM]) of 98.3% acid at 339°C (642°F).[14]

Dealloying

Dealloying is the process in which a component of a multicomponent alloy is preferentially attacked. The two common examples of this phenomenon are removal of zinc, nickel, or aluminum from susceptible types of brass, and graphitic corrosion of cast irons. Dealloying is not usually encountered in sulfuric acid because brasses are not used in sulfuric acid service and because cast irons are generally not prone to graphitic corrosion in concentrated sulfuric acid and are not generally used in weak acid.

End-Grain Attack

If the end grain (cut surface) of steel plate is exposed to sulfuric acid, the corrosion rate into the end grain can be very severe. This can happen in nozzles set into tanks or piping, at welds with incomplete penetration, and so on. Where end grain is exposed in acid, it should be overlaid with weld to prevent this type of attack.

High-Temperature Corrosion

High-temperature corrosion, in the conventional sense, is not a problem with liquid acid products. It is, however, of concern in some parts of the acid production plant, such as hot gas ducts, exchangers, and converters. Alloy selection for equipment operating at high temperature is based on resistance to oxidation/corrosion and elevated mechanical property requirements.

Inhibition

To inhibit means to retard or slow the rate of corrosion, usually by the addition of other chemicals to the system. This is normally only viable when the bulk system is contained and is circulating so that the inhibitor is not continuously lost. The addition of inhibitors must also not react with or contaminate the process acid. These exceptions mostly exclude the use of inhibitors in commercial sulfuric acid systems. However, additions of nitric acid to sulfuric acid can reduce the corrosion rate of both carbon and stainless steels. Traces of HF can inhibit corrosion of chromium- and copper-bearing nickel alloys, such as alloy G (N06007), in concentrated acid. In the copper-free grades, such as alloys 625 (N06625) and C-276 (N10276), traces of HF accelerate attack below about 80% sulfuric acid concentration; in more concentrated acid solutions they become inhibitive.

Prediction of Corrosion Rate

The correlation between laboratory-generated data and field experience or testing is not always good. A comparison of corrosion of various stainless steels in laboratory and plant exposures showed that some alloys had similar results in both test series while the laboratory results were higher than those generated in plant tests with other alloys. Of the alloys used, 304 (S30400) and the duplex alloys—255 (S32550), 2205 (S31803), and Zeron® 100 (S32760)—had similar rates in both locations, while the corrosion rates of 316L (S31603), 309 (S30900), and 310 (S31000) were substantially lower in the plant than in the laboratory. This may have been due to the slow passivating process for some alloys, as the plant immersion was approximately 10 weeks compared to 1 week in the laboratory.[15]

There is also variability between published results obtained under nominally identical conditions. Much of the problem is due to contaminants present in plant liquors being absent from the chemically pure solutions often used in laboratory corrosion tests. Even synthetic solutions intended to simulate plant conditions might not do so successfully with regard to minor, unstable, or transient contaminants.

The detailed test techniques used can have a strong influence on the results obtained. The preparation of samples, particularly in terms of their surface condition, has a big impact on short-term corrosion tests, but the effects would be negligible in long-term plant exposure. The ratio of the test volume of acid to the sample area also influences the corrosion rates measured. The critical volume-to-area ratio normally accepted for laboratory testing is 0.2 to 0.4 ml/mm^2, but this has been shown to be inadequate in hot sulfuric acid. A more appropriate ratio for hot acid is 1ml/mm^2. Using an appropriate ratio helps prevent iron concentration from becoming high enough to reduce corrosion rates in laboratory tests to much less than rates occurring in plant exposure. However, iron content should be monitored during laboratory testing and maintained at <10 mg/L and preferably at 5 mg/L to better simulate plant conditions.[16]

Corrosion is a complex process and corrosion rates are affected by many variables. Methods have been developed to predict which variables are likely to be critical in a specific situation. In many cases, a combination of rough calculations, rule of thumb, past case histories, and intuition are used to make interpretive judgments.

As understanding of corrosion mechanisms has increased, it has become possible to establish mathematical models that allow reasonably accurate predictions of corrosion rates. Corrosion of carbon steel by concentrated sulfuric acid is an example where such an approach has been tested, with positive results.[17]

Corrosion is often accelerated by increased velocity, either because of mechanical damage to the corrosion film or because of diffusion of corrosive species or products through the boundary layer. For these cases, corrosion is approximately proportional to the velocity. In other cases, there may be "breakaway" attack—a sudden increase in severity of corrosion above some critical velocity.

Carbon steel piping in concentrated sulfuric acid exhibits a corrosion rate approximately proportional to the velocity.[18] For steel in concentrated sulfuric acid at ambient temperature, the velocity at which the corrosion rate reaches a practical limit is about 3 ft/s (0.9 m/s), while for gray cast iron it is about 4.9 ft/s (1.5 m/s).[19]

There are several laboratory tests that may be employed to study velocity effects (e.g., circulating loops, rotating specimens), while electrochemical techniques can identify flow conditions responsible for diffusion limitations. For concentrated sulfuric acid, these tests for carbon steel or cast iron are not always reliable because steady-state rates are obtained only after long exposure times.

In practice, the prediction of corrosion rates in a given acid concentration at high velocities is difficult because localized turbulence (e.g., over weld irregularities, flange joints, elbows) substantially increases attack. However, calculations of corrosion rates for straight, smooth pipe are reasonably accurate. These predictions are useful for design purposes and, in specific cases, in estimating the effects of changing operating parameters on the remaining life of equipment.

Mechanisms

In electrochemical attack, either the cathodic or anodic reactions may be impaired by the accumulation of a layer of corrosion products on the metal surface. At the cathode, the layer may prevent access of the cathodic depolarizers or other reactants participating in the corrosion process. At the anode, insoluble products inhibit the diffusion of metal ions into the solution. In either case, diffusion controls the corrosion rate. The effect of velocity is to disturb the otherwise protective film or prevent its formation altogether.

Diffusion Control

In the case of carbon steel in cold, concentrated sulfuric acid, corrosion is limited by the solubility of ferrous sulfate ($FeSO_4$) in the acid.[18] Despite a high initial rate of attack, the metal quickly becomes covered with an adherent, protective film that stifles further attack. This is why carbon steel tanks are acceptable for storage of concentrated sulfuric acid. Under *flowing* conditions of low velocity (such as in piping and actively used nozzles), the fresh acid tends to dissolve ferrous sulfate, and corrosion proceeds at a rate to maintain a layer of constant thickness. It is then possible to predict the corrosion rate from the properties of the acid and the flow/geometry conditions. (Note: Under conditions of high velocity or impingement/turbulence, the fluid shear stress at the surface mechanically removes the film and the high initial rate of perhaps several hundred mils per year is maintained. In these cases, more resistant alloys are required.)

The following example illustrates the calculation of corrosion rate in metric units for some specific conditions using the equation[17]

$$CR = AT^{0.654}\ v^{0.913}\ \rho^{1.567}\ \mu^{-1.221}\ d^{-0.087}\ (w - 0.01)$$

where:
CR = corrosion rate in mm/y,
A = 1.42×10^{-4},
T = temperature in °K,
v = velocity in m/s,
ρ = density in kg/m^3,
μ = viscosity in cp,
d = pipe diameter in meters, and
w = the solubility of $FeSO_4$ in mass %.

Using a spreadsheet to facilitate the calculation of exponential functions and conversion of units, the calculation can be approached as shown in the following exercise.

Problem: Determine the corrosion rate (CR) for 95% sulfuric acid in an NPS 6 Schedule 40 pipe at 25°C (298°K) and 2 ft/s (0.6 m/s) velocity (using U.S. customary engineering units), assuming the acid contains 70 ppm iron.

Step 1: The calculation is valid only if the Reynolds number ($Re = D \times V \times d/v$) is >2100. At lower Reynolds numbers the flow may be laminar; if so, the CR should be significantly lower and decrease along (down) the length of pipe.

Determine the Reynolds number from the formula

$$Re = D/12 \times ft/s \times (62.4 \times Sp.\ Gr.) \ / \ (\mu \times 0.0000672)$$

where:

D = pipe ID (inches),
ft/s = velocity (feet/second),
Sp. Gr. = specific gravity (see Table 3.1 for concentrated acid and Table 3.3 for oleum), and
μ = viscosity (Figure 7.8[17]).

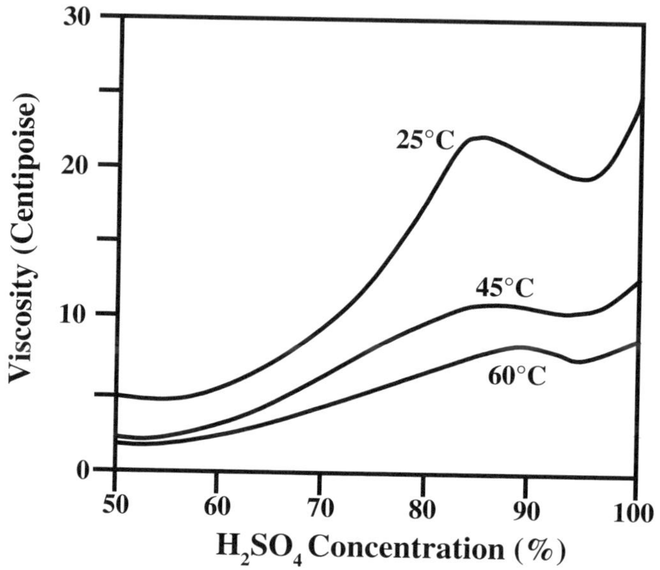

Figure 7.8 Sulfuric Acid Viscosity at Different Temperatures

$$Re = 6.06/12 \times 2 \times (62.4 \times 1.834) \ / \ (18.5 \times 0.0000672) = 9297$$

(Since Re is >2100, the flow is turbulent; proceed as below.)

Step 2: Calculate CR as follows.

$$F_{Temp} = 0.2 \times K^{0.654} = 0.2 \times 298^{0.654} = 8.302$$

$$F_{Velocity} = ft/s^{\ 0.913} = 2^{0.913} = 1.883$$

$$F_{Density} = (62.4 \times Sp.\ Gr.)^{\ 0.654}\, 1.567 = 114.4416^{0.654}\, 1.567 = 1682$$

$$F_{Viscosity} = \mu^{-1.221} = 18.5^{-1.221} = 0.02837$$

$$F_{Diameter} = d^{-0.087} = 6.065^{-0.087} = 0.8549$$

Determine solubility of ferrous sulfate (Figure 7.9[17]) as 0.05 (W_S).

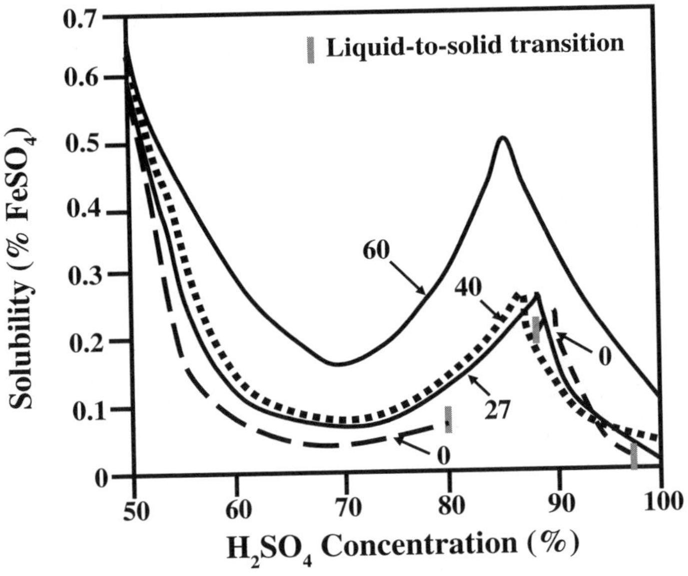

Figure 7.9 Solubility of Ferrous Sulfate in Sulfuric Acid

Convert ppm Fe to mass % $FeSO_4$:

70 ppm Fe/10000 × 151.9/55.8 = 0.0191% (W_B)

Then:

$$CR = F_{Temp} \times F_{Velocity} \times F_{Density} \times F_{Viscosity} \times F_{Diameter} (W_S - W_B)$$

$$CR = 8.302 \times 1.883 \times 1682 \times 0.02837 \times 0.8549\ (0.05–0.0191)$$

$$CR = 19.7 \text{ mpy } (0.5 \text{ mm/y})$$

For storage tanks, similar calculations may be made that predict the probable corrosion rate due to ingress of atmospheric moisture from humid air during filling and draining operations.[20] These calculations assume that the ingress of atmospheric moisture (which dilutes strong acid retained in the ferrous sulfate film) is the limiting step, and that corrosion will continue until the acid is consumed or the acid retained in the corrosion product layer becomes 50% or lower. The calculations are based on tank diameter, average dew point of the local atmosphere, scheduled loading/unloading cycles, average temperature, and so on. Calculations can also be made that take into account localized corrosion, preferential weld attack, and hydrogen grooving.

References

1. N. Sridhar, "Behavior of High Performance Alloys in Sulfuric Acid Streams," Corrosion '87, paper no. 19 (Houston, TX: NACE International, 1987): 16 pp.
2. Anon, "Corrosion Handbook—Stainless Steels" (Sandviken, Sweden: AB Sandvik Steel, 1999): pp. I:I–II:88.
3. Various authors, in Sulfuric Acid section, CD, "Dechema Corrosion Handbook" (Frankfurt, Germany: Dechema eV, 2001).
4. DuPont (1986) in C. P. Dillon, ed., "Concentrated Sulfuric Acid and Oleum," vol. MS-1, Materials Selector for Hazardous Chemicals (St. Louis, MO: MTI Inc., 1997): 212 pp.
5. C. P. Dillon, "Corrosion Control in the Chemical Process Industries," 2nd Edition, publication No. 45 (St Louis, MO: MTI Inc., 1994): 420 pp.
6. N. Sridhar, "Mechanism of Corrosion in Concentrated Sulfuric Acid," Sulphur '85 conference (London: British Sulphur Corp. Ltd., 1985): pp. 247–265.
7. G. Coates, "Corrosion of Stainless Steels in Weak Sulphuric Acid Solutions," NUCCI no. 3-83 (Degefors, Sweden: Nyby Uddeholm AB., 1983): pp. 6–10.
8. R. J. Landrum, "Designing for Corrosion Control: A Corrosion Aid for the Designer" (Houston, TX: NACE International, 1989): p. 158.
9. Anon, "The Corrosion Resistance of Nickel-Containing Alloys in Sulfuric Acid and Related Compounds," CEB-1 (Suffern, NY: The International Nickel Co. Inc., 1983): 90 pp.
10. J. R. Rodda, M. B. Ives, D. Drexler, "Anodic Protection of Sulfuric Acid Storage Tanks" (2004), unpublished draft paper.
11. J. R. Rodda, M. B. Ives, "The Corrosion Science of Sulfuric Acid Storage Tank Failures" (2004), unpublished draft paper.
12. H. Decking, W. Schalk in E. Altpeter, R. Kirchheiner, F. E. White, "Sulphuric Acid Corrosion of Some Special Stainless Steels and Nickel Alloys: Laboratory Tests and Plant Experience," NACE Conference, Grado, Italy (1995): 9 pp.
13. M. Tiivel, F. J. McGlynn, A. A. Trickett, "Carbon Steel Sulfuric Acid Storage Tank: Inspection Guidelines" (North York, ON, Canada: MARSULEX Inc., 1986): 46 pp.
14. U. Sander et al. in Sulfuric Acid section, CD, "Dechema Corrosion Handbook" (Frankfurt, Germany: Dechema eV, 2001).
15. J. R. Rodda, M. B. Ives, K. S. Coley, "Corrosion Survey of Stainless Steels in Hot, Concentrated Sulfuric Acid, Comparison of Laboratory and Plant Data" (2004), unpublished draft paper.
16. J. R. Rodda, M. B. Ives, "Determination of Corrosion Rates in Hot, Concentrated Sulfuric Acid," Corrosion 59, 4 (2003): pp. 363–370.
17. S. W. Dean, G. D. Grab, "Corrosion of Carbon Steel by Concentrated Sulfuric Acid," Corrosion/84 paper no. 147 (Houston, TX: NACE International, 1984): 11 pp.
18. B. T. Ellison, W. R. Schmeal, "Corrosion of Steel in Concentrated Sulfuric Acid," J. Electrochemical Society 125, 4 (1978): p. 524.
19. Anon, "Materials for the Handling and Storage of Concentrated (90–100%) Sulfuric Acid at Ambient Temperatures," RP0391 (Houston, TX: NACE International, 1991): 14 pp.
20. S. W. Dean, G. D. Grab, "Corrosion of Carbon Steel Tanks In Concentrated Sulfuric Acid Service," Corrosion/85 paper no. 298 (Houston, TX: NACE International, 1985): 9 pp.

8

Corrosion of Metals and Alloys in Concentrated Sulfuric Acid and Oleum

This chapter discusses the corrosion behavior of metals and alloys in nominally pure, concentrated sulfuric acid and oleum. Corrosion in contaminated acids and acid mixtures is discussed in Chapter 10. Materials of construction for specific plant items and types of equipment are described in Chapter 12.

Aluminum Alloys

Aluminum, despite its amphoteric nature, is resistant to hot 99% sulfuric acid. Laboratory tests in various concentrations of acid at 180°C (356°F) show that aluminum is the most resistant metal tested in the 98–99% range, although it is rapidly attacked below 96%.

Some experiences with aluminum are:[1]

- An experimental air-cooled heat exchanger made of aluminum showed no attack after several months cooling 99% acid in a contact plant. However, no full-sized installation was made because the plant management considered aluminum exchangers a potential safety hazard in the event of inadvertent dilution of acid.
- An aluminum heating coil (50 psig [345 kPa] steam) in a small pressure vessel was successfully used to preheat 99% acid feed for a globular bisulfate of soda (GBS) reaction. Bimetallic aluminum/copper-alloy pipe performed well in an installation in which glass-lined steel pipe suffered pinholing attack in 99% acid at 180°C (356°F).

Aluminum alloy 3003 was only slightly corroded in laboratory tests in 101, 103, 107, and 115% sulfuric acid at ambient temperature. Aluminum alloy 1100 pipe carries 99.95% acid at 45°C (113°F) in one plant. In another plant aluminum alloy 1100-H112 pipe is used to carry 98% acid at 200°C (392°F).[2]

Steels

Ferrous-based materials are commonly used in strong sulfuric acid, especially if temperatures and velocities are low. Many factors affect the corrosion resistance of steel in strong acids, as illustrated in the following sections.

Effect of Temperature

A 21°C (38°F) temperature increase, from 25 to 46°C (77 to 115°F) of concentrated acid with an iron content of 45 ppm under near ambient storage conditions increases the corrosion rate of carbon steel by three to four times, Figure 8.1. Temperature increase has a similar effect on the corrosion rate of carbon steel in 93.5% sulfuric acid, also with 40 ppm of Fe, Figure 8.2.[3] The high iron content in both of these cases is likely to have reduced the overall corrosion rate, but the temperature trends are still valid.

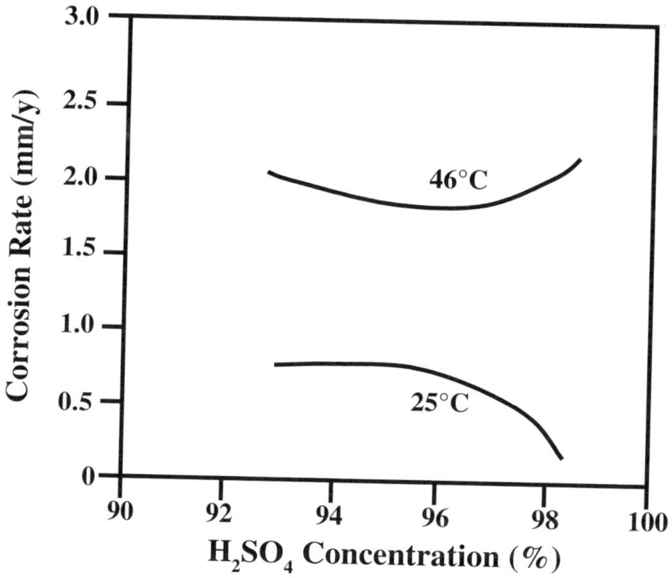

Figure 8.1 Corrosion Rate of Carbon Steel in Sulfuric Acid with 45 ppm Iron at 25 and 46°C (77 and 115°F)

Corrosion Product Film

Initial corrosion of iron is stifled by a film of ferrous sulfate. The actual corrosion rate, if the film is undisturbed by velocity or other physical disturbance, oxidation, or solubilization (thermal or chemical), is proportional to the solubility of the hydrated sulfate.[4–6] The stability and reliability of the film can be enhanced by anodic protection (AP), although this is economically feasible only for large storage tanks (see "Anodic Protection (AP)," Chapter 7).

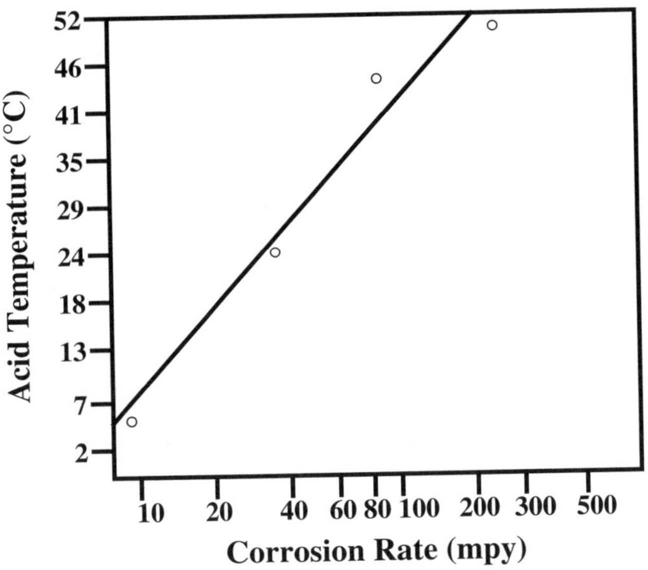

Figure 8.2 Effect of Temperature on the Corrosion Rate of Carbon Steel in 93.5% Sulfuric Acid with 40 ppm Iron

Effect of Acid Concentration

The effect of acid concentration on the corrosion of cast iron and steel is considerable, particularly in the transition between strong acid and oleum, Figure 8.3, at 27°C (81°F).[7] Inadvertent dilution of strong acid should be avoided because corrosion of carbon steel in less than around 90% acid can be severe.

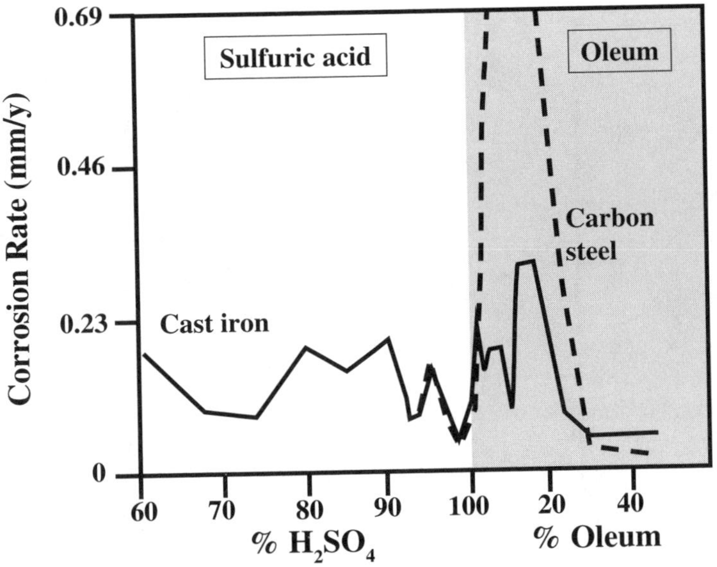

Figure 8.3 Corrosion Rate of Cast Iron and Carbon Steel in Static Sulfuric Acid and Oleum at Room Temperature

There have been failures or near failures of carbon-steel storage tanks nominally handling concentrated acid. In one case, pitting corrosion gouged a tank wall more than halfway through in a band about 15 cm high, just above the floor, in a 20,000-ton tank.[8] Localized corrosion by weak acid also led to an explosion in an alkylation waste acid tank, resulting in the loss of one life, injuring eight, and discharging 4.2 million liters of H_2SO_4.

At 25°C (77°F), 98.5% acid is less corrosive than 93% or 96% acid, but at 38°C (100°F) the rates are the same. Therefore, the heating of 98–99% grade to prevent freezing must be carefully controlled.[3] The measured corrosion rate in a steel tank holding 92% black acid (i.e., an acid recovered from a diethyl sulfate process containing aliphatic carbonaceous contaminants) was 7 mpy (0.2 mm/y), with pitting to 100 mil (3 mm) total depth over a ten-year period at 80°C (175°F).[9]

Dihydrate ($H_2SO_4 \cdot 2H_2O$) 84% acid is the second most corrosive concentration, with rates of 40 mpy (1 mm/y) at 27°C (80°F) and 135 mpy (3.4 mm/y) at 66°C (150°F), even under static conditions. In general, steel may be excessively attacked in the 80–88% range and at temperatures above about 50°C (120°F).[3] Temperature has a strong effect on corrosion of carbon steel in strong acid and oleum, Figure 8.4.[10] See Figure 8.1 for corrosion rates in 93–99% acid with 45-ppm iron content.

The outstanding characteristic of corrosion in oleum is the high rate of attack on steel in the range 101–105% sulfuric acid (i.e., around 5 to 20% oleum), especially at temperatures of about 50°C (120°F), see Figure 8.3. Small amounts of nitric acid, sometimes added as an antifreeze measure, may reduce corrosion of steel in the oleum ranges below 30% and above 65%. Steel is preferred for shipment and storage. Typically, corrosion rates are about 3–5 mpy (0.08–0.13 mm/y) at ambient temperatures. Interestingly, the corrosion rate of steel in 5–30% oleum *decreases* with increasing temperature, Figures 8.5 (modified slightly from the original figure) and 8.6.[1] Figure 8.6 also shows that the corrosion rate increases substantially with increasing acid velocity from 5.6 to 9.4 ft/s (1.7 to 2.8 m/s).

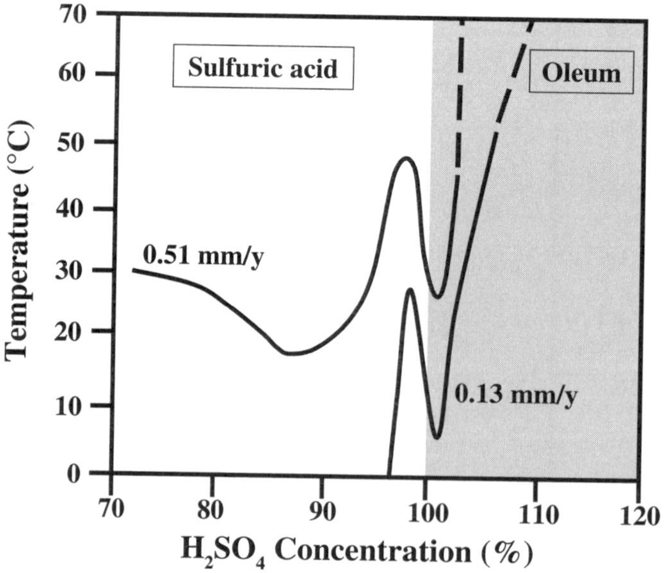

Figure 8.4 Part of Isocorrosion Curve for Corrosion Rate of 5 mpy (0.13 mm/y) and 20 mpy (0.5 mm/y) for Carbon Steel in Sulfuric Acid and Oleum with 45 ppm Iron

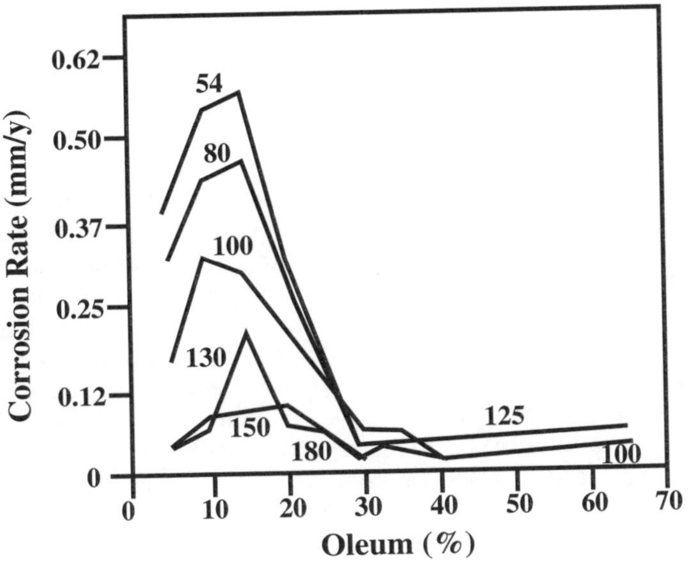

Figure 8.5 Effect of Temperature (°F) on Corrosion Rate of Tank Car Carbon Steel in Oleum

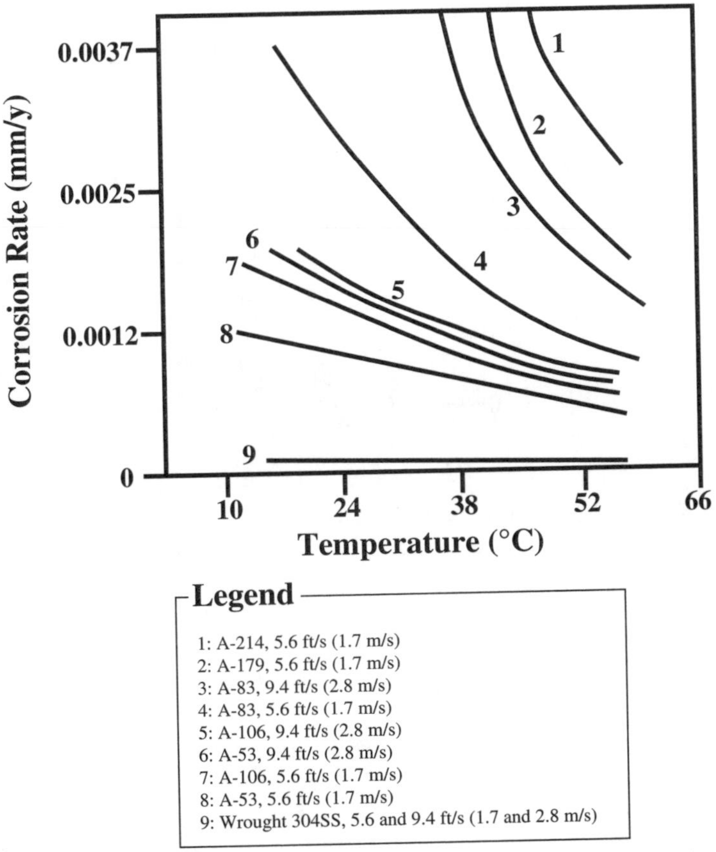

Figure 8.6 Corrosion Rate of Various Grades of Carbon Steel Compared to Type 304 Stainless Steel in 20% Oleum at 5.6 and 9.4 ft/s (1.7 and 2.8 m/s)

Effect of Velocity

Erosion and abrasion effects remove the normally protective ferrous sulfate film. Acid velocity has a strong influence on the corrosion rate of carbon steel and cast iron, e.g., in 95% sulfuric acid at 50°C (122°F), Figure 8.7.[11] At 4 ft/s (1.2 m/s) at 40°C (105°F), the corrosion rate measured in a 3-in (7.6-cm) steel pipe was 400 mpy (10 mm/y). At 80 to 90°C (175 to 190°F), the rate was 350 mpy (9 mm/y) at 1.6 ft/s (0.5 m/s) but 1,200 mpy (30 mm/y) at 6.5 ft/s (2 m/s).[9] The expected corrosion rate in a specific acid concentration under stated conditions of temperature, velocity, iron content, etc., can be calculated (see Prediction of Corrosion Rate, Chapter 7).

Because of the difficulty in controlling localized velocity in certain installations (e.g., at elbows and ells and downstream of irregularities such as welds), some users have completely eliminated carbon-steel piping in strong acid service. In complex small-diameter control loops, stainless steel is normally preferred. Similarly, excessive velocity at inlet and outlet nozzles can cause accelerated local corrosion. Siting of nozzles and the use of wear plates or sleeves should be considered.

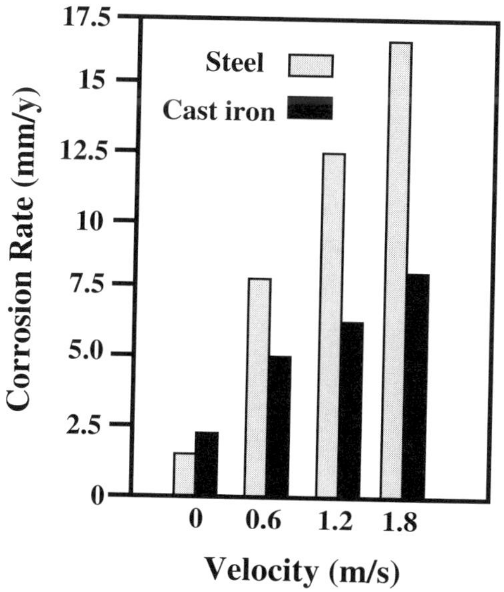

Figure 8.7 Effect of Velocity on Corrosion Rate of Carbon Steel and Cast Iron in 95% Sulfuric Acid at 50°C (122°F)

Hydrogen Grooving

Hydrogen grooving, sometimes called "tiger claws," may occur where the movement of hydrogen gas disturbs the otherwise protective ferrous sulfate film. This occurs notably in the upper hemisphere within the liquid phase of cylindrical components (e.g., lines, horizontal tanks, piping, manways, and nozzles) exposed to solar heating, and on the walls and tops of tank cars. It may occur even on horizontal welds with excess build-up. Hydrogen grooving on a sulfuric acid tank car is shown in Figure 8.8.

Figure 8.8 Hydrogen Grooving at the Top of a Sulfuric Acid Tank Car

During the corrosion of carbon steel by concentrated H_2SO_4, a soft, ferrous sulfate ($FeSO_4$) film forms on the steel surface and hydrogen is evolved as a gas (H_2):

$$Fe + H_2SO_4 = FeSO_4 + H_2$$

The $FeSO_4$ acts as a corrosion barrier and significantly reduces the corrosion rate. The H_2 collects as small bubbles. If in sufficient quantity, they cause disruption of the protective $FeSO_4$ film, either by mechanical means (erosion) or by increasing chemical dissolution as they flow upward along the film on vertical and inclined surfaces. The disruption is not uniform on the film but occurs along preferred paths. The paths generally are parallel to each other but not necessarily uniformly spaced. The steady stream of hydrogen bubbles follows the established vertical paths, gradually deepening them into channels or grooves, Figure 8.9.[12] This figure also includes a cross section of the plate illustrating the local nature of this deep, sharp grooving. Corrosion rates as high as 300 mpy (7.6 mm/y) have been reported on the sides of storage tanks in these grooves. In time, the grooves become deep and a hole can form in the wall, causing the vessel or piping to fail.

Inward-sloping tank walls are very prone to hydrogen grooving. A simple inspection method to detect if a tank is not plumb is to use a 3 or 4 ft (0.9 to 1.2 m) spirit level. Walls found to be inward sloping then are inspected using ultrasonic techniques to look for loss in wall thickness caused by corrosion.

For storage tanks, replacing badly corroded wall sections with new carbon-steel plates often is the most economical solution. If grooving damage is not too severe, "wallpapering" the affected areas of the tank sidewall with stainless-steel sheet can

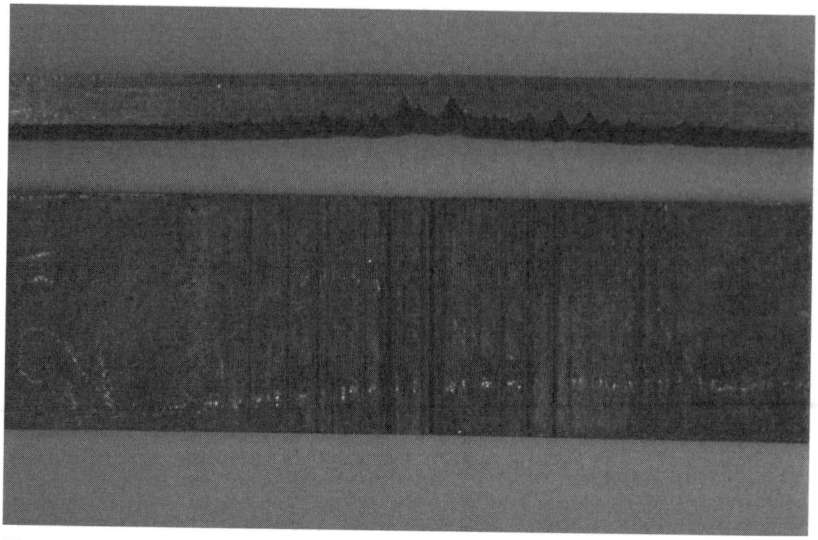

Figure 8.9 Hydrogen Grooving in a Sulfuric Acid Storage Tank Wall Plate

be successful. However, one must be careful because H_2 and sulfates can build up between the stainless-steel sheets and the tank wall. Repairs should be made in accordance with API standard 653.[13] Baked phenolic linings have been used to prevent hydrogen grooving. If a lining is applied, it is recommended that any stainless-steel accessories (such as dip tubes) also be coated because the corrosion potential differential between the stainless steel and steel exposed at pinholes in the coating may accelerate the corrosion of the steel. For large-diameter nozzles and manways, weld overlaying or cladding with type 316 (S31600) stainless steel or alloy 20 (N08020) has been used to avoid hydrogen grooving.

Another solution to minimize erosion-corrosion and hydrogen grooving is to apply anodic protection,[12] a method commonly used to reduce iron contamination of the acid. If the acid inlet nozzle in the roof of a storage tank is located near the shell, the wall area directly beneath the inlet is very susceptible to erosion-corrosion and hydrogen grooving. To avoid this corrosion, the inlet tube should be located away from the wall, near the center of the roof, or a stainless-steel inlet dip tube should be used to route the incoming acid away from the wall.

Hydrogen grooving and erosion-corrosion on the sidewalls beneath acid inlets have caused catastrophic failures of storage tanks. For example, a 3,000-ton storage tank in Canada containing 2,800 tons of 93% acid ruptured suddenly.[14] The damage caused was extensive due to the mass of the acid released but, fortunately, in this incident there were no fatalities. As acid tanks have become larger, the danger of this type of incident has increased, and prevention of localized damage and monitoring to ensure sound tanks has become normal procedure for all storage facilities.

Hydrogen grooving has caused failures in acid piping, such as in the top section of a 3-in (7.6-cm) diameter carbon-steel pipe elbow in contact with 93% H_2SO_4, Figures 8.10 and 8.11.[15–17] The corrosion occurred during periods when the acid was not being pumped; that is, it was static in the line. Note the deep groove at the 12-o'clock position of the pipe. Eventually the hydrogen grooving caused a hole and failure of the pipeline. Stagnant conditions, often combined with solar heating, promote hydrogen

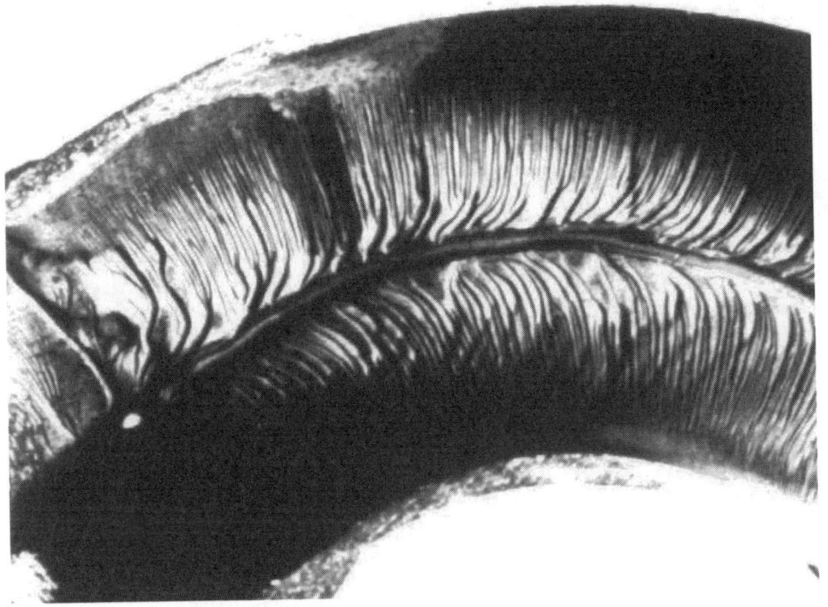

Figure 8.10 Hydrogen Grooving in Pipe

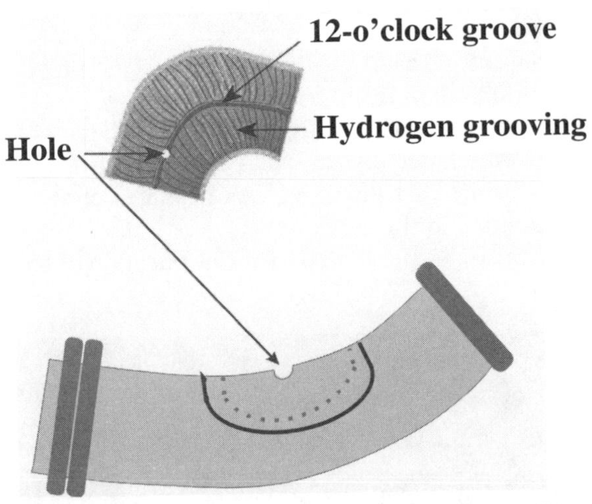

Figure 8.11 Sketch of Hydrogen Grooving in Pipe

grooving in piping. A minimum flow velocity of 1 ft/s (0.3 m/s) is often recommended to prevent this attack.

At ambient temperatures, often the most economical and practical remedy to prevent hydrogen grooving of piping less than 6 in (15 cm) in diameter is to use type 300-series stainless steel (SS) as the material of construction; Type 316 SS (S31600) is usually recommended for 93% H_2SO_4 concentration; alloy 20 (N08020) generally is used at H_2SO_4 concentrations below 93%. Either type 304 (S30400) or 316 SS is suitable at concentrations >93%. Low-carbon grades, such as type 304L (S30403) and

316L (S31603), are not required because concentrated H_2SO_4 does not cause intergranular corrosion at ambient temperatures.

Hydrogen Blistering

Hydrogen blistering may occur in steel plate or pipe that contains significant manganese sulfide (MnS) inclusions. This phenomenon is most likely to occur when there are minor dilution effects of the acid, e.g., at an area where the liquid level changes in storage tanks (in which there may be ingress of atmospheric moisture). Atomic (nascent) hydrogen results from corrosion, enters the steel, and dimerizes at the nonmetallic inclusions to form blisters, Figure 8.12.[12]

One company reports that most of the API 650[18] tanks it inspected had blisters 2 to 12 inches (5 to 30 cm) in diameter (many of which had ruptured and suffered accelerated corrosion) on floor plates and sidewalls.[19]

Blisters in tanks are largely ignored, and no failures have been reported. Blisters may be vented by drilling a small hole from the inside and plugging with a polytetrafluoroethylene (PTFE) rod. (Note: Hydrogen gas becomes *hot* on expansion and may be self-igniting.)

Carbon steels made to fine-grain practice are blister-prone because they contain layers of flattened sulfide inclusions as well as galaxies of alumina inclusions oriented as bands toward the center of the plate. Blisters are seldom found in tanks constructed from semikilled steels because (1) sulfide inclusions in these steels are rounded in shape and (2) there are no alumina inclusions. However, many plate mills now kill all their steels, and ASTM A 285[20] and A 293[21] steels can no longer be depended on to be in suitable condition to resist blistering.

Blistering can be largely avoided by using cleaner steel. Most steel manufacturers can supply common grades (such as ASTM A 516[22]) with low sulfur content (i.e., 0.010% or 0.005% maximum) and also "inclusion shape control," which produces fewer and rounder inclusions in the steel.

Hydrogen blistering can also be prevented by the application of anodic protection (AP).

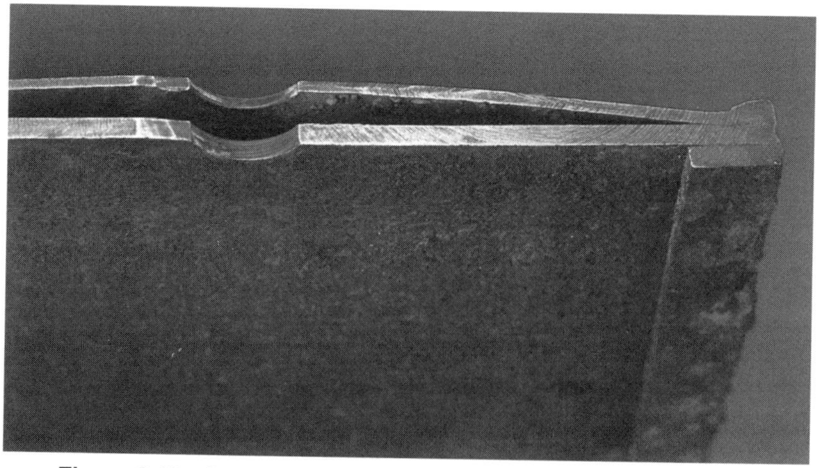

Figure 8.12 Section Through a Hydrogen Blister in a Base Plate of a Sulfuric Acid Storage Tank

Weld Corrosion

Rough welds, particularly those associated with mill-scale and/or slag deposits, may suffer accelerated corrosion. This is probably due to acid retention and subsequent dilution when the liquid level is lowered in a vessel or piping is drained.[23] The problem may be prevented, or at least alleviated, by grinding welds smooth when possible.

Defects associated with welds in acid tanks include the following:[12]

- Residual slag inclusion and porosity—can be leached out in acid service and lead to leaks
- Pitting—weld surfaces are often subject to pitting corrosion
- Lack of penetration of butt welds—corrosion exposing defects and lack of fusion can lead to weld removal
- Galvanic effects—corrosion between the weld and adjacent plate caused by differences in composition (see Figure 7.7)
- Weld cracking—improper joint design, poor workmanship, hydrogen embrittlement, and external forces that were not designed for have caused cracked weld

Brittle Fracture

For service below 40°C (104°F), steels should be selected to avoid brittle fracture. Several large storage tanks have failed by brittle fracture at low ambient temperatures with catastrophic results. Shells of brick-lined towers in contact acid plants have also failed by this mechanism.

Brittle fracture occurs as a result of the combination of a crack-like flaw and near-yield point tensile stresses in a steel that is near or below its nil ductility transition temperature (NDTT), also called ductile-to-brittle transition (DBT) temperature. The latter term is favored by some to avoid confusion of NDT terms with the abbreviation for non-destructive testing.

The DBT of steel can be lowered by increasing the manganese-to-carbon ratio, by austenizing in the steel-making practice and a normalizing heat treatment. ASTM A 516[22] steels have improved DBT, as compared with, for example, ASTM A 515[24] grade, and are commonly employed for ambient-temperature service in temperate climates.

Steel selection depends upon metal thickness, the thicker plates having a higher DBT at the same chemical composition. Plates over 1.5-in (38-mm) thickness are usually normalized. In the case of unrefrigerated tanks, selection is usually based on the lowest one-day mean temperature in the geographical area of concern. The design metal temperature is taken as 8.33°C (15°F) above the lowest one-day mean ambient temperature for a given location, or from another reliable meteorological data source.

The American Petroleum Institute (API) Standard 650[18] provides information concerning minimum permissible design metal temperatures for plates used in tank shells without impact testing, a table of materials groups, and relevant piping/forging specifications. The British Standards Institution (BSI) Standard BS 7910:1999[25] is the current protocol for the evaluation of equipment and has replaced the previous British standard[26] on this topic.

Effect of Contaminants in the Acid

The purer the acid, the higher the corrosion rate; dissolved iron (i.e., ferrous sulfate) retards the corrosion of steel.[5,27] In general, an iron content of less than 25 ppm is considered corrosive.[12] Carbon steel is protected by the formation of a ferrous sulfate film as long as it is neither disturbed nor solubilized. On the other hand, ferric ions are not protective, being capable of reduction to the ferrous state by corrosion processes.

Because alloy materials are becoming more commonly used in the production of sulfuric acid, iron levels in the acid are reduced and modern acid is probably more corrosive to carbon steel tanks, etc., than acid made from older, largely steel plants.

Nitric acid additions in the range 0.5 to 2.0% and > about 8% are inhibitive in 70% sulfuric acid. Below about 1,000 ppm, a purple complex of $Fe(NO)^{++}$ is formed, which is nonprotective. Nitric acid contamination is harmful to steel in the range of 2 to 8%. Nitrate and nitrite ions oxidize the otherwise protective ferrous sulfate film. Increased corrosion then occurs if these ions are not present in sufficient amounts to passivate the steel surface by the formation of a ferric oxide film.

Chloride or fluoride contamination in concentrated sulfuric acid has little effect on corrosion of steel.

Traces of sulfur dioxide have no obvious effect on the corrosion of iron or steel, except indirectly by the reduction of oxidizing agents such as nitrates, nitrites, or chromates.

Cast Irons

Gray Cast Iron

Gray cast irons are more resistant than carbon steel, due to either a more adherent iron sulfate film or electrochemical effects of contained graphite. Temperatures up to 100°C (212°F) are tolerated, unlike in carbon steel. Cast iron has been traditionally used for pumps and serpentine cascade coolers. Cast-iron coolers using air for heat removal have also been used in the past.

Cracking of cast iron, when connected to a stainless-steel cooler protected by anodic protection, has occurred in concentrated acid. The AP increases the oxidizing capacity of the acid, which then acts like oleum (see below) in oxidizing the graphite flakes exposed in the cast-iron matrix. The problem can be prevented by placing a ductile iron spool piece between the gray cast-iron piping and the anodically protected stainless-steel exchanger.

Under static conditions, the corrosion of cast iron is usually less than that of steel. In laboratory tests in unstirred solutions, however, carbon steel was more resistant than cast iron in some acid strengths at ambient temperature (see Table 8.4 below). These were short duration tests, only 2 weeks, so initial surface condition may have been a factor.

Cast iron was about ten times as resistant as steel at 4 ft/s (1.2 m/s) in 92% black plant acid at 40°C (105°F).[28] Hydrogen grooving, however, can occur in piping and tees under static conditions. In 41-day field tests in a pipeline containing 99% acid with 28–54 ppm NO_x and 2–19 ppm SO_2 at 30°C (86°F), cast iron corroded at 18 mpy (0.46 mm/y), while the corrosion rate for steel was 40 mpy (1 mm/y).[29]

The superiority of cast iron to carbon steel under velocity conditions has been investigated and confirmed in laboratory tests, using rotating cylinder electrodes.[30] The reported corrosion rates are unusually high, due either to the use of low-iron-content, analytical-grade acid or to the brevity of the test exposure. Initial corrosion rates tend to be high until the protective film of corrosion products is formed.

Cast iron is not suitable for oleum service, due to attack by sulfur trioxide on the continuous network of graphite flakes and/or silicon-rich constituents, and it can fail with explosive force. Process iron, which has a controlled microstructure consisting of small graphite flakes in isolated clusters, and ductile iron, which has individual graphite nodules, are not affected by AP or oleum in this way.

Ductile Iron

Ductile iron is treated so that the graphite coalesces into spheroidal particles instead of forming flakes. Ductile iron is routinely substituted for cast iron to take advantage of the better ductility and higher margin of safety. The corrosion rate of ductile iron is slightly higher than that for cast iron, but the difference is usually acceptably small. Ductile iron is used in both concentrated acid and oleum. It is free of oleum and AP problems associated with cast iron. As with other cast products, there are possible quality-assurance problems relative to proper alloying and heat treatment. A poor-quality ductile iron poses the same safety problems as cast iron.

Mondi® is a proprietary grade of ductile iron piping developed for strong sulfuric acid (>92% H_2SO_4) piping. It has improved corrosion resistance to 92–99% acid at temperatures up to 149°C (300°F). It has 95% ferrite content so it has excellent performance around anodically protected acid coolers. Piping systems made with this grade of iron operate at higher velocities than standard ductile and gray cast iron. Velocities range from 1.2 to 3.0 m/s (4 to 10 ft/s) for 3-in to 30-in line sizes.[31]

Silicon Cast Irons

By far the most resistant cast iron for strong sulfuric acid service is the 14.5% silicon iron (F47003) that is considerably more resistant than conventional gray cast iron, Figure 8.13.[32]

This material is used to make various cast pumps, valves, tubes, and pipe used in concentrated acid. The siliceous protective film is highly resistant, even to abrasion. This alloy will withstand 100% acid to atmospheric boiling point, usually with corrosion rates of less than 5 mpy (0.12 mm/y). Unfortunately, it is brittle and difficult to machine and weld. New cast pipe (but not pumps or valves) can be welded after preheating to 870°C (1,600°F) using the oxyacetylene process and a flux.[32] These castings cannot, however, be welded after service exposure.[23]

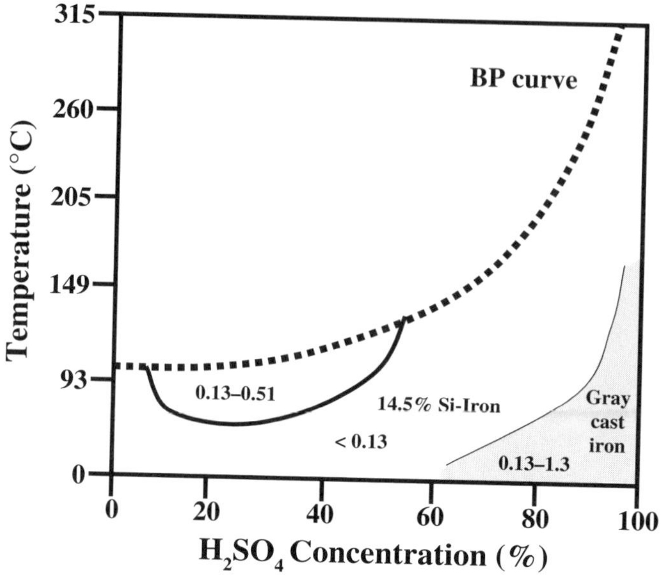

Figure 8.13 Corrosion Behavior in Sulfuric Acid of High-Silicon Cast Iron Compared with Gray Cast Iron

There are three high-silicon iron grades in ASTM A 518:[33]

- Grade 1 is conventional F47003 material with 0.5% molybdenum maximum.
- Grade 2 contains the addition of 3.25 to 5% chromium, with 0.4–0.6% Mo. It can be substituted for Grade 1, having equivalent resistance to sulfuric acid plus resistance to ferric chloride. Grade 2 was developed to resist HCl contaminated with ferric ions; it supersedes an older 3% molybdenum alloy. A proprietary variant of Grade 2 is vacuum-treated to achieve somewhat higher strength level.[32]
- Grade 3 is conventional 14.5% silicon iron with molybdenum limited to 0.2% maximum.

There is also a proprietary grade of slightly higher silicon content (15–16%) available for shaft sleeves and pumps. In Europe, other national specifications for silicon cast irons are used.[34]

In concentrated sulfuric acid, corrosion resistance can be adversely affected by sulfur trioxide, sulfur dioxide, and fluorides. These silicon cast irons are unsuitable for oleum service because of sulfur trioxide attack. The effect is to oxidize silicon particles in the iron to produce silicon dioxide that has a bigger volume than the silicon. This can cause sudden cracking of silicon-containing irons in fuming acid.[35] A similar effect can also occur in gray cast iron containing silicon.

The lower-silicon-containing Durcomet® 5 with 4–6% silicon has a corrosion rate of <1 mpy (<0.025 mm/y) in 95% acid at 90°C (194°F) and <20 mpy (<0.50 mm/y) at up to 110°C (230°F).[36] Test alloys with different silicon contents were tested in 95% sulfuric acid at 90°C (194°F) and corrosion rates were found to vary from 12.3 mpy

(0.31 mm/y) at 4.3% Si to 19.3 mpy (0.49 mm/y) at 4.8% Si to 0.8 mpy (0.02 mm/y) at 5.2% silicon.[37]

Silicon cast irons must be protected from thermal and/or mechanical shock because of their low ductility and mechanical strength.

Nickel Cast Irons

The austenitic nickel cast irons, Ni-Resist®, such as Grade 1 (F41000) and Grade 3 (F43000), offer little advantage over gray cast iron in sulfuric acid, although laboratory data suggest useful resistance in 15% oleum. These alloy cast irons are suitable in abrasive service in oleum at ambient temperature. A 30% nickel alloy (F41004) is superior to the 15% nickel grade (F41000), with a rate of the order of 4 mpy (0.1 mm/y) at 260°C (500°F).[38] In plant practice, F41000 has been used to pump strong acid at temperatures up to 38°C (100°F). The corrosion resistance of some nickel cast irons (Ni-Resist®) has been compared with other ferrous alloys in hot 2–7% oleum, Table 8.1.[38] These data show that this range of oleum strengths is very aggressive to all of these materials and that the 300-series stainless steels are preferred.

Table 8.1 Corrosion Rates of Various Ferrous Materials in 100.5–101.5% H_2SO_4 (2–7% Oleum) at 149–163°C (300–325°F)

Alloy	Corrosion Rate mm/y (mpy)
302 SS (S30200)	0.05 (2)
304 SS (S30400)	0.15 (6)
Ductile Iron—Annealed	0.33 (13)
Ductile Iron—As Cast	0.53 (21)
Carbon Steel	0.58 (23)
Gray Cast Iron	0.89 (35)
Ni-Resist® Type 3	1.85 (73)
Ni-Resist® Type 2	1.98 (78)
Ni-Resist® TypeD2	2.13 (84)

Stainless Steels

All stainless steels can be classified into three groups according to metallurgical structure and response to heat treatment. These are the martensitic, ferritic, and austenitic groups. Further subdivisions include duplex alloys with austenitic/ferritic microstructures and precipitation-hardening (PH) grades strengthened by an age-hardening treatment.

Stainless steels are commonly used in concentrated sulfuric acid, with or without anodic protection depending on the application.

Any cross-contamination between strong sulfuric acid and water (as in shell and tube exchangers) will result in rapid attack by the hot dilute acid formed by the exothermic reaction. To prevent through-wall penetration of exchanger tubes by pitting or SCC from the water side, proper design is required to prevent deposits and to minimize wall temperature. In high-chloride waters, high-performance stainless steels may be required.

Surface contamination with chlorides prior to service (e.g., piping shipped or stored under marine atmospheres) will result in formation of HCl *in situ* upon exposure to sulfuric acid. A preliminary wash with potable or other low-chloride water is mandatory under these conditions. Stainless-steel equipment under AP will tolerate some chloride contamination, probably up to about 250 ppm.

Ferritic Grades

Ferritic stainless steels show a transition from ductile to brittle behavior over a narrow temperature band. This transition can occur above room temperature in steels with high levels of carbon, nitrogen, or chromium. This effect, combined with a tendency to sensitize from heating, limited the usefulness of the early ferritic stainless steels and restricted them to thin sections. The modern low-carbon, low-nitrogen grades (with or without stabilizing additions) have limited toughness and are still usually restricted to thin sections. The ferritic stainless steels are resistant and sometimes immune to chloride stress corrosion cracking. This type of alloy can be subject to 475°C (887°F) embrittlement caused by precipitation of a chromium-rich phase. The ferritic grades are also particularly prone to σ-phase precipitation because of their high chromium and molybdenum contents. Both types of embrittlement can be removed by heating and rapid cooling.

The standard ferritic grades of stainless steel, such as 409–430 (S40900–S40300) and 434 and 444 (S43400 and S44400), contain 11–17% chromium with carbon and nitrogen levels kept low to avoid embrittlement. These grades of stainless steels find little application in strong sulfuric acid. The more highly alloyed superferritic stainless steels do find some uses in sulfuric acid; these will be discussed later in the section on high-performance alloys.

Duplex Stainless Steels

Duplex stainless steels have a controlled balance between austenite- and ferrite-bearing constituents. The duplex structure contains approximately 50/50 austenite and ferrite phases, resulting in higher strength as compared to the 18–8 grades, as well as improved corrosion resistance in some environments (e.g., chloride-bearing aqueous solutions). The original duplex stainless steels were typified by type 329 (S32900) with 26% Cr, 4.5% Ni, and 1.5% Mo. Modern duplex or superduplex grades typically contain nitrogen that permits the use of this type of steel, often without the need for heat treatment.

Grades such as 2205 (S31803), 2304 (S32304), and 2507 (S32750) find some applications in weak and occasionally in concentrated sulfuric acid, in which they have reasonable resistance at moderate temperatures, Figure 8.14.[39–41] This figure shows the isocorrosion curves at 0.1 mm/y (3.9 mpy) for 2205 (S31803), 2507 (S32750), and Zeron 100® (S32760) in comparison with some austenitic alloys in naturally aerated 70–100% sulfuric acid.

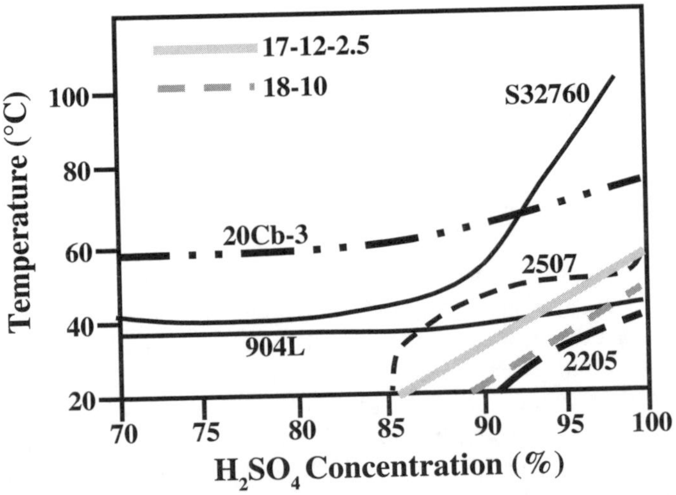

Figure 8.14 Isocorrosion Curves at 0.1 mm/y (3.9 mpy) of Various Duplex and Austenitic Alloys in 70–100% H_2SO_4

There are also a number of proprietary grades, including 7-Mo PLUS® (S32950) and Ferralium 255® (S32550), that have good resistance to concentrated sulfuric acid, e.g., in 98% acid up to about 150°C (302°F).

Austenitic Stainless Steels (Based on 18Cr, 8Ni)

The austenitic stainless steels constitute a large, diverse body of alloys developed from the original 18%-chromium, 8%-nickel stainless steel.

This basic austenitic stainless steel, e.g., type 302 (S30200), has a high carbon content that precludes its use because of excessive corrosion associated with sensitization. Fortunately, suitable grades of this type of stainless steel are available in both wrought and cast forms.

The standard commercial grade is type 304 (S30400) with approximately 18% Cr, 8% Ni, <0.08% C. Welding can impair its corrosion resistance due to sensitization, especially in the heat-affected zone (HAZ). Stabilized versions of type 304 (S30400) are available. Additions of titanium or niobium (also known as columbium) in amounts equal to 5 or 10 times the carbon content in types 321 (S32100) and 347 (S34700), respectively, protect against IGA by precipitating carbon as titanium or niobium carbides instead of chromium carbides, thus preventing chromium depletion.

A molybdenum-bearing variant is type 316 (S31600), containing about 17% Cr, 12% Ni, and 2–3 % Mo. The molybdenum addition improves corrosion resistance in many environments, e.g., in chloride-containing solutions. There is also a stabilized version of the molybdenum-containing grade that is commonly used in Europe and is becoming more common in North America. This grade, 316Ti (S31635), is a high-carbon stainless steel with the carbon stabilized by the addition of a titanium addition equal to at least five times the carbon content. Its main application is in situations where it is exposed to temperatures between 550 and 800°C (1,022 and 1,471°F) for prolonged periods.

Another standard stainless steel with even higher molybdenum content is type 317 (S31700). This stainless steel also has slightly higher levels of chromium and nickel and has improved corrosion resistance in many environments, including weak and intermediate-strength sulfuric acid.

The more usual way to guard against carbide effects is to use low-carbon L grades with <0.03% carbon. These low-carbon grades, 304L (S30403), 316L (S31603), and 317L (S31703), have largely replaced the original higher-carbon and stabilized grades except for very specific applications.

Modern steel-making practice has made it possible to produce L-grade stainless steels with controlled nitrogen additions that have the mechanical properties of the equivalent high-carbon, non-L grade. Dual-grade stainless steels, 304/304L and 316/316L, are now available and permitted for use in many applications. ASME has ruled that dual-grade steel may use the straight non-L-grade allowable stresses for all forms up to 540°C. If this trend continues, the incidence and likelihood of intergranular attack from carbide precipitation will diminish, and demand for 321/347 will probably decrease further. Improved steel-making control also means that 316 can be made more cheaply with the molybdenum content at the low end of the permitted range of 2–3%. This can have a serious effect on the corrosion resistance of this type of steel. In applications where molybdenum is a key factor in corrosion resistance, a minimum level should be specified or type 317 or 317L, with 3–4% molybdenum, should be used.[42]

Type 304L (S30403) and 316L (S31603) are the commonly used stainless steels for service in concentrated sulfuric acids. The regular-carbon grades (S30400 and S31600) are sometimes used for threaded and screwed connections (but only in small-diameter pipe and fittings where welded construction is impractical), while the low-carbon or stabilized grades, types 321 and 347 (S32100 and S34700), are used in the preferred welded construction. This is a precautionary concept, as sensitization and intergranular corrosion attack (IGA) are not usually a problem in strong acid or oleum. IGA can occur, however, if there is inadvertent dilution.

Types 304 and 304L (S30400 and S30403) and the molybdenum-bearing grades, types 316 and 316L (S31600 and S31603), are generally resistant to strong sulfuric acid at ambient temperature and, unlike carbon steel and cast iron, resist erosion-corrosion. Corrosion resistance of these grades is good up to 18–20 ft/s (about 6 m/s) in the absence of abrasive particles (although normal design would be limited to about 6–8 ft/s [2–3 m/s]). They also resist oleum; see Table 8.7 below.

Increased velocity and temperature usually increase corrosion rates of ferrous alloys in 93.2% acid, Table 8.2.[1] These data show that cast iron is relatively insensitive to velocity, while type 316 (S31600) and alloy 20 (N08020) are not corroded in the range of velocities and temperatures tested. Type 304 (S30400) is strongly corroded at the highest velocity (35 ft/s; 10.7 m/s) and temperature (79°C; 175°F). Similar data for 99.3% acid are shown in Table 8.3. In this stronger acid, type 316 (S31600) and alloy 20 (N08020) were corroded at the most severe test conditions used (35 ft/s; 2.1 m/s and 102°C; 215°F). Carbon steel and ductile iron were badly attacked under all test conditions.

Common iron-based alloys were exposed to a range of acid concentrations in unstirred, short-duration (2 weeks) laboratory tests at ambient temperature. These tests showed that only alloy 20 (N08020) was resistant to the weaker acid strengths (down to 73%) even at room temperature. All of the alloys tested were acceptable in

Table 8.2 Effect of Velocity on Corrosion Rate mpy (mm/y) of Ferrous Alloys in 93.2% Sulfuric Acid

	5 ft/s (1.5 m/s)		15 ft/s (4.5 m/s)		35 ft/s (10.7 m/s)	
Alloy	46°C (115°F)	79°C (175°F)	46°C (115°F)	79°C (175°F)	46°C (115°F)	79°C (175°F)
Carbon steel	1,000 (25)	—	>2,000 (>50)	—	—	—
Ductile iron	100 (2.5)	—	120 (3)	—	—	1,500 (37.5)
Cast iron	65 (1.6)	120 (3)	70 (1.8)	120 (3)	90 (2.3)	120 (3)
Type 304	<1 (0.03)	4 (0.1)	<1 (0.03)	6 (0.15)	<1 (0.03)	75 (1.88)
Type 316	<1 (0.03)	3 (0.8)	<1 (0.03)	4 (0.1)	<1 (0.03)	4 (0.1)
Alloy 20	<1 (0.03)	2 (0.05)	<1 (0.03)	3 (0.08)	<1 (0.03)	3 (0.8)

Table 8.3 Effect of Velocity on Corrosion Rate mpy (mm/y) of Ferrous Alloys in 99.3% Sulfuric Acid

	5 ft/s (1.5 m/s)		15 ft/s (4.5 m/s)		35 ft/s (10.7 m/s)	
Alloy	79°C (175°F)	102°C (215°F)	79°C (175°F)	102°C (215°F)	79°C (175°F)	102°C (215°F)
Ductile iron	110 (2.75)	100 (2.5)	150 (3.75)	140 (3.5)	180 (4.5)	180 (4.5)
Cast iron	12 (0.3)	14 (0.35)	16 (0.4)	16 (0.4)	23 (0.58)	23 (0.58)
Type 304	6 (0.15)	7 (0.18)	6 (0.15)	7 (0.18)	6 (0.15)	8 (0.2)
Type 316	2 (0.05)	45 (1.13)	2 (0.05)	50 (1.25)	3 (0.08)	120 (3)
Alloy 20	2 (0.05)	22 (0.55)	2 (0.05)	30 (0.75)	2 (0.05)	90 (2.25)

the stronger acid concentrations, Table 8.4.[43] The data for 93% and 98% are very similar to ones generated in plant trials on pipelines.

The upper temperature limit for stable passivity of type 304 (S30403) is about 40°C (104°F) in 93% acid, 70°C (158°F) in 98.5% acid, and over 100°C (212°F) in concentrations above 99% in most applications.[17]

Table 8.4 Corrosion Rates in mpy (mm/y) of Ferrous Alloys in Sulfuric Acid at 25°C (77°F)

% H_2SO_4	Carbon steel	Cast iron	Type 304	Type 316	Alloy 20
73	26 (0.7)	9 (0.2)	67 (1.7)	5 (0.1)	<1 (0.03)
78	6 (0.2)	15 (0.4)	13 (0.3)	25 (0.6)	<1 (0.03)
85	19 (0.5)	20 (0.5)	2 (0.1)	8 (0.2)	<1 (0.03)
93	25 (0.6)	10 (0.3)	4 (0.1)	<1 (0.03)	<1 (0.03)
98	1 (0.03)	2 (0.1)	<1 (0.03)	<1 (0.03)	<1 (0.03)

There is a well-defined area in 97–99% acid where the 5 mpy (0.12 mm/y) isocorrosion lines reappear at higher temperatures, permitting operation in the 100 to 200°C (212 to 392°F) range under carefully controlled conditions.[44] Type 316 is somewhat less resistant in these higher concentrations. This high-temperature resistance is observed also in stainless steels other than type 304 (S30403), Figure 5.5. An empirical equation has been derived for a Corrosion Index (CI) relating to this resistance, where:

$$CI = 0.35\,(Fe + Mn) + 0.70\,Cr + 0.30\,Ni - 0.12\,Mo$$

Alloys with a CI greater than 39 have the best possibility of resistance.

The effect of nickel content on the corrosion resistance of stainless steels with 18% chromium and 2.5% molybdenum was studied in 97% sulfuric acid. It was found that the corrosion rate decreased as nickel content increased below 60°C (140°F), but that at higher temperatures the trend reversed and increasing nickel content increased the corrosion rate, Figure 8.15. (Corrosion rates in this figure are in g/m^2h, which is approximately equal to mm/y; accurate conversion is not possible because of the changing nickel content). This change in behavior as temperature increased was explained in terms of the formation of water on the passivated surfaces of nickel-containing steels leading to dissolution of the passive film at the higher temperature.[45]

Freshly mixed C.P. sulfuric acid is autopassivating to both types 304 (S30400) and 316 (S31600) above 90% concentration at 38°C (100°F), even with a nitrogen purge.[38] Aerated acid at 50°C (122°F) is resisted by type 316 (S31600), but type 304 (S30400) requires at least 93% concentration. In the 90–92% range, type 316 (S31600) is preferred because type 304 (S30400) could become active and not repassivate. In potentiodynamic scanning tests, the potential of type 316 (S31600) cycles between the active and passive states in 93% acid at 70°C (158°F) to produce low to moderate *average* corrosion rates.[46]

Type 304L (S30403) is more resistant than type 316L (S31603) in concentrations above 94% acid up to about 50°C (122°F) in the 2 ft/s (0.6 m/s) velocity range.[11] The

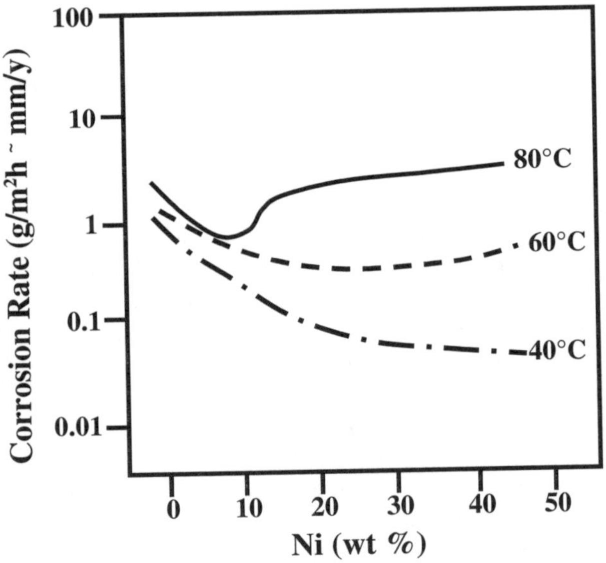

Figure 8.15 Effect of Nickel Content on the Corrosion Rate of an 18% Cr, x% Ni, 2.5% Mo Steel in 97% Sulfuric Acid at Different Temperatures

advantage of type 304 (S30400) is not maintained at 70°C (158°F), where type 316 (S31600) is superior in some laboratory tests for reasons that were not determined. Some investigators have independently confirmed the oscillation between active and passive states and report about three orders of magnitude difference between rates in these states, e.g., from <0.01 to >100 mpy (<0.03 to >2.5 mm/y) at 70°C (158°F). Acid velocity in the 1.5 to 18 ft/s (0.5 to 6 m/s) range did not affect passivity but did influence frequency of oscillation.[47] Chloride contamination of 100 ppm also increased the frequency of oscillation. These investigators report type 316 (S31600) and alloy 255 (S32550) to be superior to type 304 (S30400), suggesting an absence of oxidizing species in their laboratory tests (because field experience shows the converse to be true at about 93% acid concentration).

In 93% acid contaminated with sulfur dioxide, type 316 (S31600) had a corrosion rate of 5 mpy (0.12 mm/y) at 25°C (77°F), but an average of 50 mpy (1.2 mm/y) at 75°C (167°F), going active briefly every few minutes.[38]

From a practical standpoint, stainless steels should not be used under conditions in which they are on the active/passive boundary. Corrosion rates can be very severe in this region.

Both types 304 and 316 show breakaway corrosion under hot-wall conditions in 93% acid at slightly above 60°C (140°F), although control coupons (unheated specimens) are resistant to 75°C (167°F). Care must be taken in heat-tracing strong acid lines to prevent freezing. Self-limiting electric tracing or "stand-off" low-pressure steam tracing must be used to avoid localized hot spots.

Type 304 (S30400), type 316 (S31600), and alloy 20 Cb-3 (N08020) suffered general corrosion in 99% sulfuric acid at room temperature. Increasing the temperature to 74°C (165°F) resulted in intergranular attack as well as general corrosion in all three alloys. Further increase in temperature to 127°C (261°F) resulted in pitting corrosion together with unacceptably high general corrosion rates.[48]

Data from tests in 98.7% acid at 100°C (212°F) showed that type 304 (S30400) was considerably more resistant than type 316 (S31600) or alloy 904L (N080904), Table 8.5.[49]

In higher-strength acid (>99%), higher-chromium alloys are generally superior to the standard 304 (S30400) and 316 (S31600) stainless steels, Table 8.6.[17] The nickel-rich alloy 825 (N08825), however, was less resistant even than type 304 (S30400), in spite of its high chromium content. These data were generated from tests in 99.8% acid at 194°C (380°F) in a pump tank and in 99.0% acid at 172°C (340°F) below the distributor.

A modified form of 310S (S31008) has been designated 310M and is being used as a proprietary alloy in strong sulfuric acid service. At a temperature of 180°F (82°C), this stainless steel has a corrosion rate of 2.5 mpy (0.064 mm/y) in 97% acid, approximately 1 mpy (0.025 mm/y) in 98%, and 0.5 mpy (0.013 mm/y) in 99% sulfuric acid.[50]

At ambient temperature, all stainless-steel grades are suitable for oleum service. At higher temperatures, e.g., >90°C (194°F), the molybdenum-free grades are better than the molybdenum-containing grades. Type 316 (S31600) is at least an order of magnitude less resistant, Table 8.7.[43] This table shows data obtained from tests in a line from a steam still outlet to a preheater. Type 316 (S31600) has failed in oleum reboilers and type 316 (S31600) rivets, inadvertently used in a type 304 (S30400) spiral heat exchanger, suffered accelerated attack.

Below 14% oleum, flow rates may increase corrosion, even at moderate temperatures, e.g., 60°C (140°F). These problems may be corrected by proper design, reducing velocity, and impingement at susceptible points. Higher alloys, such as type 309 (S30900), have been used for tubes in reboilers.

However, types 309 (S30900) and 310 (S31000) are not generally available as piping, except in mill lots.

Table 8.5 Corrosion Rate of Stainless Steels in Flowing 98.7% Sulfuric Acid at 100°C (212°F)

Alloy	mpy (mm/y)
Type 304	19.7 (0.5)
Type 316	135 (3.44)
Type 904L	90.6 (2.3)

Table 8.6 Corrosion Rate mpy (mm/y) of Various Alloys in a Sulfuric Acid Pilot Plant at 172–194°C (340–380°F)

Material	Pump Tank	Below Distributor
Type 304	2 (0.1)	2 (0.1)
Type 316L	6 (0.2)	5 (0.1)
Type 317L	5 (0.1)	16 (0.4)
Type 309S	<1 (0.03)	<1 (0.03)
Type 310S	<1 (0.03)	<1 (0.03)
Alloy 255	<1 (0.03)	<1 (0.03)
Alloy 825	4 (0.03)	21 (0.5)

Table 8.7 Corrosion Rates in mpy (mm/y) of Various Stainless Steels in 10–12% Oleum at 160°C (320°F)

Alloy	mpy (mm/y)
Alloy 26-1	0.4 (0.01)
Type 309	0.6 (0.02)
Type 304	1.4 (0.04)
Type 316	20 (0.5)
Alloy 20Cb-3	80 (2)

Intergranular Attack (IGA) Normally, there is no intergranular corrosion attack of sensitized stainless steel at and above about 78% sulfuric acid.[51] However, this can be caused by oxidizing contaminants. The low-carbon or L-grades (S30403 and S31603) are preferred as insurance against dilution or contamination. In piping particularly, the L-grades demand little or no additional cost over regular-carbon grades.

Anodic Protection (AP) Anodic Protection (see Anodic Protection [AP], Chapter 7) is effective for type 304 (S30400) and 316 (S31600) and other stainless steels and can extend the temperature of application in all strong acid solutions. It can ensure that the stainless steel remains in a stable, passive condition, effectively noncorroding, at temperatures and acid strengths at which it would otherwise be in an active or active/passive corroding condition. The electrical-current requirements to apply AP increase as acid strength decreases or acid temperature increases.

In laboratory tests, a rate of 29 mpy (0.75 mm/y) for freely corroding type 316 (S31600) in 93% acid (plus 4 ppm iron and 800 ppm SO_2) at 70°C (158°F) was reduced to less than 2 mpy (0.05 mm/y) by AP.[52]

At 100°C (212°F), the corrosion rate is reduced from about 200 mpy (5.0 mm/y) to less than 2 mpy (0.05 mm/y).[46] Type 316L (S31603) shell and tube coolers have been routinely protected by AP. The limiting temperature, however, for unprotected coolers is about 75°C (167°F), while anodically protected coolers are used as high as 125°C (257°F).[53]

Silicon-Containing Stainless Steels

Chromium is an effective alloying element in strongly oxidizing environments such as concentrated sulfuric acid and nitric acid. The other alloying element that improves corrosion resistance in these environments is silicon.

Low levels of silicon (and phosphorus) segregate in the grain boundaries of stainless steels and promote intergranular attack. This preferential attack is produced because the segregates render the grain boundary anodic to the bulk structure. However, as the silicon content is increased to 3 or 4%, the concentration gradient disappears and silicon is uniformly distributed throughout the structure. Although this increased silicon increases the rate of dissolution attack, it is now spread throughout the structure and not concentrated at the grain boundaries. This type of alloy with high chromium, nickel, and silicon is very resistant to transpassive corrosion. Very low carbon levels are needed to prevent chromium carbide precipitation, which is even more detrimental under transpassive conditions than in passive regions.[54]

High-silicon austenitic stainless steels were originally developed for use in concentrated nitric acid. Interestingly, substantially equivalent compositions (e.g., alloys 1815LCSi and A610, both designated S30600) are reported to have different active/passive transition temperatures in 93–98% sulfuric acid, the very low-molybdenum alloy A610 being more resistant.[11] Alloy 1815LCSi (S30600), with up to 0.2% molybdenum, was more easily activated.

More recently, alloys of approximately 18% Cr, 16% Ni, and 5–6% Si have been produced for use in concentrated sulfuric acid. Alloys with nominally 5.5% Si include ANTINIT A614® (S32615; Böhler), SX® alloy (S32615; Edmeston), the SARAMET® alloys (Chemetics), and a 6% silicon alloy, ZeCor® (S38815; EnviroChem).

There is also Cronifer® 2509Si7 (S70003; Thyssen Krupp VDM), which has 7% silicon, the highest silicon content in iron-based alloys.[55] It is not readily available and is only used in thin sections, e.g., plate heat exchangers, or as cladding. The corrosion rate is <0.1 mm/y (<4 mpy) in 96% sulfuric acid up to about 150°C (302°F) or in 98.5% sulfuric up to about 200°C (392°F). It is also able to withstand short-term excursions to lower acid concentrations or higher temperatures without suffering major corrosion.[56] Welding requires low heat input and solution annealing of the fabricated equipment for some applications.

Fabrication of all of these high-silicon stainless steels by fusion welding is difficult and requires exacting controls. Improper welding techniques can adversely affect corrosion resistance and toughness by causing precipitation of carbides and silicon-rich π phase. In some cases, solution heat treatment and water quench of the equipment after fabrication has been used to avoid possible intergranular corrosion problems.

A recent study found that when sufficient silicon was added to an alloy (based on 310 [S31000]) to achieve passivity and provide satisfactory corrosion resistance in the most corrosive conditions, the alloys were extremely brittle. Further alloying with nickel, copper, and/or molybdenum was beneficial, providing good corrosion resistance, with alloys with a silicon level that was low enough to provide good ductility after heat treatment. The base alloy 310 (S31000) did not passivate in 98% sulfuric acid at 160°C (320°F) but did passivate under these test conditions if 6% silicon and 3% copper were added to the alloy.[57]

Most of these alloys were developed as cast products, intended for service in concentrated oxidizing acids, such as strong nitric or sulfuric contaminated with oxidants (e.g., nitrates) or reducing agents (e.g., hydrogen sulfide). Such products offer extended service life, higher temperature possibilities, and increased resistance to erosion-corrosion or abrasion. Although historically developed for valves and pumps, the modern silicon-rich varieties are provided in wrought form, such as plate or pipe.

Many of these specialty alloys are proprietary and have trademarked names; some do not have UNS numbers. Of the high-iron alloys, the 5% silicon materials have largely replaced other alloys that originally offered a better resistance than alloy 20 (CN-7M [N26455]). An example of this type of silicon-containing austenitic alloy is the family of alloys using the name SARAMET®. This alloy has good resistance to concentrated acid and better pitting resistance than 316L (S31603) in chloride solutions. It also has a good response to anodic polarization with a polarized corrosion rate of only 0.26 mpy (<0.01 mm/y) in 98.2% sulfuric acid at 115°C (239°F).[58] The original SARAMET® alloy is now called SARAMET® 23 (S30601) and it has been successfully employed for many years in strong sulfuric acid service in towers, pump tanks, piping, coolers, etc. SARAMET® 21 has a reduced alloy content so its resistance to strong acid is less than the other grades, but it can offer an economic solution with acceptable performance in many applications.

SARAMET® 35 is the latest alloy in the range and offers better performance in weaker acid than previous versions of this alloy, Figure 8.16. It is being used in applications where weak acid might form from process upsets, moisture ingress, or water additions into strong acid. Examples of use of this grade of SARAMET® include gas inlet regions in sulfur burner drying towers and strong acid pump tanks under suction with top-entry dip-tube water dilution systems.[59] These alloys are produced by a special melting and casting technique and then worked to produce the desired wrought forms. SARAMET® 21 will withstand up to 125°C (257°F) in 98% acid. Long-term field corrosion testing has shown corrosion rates of <1 mpy (<0.025 m/y) for all 98% applications at velocities up to 16 ft/s (4.9 m/s).[60]

In tests of four-day duration in mildly agitated 93% sulfuric acid at 105°C (221°F), SARAMET® 35 corroded at 4 mpy (0.10 mm/y) while SARAMET® 23 had a corrosion rate of 200 mpy (5.1 mm/y).[60] In similar tests in 90% acid at the same temperature, the rates were 15 mpy (0.38) and >1,000 mpy (>25 mm/y), respectively. These data show that this latest grade has excellent resistance to the weaker end of the strong acid range.

The SX® alloy (S32615) with 5% silicon is superior to alloy C-276 over a range of acid strengths, as shown by the results of static laboratory tests in Table 8.8.[61] These results also show that sulfur dioxide contamination has little effect on the corrosion rate of this alloy in 98% acid. In dynamic testing it was shown that SX® alloy had zero corrosion in 98% H_2SO_4 at 70°C (158°F) at 25 m/s and also in 98% H_2SO_4 at 10m/s at 115°C (239°F). Data on alloy 33 under similar conditions has been included for comparison.[62]

The corrosion rate of SX® alloy (S32615) versus temperature in 98% sulfuric acid is more resistant than some of the cast alloys traditionally used in this acid, Figure 8.17[1,61,63] and Figure 8.18.[64]

Another similar alloy with slightly higher silicon content is ZeCor® (S38815), which has excellent resistance to strong acid corrosion, Figure 8.19.[65] It should be noted that these two isocorrosion curves (Figures 8.18 and 8.19) are for 1 mpy (0.025 mm/y) corrosion rate rather than the more usual 5 mpy (0.13 mm/y) rate. ZeCor® is being used for towers, pump tanks, piping, distributors, strainers, mesh pads, etc., in sulfuric acid plants.

Silicon is well known to be attacked by acidic fluorides, and these silicon containing alloys can be susceptible to this type of attack (see Fluoride Ion Contamination in Chapter 10).

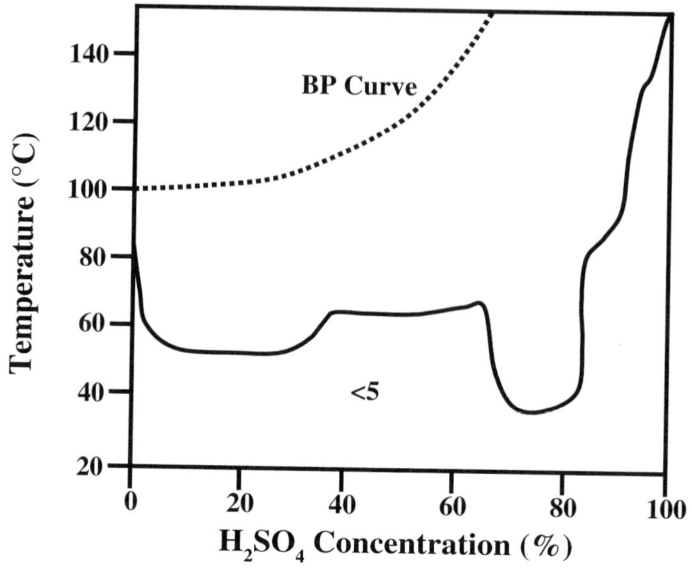

Figure 8.16 Isocorrosion Curve at 5 mpy (0.13 mm/y) for SARAMET® 35

Table 8.8 Corrosion Rates mm/y (mpy) of SX® Alloy and Alloy C-276 in Sulfuric Acid

% H_2SO_4	Temperature °C (°F)	SX® Alloy(S32615)	Alloy C-276 (N10276)	Alloy 33 (R20033)
95	110 (230)	0.1 (3.9)	2.16 (85)	—
96	100 (212)	0.06* (2.4)	1.10 (43)	—
	100 (212)	0.05** (2)	—	—
98	110 (230)	0	0.49 (19)	0.05 (2)
	110 (230) +SO2	0.02 (0.8)	—	—
	130 (266)	0.01 (0.4)	1.43 (56)	0.07 (2.7)
	150 (302)	0.08 (3.1)	—	0.08 (3.1)

* Cast
** Weld metal

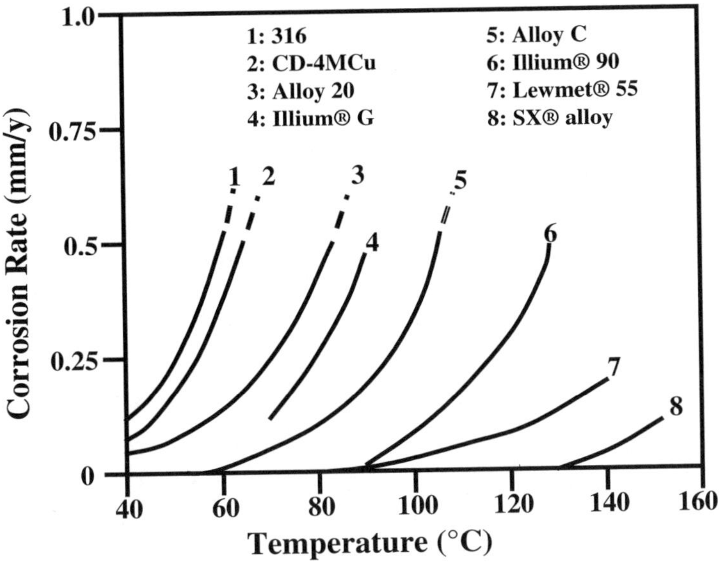

Figure 8.17 Effect of Temperature on the Corrosion Rates in 98% H_2SO_4 of SX® Alloy and Traditional Sulfuric Acid Alloys

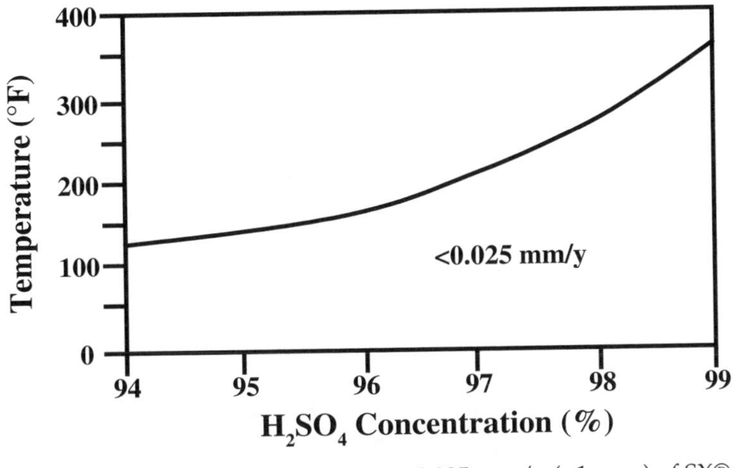

Figure 8.18 Isocorrosion Curve at <0.025 mm/y (<1 mpy) of SX® in Concentrated Sulfuric Acid

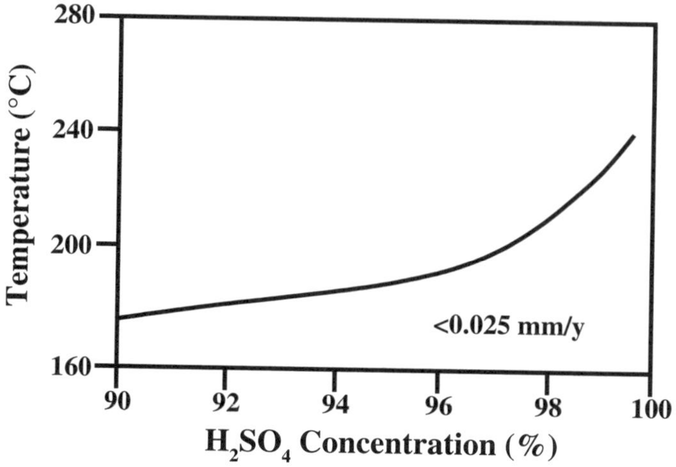

Figure 8.19 Isocorrosion Curve at 0.025 mm/y (1 mpy) of ZeCor® in Concentrated Sulfuric Acid

High-Performance Alloys

There is a group of high-performance alloys, sometimes called corrosion-resistant alloys (CRAs), which consists of both stainless steels (i.e., containing >50% iron) with high chromium and nickel content and alloys with nickel as the predominant alloying element (these are not true nickel-based alloys because they contain <50% nickel). In the development of alloys to resist weak or intermediate strengths of acid (i.e., < 70%), first castings, then wrought materials, were produced of approximately 20% Cr, 29% Ni, 3% Mo, and 3% Cu (CN7M [N08007]) and designated "alloy 20." Modern variants of the original alloy, e.g., alloy 20Cb-3 (N08020), are used in concentrated sulfuric acid primarily to resist velocity or turbulence. A slight increase in acceptable service temperature is demonstrated, compared with the lower alloy type 304 (S30400) stainless steel. The high-alloy grades are equally amenable to AP as are the lower alloyed stainless steels.

The corrosion rate of alloy 20 (N08020) strip increases with temperature in 90% acid, Table 8.9. This table shows that this alloy has good resistance to the higher-strength acid only at moderate temperatures.[66]

Table 8.9 Average Corrosion Rate of Alloy 20 in 90% H_2SO_4 at Different Temperatures

% H_2SO_4	Temp. °C (°F)	Average Corrosion Rate mpy (mm/y)
90	60 (140)	4 (0.102)
	85 (185)	15 (0.381)
	95 (203)	27 (0.686)

There is a series of nickel-rich alloys that fall between the highly alloyed austenitic stainless steels and the nickel-based alloys. Titanium-stabilized alloy 825 (N08825) has substantially the same isocorrosion chart as alloy 20Cb-3 (N08020) in concentrated acid. It is less expensive than alloy 20Cb-3 and has been widely used in the United Kingdom, where alloy 20Cb-3 is used primarily in cast form.

Low-carbon alloy G (N06007) and its variants, G-3 (N06985) and G-30 (N06030), offer little advantage over alloy 20Cb-3 (N08020) or alloy 825 (N08825). They cost more and increase the service-temperature limitation only above 93%, and then only slightly. However, alloy G-3 is substantially unaffected by up to 200 ppm chloride ion contamination in concentrated acid and is less expensive than nickel-chromium-molybdenum alloys, namely alloy C-276 (N10276) and its several variants.[67] Acid velocity increases the corrosion rates of these alloys in concentrated acid, Figure 8.20.[11]

Alloys of 20% Cr, 25% Ni, and 4% Mo variety (e.g., N08320), as well as those with added molybdenum and copper (such as alloy 904L [N08904]), resist concentrated sulfuric acid to about 50°C (122°F). Alloys with copper have somewhat better resistance than copper-free grades.[17] The presence of copper in stainless steels has long been known to improve their resistance to sulfuric acid solutions.[68]

Among the high-performance stainless steels, 6% molybdenum compositions, e.g., alloy AL6XN (N08367), alloy 926 (N08926, e.g., 1925hMo [VDM], 25-6Mo [SMC], and UR B26 [Acelor]), and 254SMO (S31254), are preferred over 4% molybdenum grades for resistance to high-chloride waters.[69] Such compositions are also more resistant to 70–80% sulfuric acid, but their resistance diminishes above that concentration to less than that of type 316L (S31603) and 904L (N08904) at 90% concentration and above, Figure 8.21 (data taken from various sources). Presumably this lack of resistance is

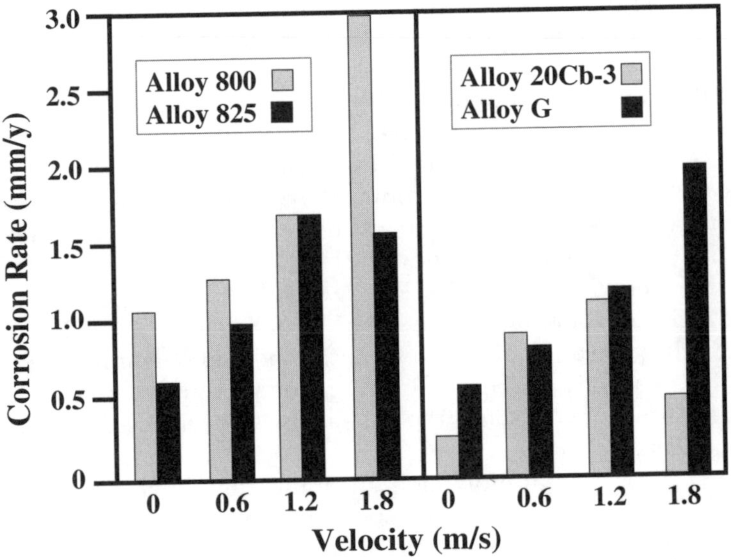

Figure 8.20 Effect of Velocity on High-Performance Alloys in 95% Sulfuric Acid at 50°C (122°F)

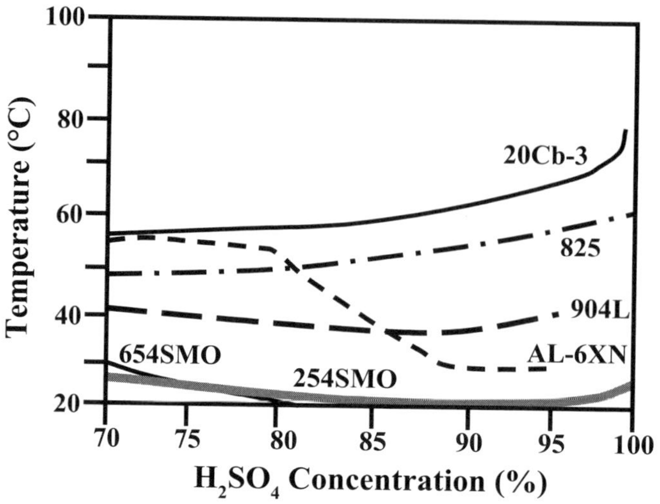

Figure 8.21 Isocorrosion Curves at 0.1 mm/y (3.9 mpy) for Various High-Performance Alloys in Concentrated Sulfuric Acid

due to inadequate chromium content to offset the readily oxidizable molybdenum content. Note that alloy 20Cb-3 (N08020) maintains its resistance above 80% concentration. Note also that the data for similar alloys, e.g., N08367 and S31254, are not in good agreement. This may be due to differences in performance of these alloys or, more likely, to differences in test technique, acid used, etc.

These high-molybdenum stainless steels, however, show good resistance to intermediate-strength sulfuric acid containing chlorides. They are also used in anodically protected coolers using seawater or other high-chloride water as the cooling medium.

There are also now austenitic alloys with even higher chromium and molybdenum contents, such as alloy 654SMO (S32654), alloy 31 (S08031), and Incoloy® alloy 27-7MO (S31277) with 7% molybdenum. This type of alloy was developed for better resistance to aggressive chloride environments, e.g., to resist crevice corrosion in hot seawater. Some of these alloys are even less resistant in concentrated acid, i.e., above about 60%, than 904L (N08904) or alloy 20Cb-3 (N08020), while others are almost as good until acid strength exceeds 90%, Figure 8.21.[70–73]

The two high-molybdenum alloys 25-6MO (alloy 926 [N08926]) and 27-7MO (S31277) are better than alloy 825 (N08825) until the acid strength exceeds about 80% and about 88%, respectively, Figure 8.22.[74] In the more concentrated acid the low-molybdenum 825 is more resistant. Type 316 stainless steel is also included in this figure for comparison.

Corrosion rates of several molybdenum-containing alloys are compared in 95% sulfuric acid in Table 8.10.[75] These data show that the 6 and 7% molybdenum alloys are marginal under these conditions and are inferior to alloy C-276 (N10276).

Other high-performance alloys include so-called superferritic stainless steels such as E-Brite® (S44627) and superduplex stainless steels such as alloy 255 (S32550). The behavior of these alloys is compared with that of some of the 300-series austenitic stainless steels and various other alloys in a range of acid concentrations and temperatures,

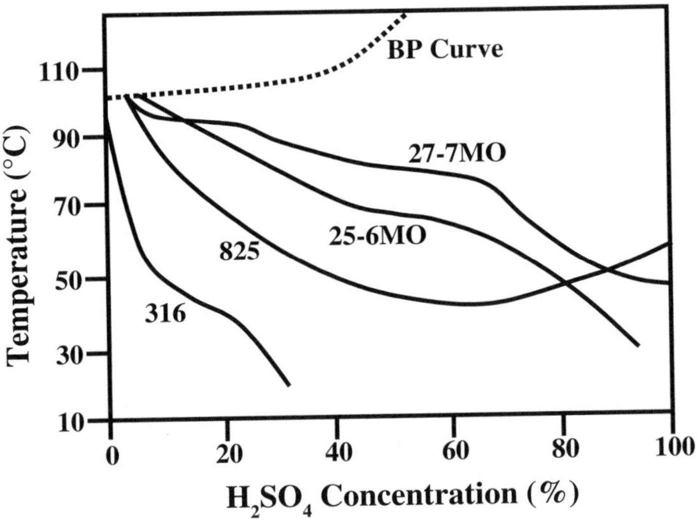

Figure 8.22 Isocorrosion Curve at 5 mpy (0.13 mm/y) for High-Molybdenum Stainless Steels in Sulfuric Acid

Table 8.10 Corrosion Resistance of Various Alloys in 95% Sulfuric Acid at 50°C (122°F)

Material (UNS No.)	Corrosion Rate mpy (mm/y)
27-7MO (S31277)	14 (0.36)
25-6MO (N08926)	18 (0.46)
254SMO (S31254)	26 (0.66)
AL6XN (N08367)	22 (0.56)
Alloy 31 (N08031)*	0.31 (0.008)[76]
Alloy 625 (N06625)	48 (1.22)
Alloy C-276 (N10276)	0.1 (0.0025)

*Tested at 60°C.

Table 8.11.[77] These data show that many of these alloys have good resistance to hot, strong acid and that type 310 (S31000) stainless, alloy 255 (S32550) duplex, and E-Brite® (S44627) and 29-4-2 (S44800) ferritics are particularly resistant over the range of conditions tested.

Alloy 33 (R20033) is not a stainless steel but is an austenitic alloy based on chromium with a nominal analysis of 33% Cr, 32% Fe, 31% Ni, 1.6% Mo, 0.6% Cu, and 0.4% N. It has excellent resistance to concentrated sulfuric acid and oleum at elevated temperatures. The corrosion behavior of alloy 33 has also been demonstrated in various field tests: In a plant stream of 96–98.5% sulfuric acid at a velocity of ≥ 1 m/s and 135–140°C, the corrosion rate of alloy 33 after 14 days was <0.4 mpy (<0.01 mm/y), compared to type 304 at 7.0 mpy (0.18 mm/y), alloy G-30 at 3.1 mpy (0.08 mm/y), and alloy A611 at 1.2 mpy (0.03 mm/y).[78] This alloy has been compared with a

Table 8.11 Corrosion Rate mpy (mm/y) of High-Performance Alloys in 97.65 to 99.2% Sulfuric Acid

	99°C (210°F)		116°C (240°F)		143°C (290°F)	199°C (390°F)	
Material	97.65%	97.65%	98.45%	99.2%	100.00%	99.20%	100.00%
Type 304L	15.3 (0.39)	29.4 (0.75)	12.5 (3.2)	1.0 (0.03)	2.3 (0.06)	3.7 (0.09)	1.9 (0.05)
Type 316L	3.8 (0.10)	31.6 (0.80)	2.9 (0.07)	3.2 (0.08)	3.2 (0.12)	4.6 (0.12)	1.6 (0.04)
Type 317	2.1 (0.05)	22.5 (0.57)	—	6.4 (0.16)	—	3.6 (0.09)	0.8 (0.02)
Type 310	3.3 (0.08)	22.2 (0.56)	0.9 (0.02)	0.7 (0.02)	0	1.1 (0.03)	0.3 (0.01)
Nitronic® 50	2.5 (0.06)	20.1 (0.51)	2.8 (0.07)	2.8 (0.07)	8.6 (0.22)	—	—
Nitronic® 60	3.5 (0.09)	23.3 (0.59)	0.7 (0.02)	1.6 (0.04)	2.5 (0.06)	—	—
Alloy 825	4.0 (0.10)	16.9 (0.43)	4.3 (0.11)	4.3 (0.11)	2.3 (0.06)	3.5 (0.09)	0.7 (0.02)
Alloy 1815	0.70 0.02)	57 (1.45)	0.7 (0.02)	7.7 (0.20)	1.1 (0.03)	5.7 (0.14)	2.8 (0.07)
Alloy 28	3.0 (0.08)	17 (0.43)	5.9 (0.15)	4.2 (0.11)	28.5 (0.72)	—	0.8 (0.02)
Alloy 20Cb-3	—	21.1 (0.54)	4.3 (0.11)	2.1 (0.05)	31.4 (0.80)	5.1 (0.13)	1.3 (0.03)
904L	—	25 (0.64)	14.6 (0.37)	7.5 (0.19)	28.9 (0.73)	—	0.9 (0.02)
Alloy 254 SMO	5.1 (0.13)	24.7 (0.63)	—	7.8 (0.20)	40.2 (1.02)	—	0.8 (0.02)
Alloy G	20.3 (0.52)	23.7 (0.60)	5.6 (0.14)	7.2 (0.18)	37.3 (0.95)	—	3.7 (0.09)
Alloy 255	2.9 (0.07)	4.3 (0.11)	—	1.1 (0.03)	—	1.2 (0.03)	0.3 (0.01)
E-Brite® 26-1	2.3 (0.06)	2.6 (0.07)	0.8 (0.02)	0	0	1.1 (0.03)	0.7 (0.02)
29-4-2	3.1 (0.08)	4.3 (0.11)	0.8 (0.02)	0.4 (0.01)	0	0.9 (0.02)	0

high-silicon austenitic alloy (A611 [S30601]), a high-chromium austenitic (310L [S31050]), and a superferritic (28-4-2) stainless steel in 7-day tests in 98% H_2SO_4 at various temperatures, Table 8.12.[79]

Another austenitic alloy that has good resistance to sulfuric acid is the proprietary MC® alloy from Mitsubishi Metal Corp. This is an austenitic alloy with between 40 and 80% chromium, 5% maximum iron, 5% maximum molybdenum, and the remainder nickel. High price and problems with workability have prevented a broad application of this alloy in the chemical process industry. This alloy has lower corrosion rates in 96% sulfuric acid at 200°C (392°F) than does alloy 33 (R20033), Figure 8.23.[80]

Exposure to concentrated sulfuric acid at 100°C (212°F) and higher temperatures caused a breakdown of the passivation in both alloys. The corrosion rate at 200°C (392°F) decreased for MC® alloy from 3.4 mm/y (133 mpy) at the beginning to 0.4 mm/y (16 mpy) after one month, a corrosion rate still well above an acceptable value

Table 8.12 Corrosion Rate mm/y (mpy) of Alloy 33 and Other Alloys in 98% Sulfuric Acid

Material	100°C (212°F)	125°C (257°F)	150°C (302°F)	175°C (347°F)	200°C (392°F)
Type 310L	0.38 (15)	0.43 (17)	0.98 (39)	0.38 (15)	0.07 (2.8)
Alloy 28-4-2	0.03 (1.2)	0.06 (2.4)	0.53 (21)	0.04 (1.6)	0.07 (2.8)
Alloy A611	0.02 (0.79)	0.36 (14)	0.81 (32)	0.70 (28)	0.61 (24)
Alloy 33	0.04 (1.6)	0.07 (2.8)	0.08 (3.1)	0.16 (6.3)	0.04 (1.6)

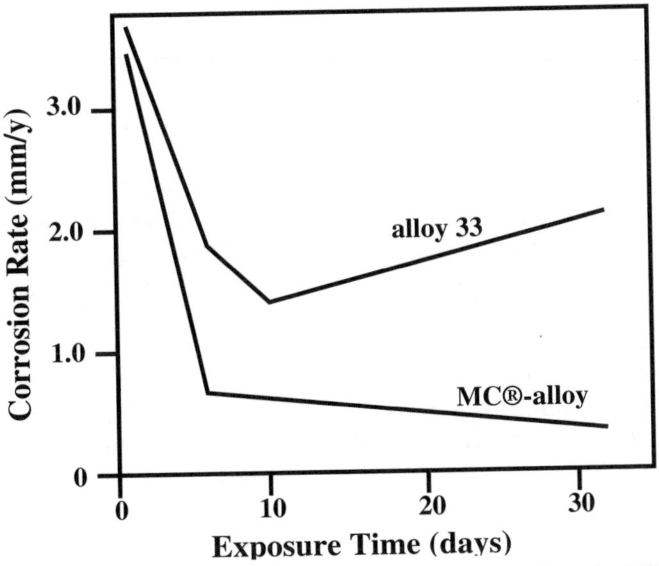

Figure 8.23 Corrosion Rates of Chromium-Based Austenitic Alloys in 96% Sulfuric Acid at 200°C (392°F)

of 0.1 mm/y (4 mpy). Alloy 33 showed a similar but less dramatic decrease in the corrosion rate within the first week, followed by an increase at longer times. These differences in behavior were attributed to changes in composition of the surfaces with time.

Cast Stainless Steels and Related Alloys

The equivalent cast version of types 304 (S30400) and 304L (S30403) are CF-8 (J92600) and CF-3 (J92700), respectively, and they exhibit approximately the same corrosion response as the wrought alloys. However, castings can have surface layers containing more than the maximum allowable carbon content of 0.08%, which can significantly reduce corrosion resistance of the surface.[81]

In the cast form, the difference between CF-8 (J92600) or CF-3 and the corresponding molybdenum-bearing grades CF-8M (J92900) and CF-3M (J92800) is insignificant. CF-3 and CF-8 are not particularly common and manufacturers of cast pumps and valves tend to standardize on CF-8M, which has a broader range of applications. The molybdenum-grade castings are often more available and cheaper than the wrought-grade equivalents and can be used interchangeably for most sulfuric acid duties. Since the cast version of these alloys is unlikely to be welded, there is rarely a justification for specifying the low-carbon grades in this case. This assumes that the valves or pumps, if weld-repaired by the manufacturer, are properly reheat-treated (solution-annealed) to restore optimum corrosion resistance. Again, availability and price are likely to favor the non-L grade and a properly heat-treated casting in CF-8 or CF-8M is likely to be as corrosion resistant as its low-carbon cast or wrought equivalents. The molybdenum grades are used for shutoff valves in strong sulfuric but are not recommended for throttling duties.

The cast form of alloy 20, unstabilized and with lower nickel content, CN-7M (N08007), is favored for throttling valves and centrifugal pumps. A solution anneal is required to maintain chemical resistance and to restore resistance after weld-repair of castings. The cast CN-7M (N08007) is more resistant than the wrought alloy 20Cb-3 (N08020) compositions under velocity conditions, Figure 8.24.[11]

The ferrite phase in CF grades suffers slightly accelerated attack in 93% acid at 80°C (176°F). However, the cast duplex alloy CD-4MCu (J93370) may offer improved resistance in 78–93% acid at 80°C (176°F) despite its 50-50 austenite-ferrite structure.[82] At acid velocities below 6 ft/s (1.8 m/s), alloy CD-4MCu (J93370) is slightly more resistant than CN-7M (N08007); at 6 ft/s (1.8 m/s), the reverse is true, Figure 8.25.[11] There are a number of commercial cast alloys that have good resistance over the whole range of acid strengths at moderate temperatures, Figure 8.26.[83–85]

In the absence of oxidizing contaminants or anodic protection, the limiting temperature for these alloys is about 50°C (122°F) in concentrated acid, but oxidizing contaminants, such as sulfur dioxide, can adversely affect the corrosion resistance of these alloys.[51] Austenitic cast alloys CN-7M and duplex CD-4MCu are substantially

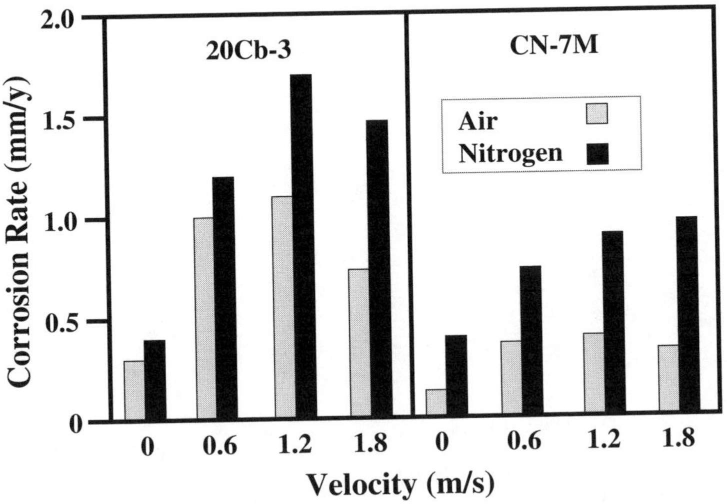

Figure 8.24 Effect of Velocity on Alloy 20Cb-3 and Cast CN-7M in 95% Sulfuric Acid at 50°C (122°F)

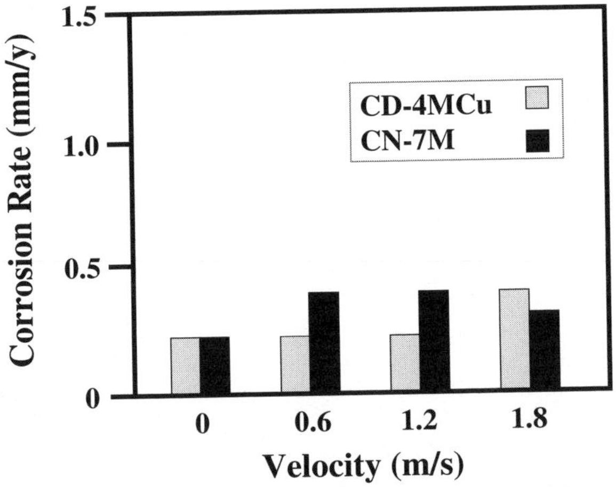

Figure 8.25 Effect of Velocity on Cast CD-4MCu and CN-7M in 95% Sulfuric Acid at 50°C (122°F)

equivalent at this temperature, although not as resistant as the nickel-chromium-molybdenum casting CW-12MW (N30002 [ASTM A 494][86]).[38] However, CN-7M is superior to CD-4MCu in the range of 78–93% acid at 80°C (176°F), although the latter is better than CF-8M. An active/passive cycling is reported in 93% acid at 80°C (176°F), analogous to that observed in type 304 (S30400) grades.[82] Isocorrosion curves

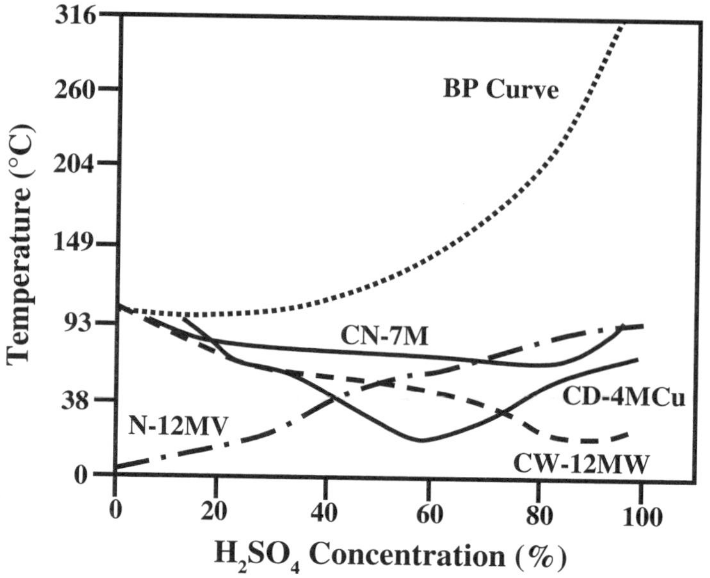

Figure 8.26 Isocorrosion Curves at 5 mpy (0.13 mm/y) of Various Cast Alloys in Sulfuric Acid

for CN-7M and CD-4MCu show that they have similar resistance in concentrated acid, but that CN-7M is superior in intermediate-strength acid, Figures 8.27 and 8.28.[87]

Usually castings are more resistant to erosion-corrosion than their wrought counterparts. An exception has been reported in the case of wrought type 304 versus cast CF-8 in deaerated 95% acid at 50°C (122°F), Figure 8.29.[11] This figure includes data for the effect of testing in an air or nitrogen atmosphere that show the beneficial effect of oxidizing conditions (air) on the corrosion of stainless steel in concentrated sulfuric acid.

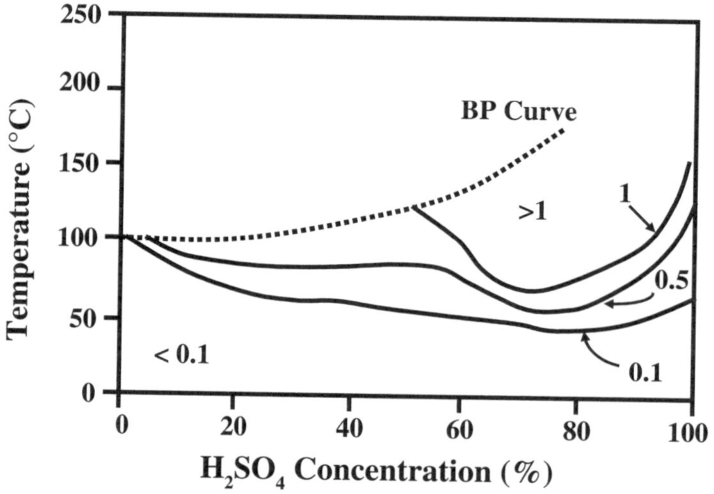

Figure 8.27 Effect of Temperature on the Corrosion Rate (mm/y) of Austenitic Cast Alloy CN-7M in Sulfuric Acid

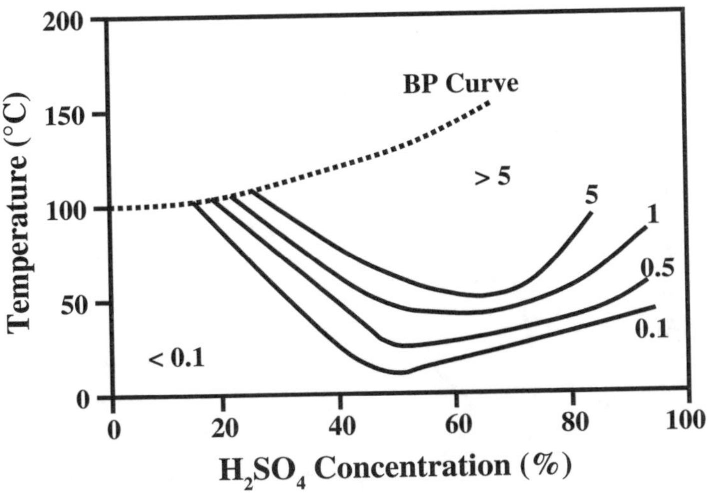

Figure 8.28 Effect of Temperature on the Corrosion Rate (mm/y) of Duplex Cast Alloy CD-4MCu in Sulfuric Acid

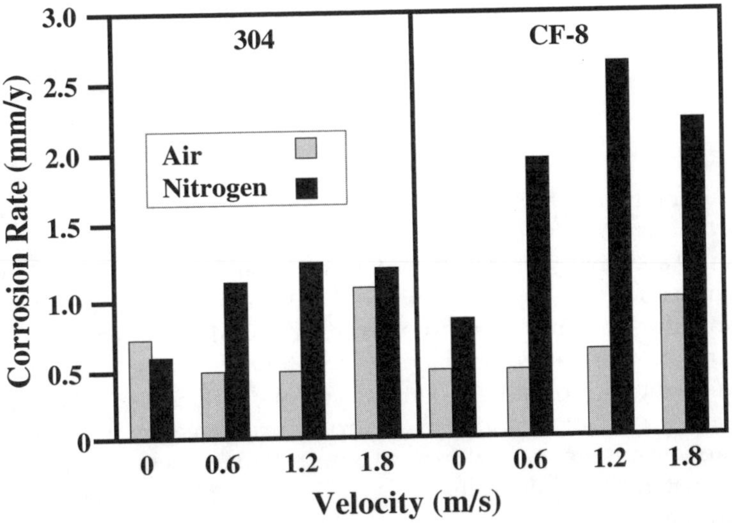

Figure 8.29 Effect of Velocity on Cast CF-8 and Wrought Type 304 in 95% Sulfuric Acid at 50°C (122°F)

Alloys for use in pumps and valves for handling hot, concentrated sulfuric acid at high velocity and with abrasive particles have been undergoing development since at least 1925. A series of casting alloys, including the Illium® range, led to the development of Lewmet® alloys.[63] Lewmet® 66 (cast or wrought austenitic nickel-based alloy) and Lewmet® 55 (an age-hardenable casting alloy) are superior to the earlier alloys of this type. Lewmet® 66 will withstand acid above 80% to 120°C (248°F), while Lewmet® 55 will also handle 70–80% acid at this temperature, Figure 8.30. This figure compares several high-nickel alloys in concentrated sulfuric acid (containing 45 ppm iron) with alloy SX® (S32615).[88]

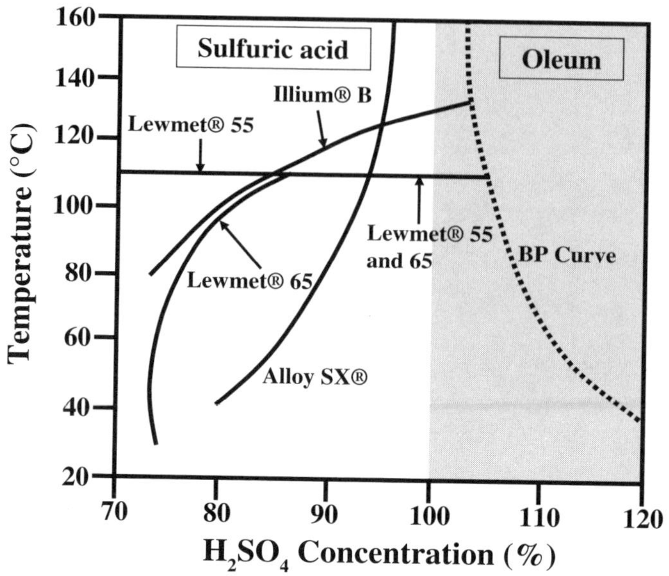

Figure 8.30 Isocorrosion Curves at 5 mpy (0.13 mm/y) for Various Casting Alloys and SX® in Concentrated Sulfuric Acid with 45 ppm Iron

In the hardened condition at 500 HB (Brinell hardness), Lewmet® resistance is comparable to that of the annealed material at 250 HB.[89] Because acid-tower pumps often encounter abrasive particles from deterioration of brick linings, Lewmet® 55 is frequently chosen for this application as well as for gate valves.[90] Wrought Lewmet® 66 is used for plugs, stems, butterfly valves, and piping. Lewmet® alloys are also used for inline strainers at pump suctions and tower outlets; dilution quills for the introduction of water into acid circulation systems; orifice plates for tower distribution, pressure measurement, and flow control; vortex breakers in vessel outlets; and thimbles in tower outlets, gas inlets, and pump tank outlets.[91] Cast versions of the 5–6% silicon austenitic stainless steels, e.g., S32615, are also now being used for these hot, strong acid applications.

Nickel and Nickel Alloys

Nickel (N02200) and nickel-copper alloys (such as alloy 400 [N04400]) are not used in oxidizing concentrated sulfuric acid applications. Alloy 400 (N04400) is, however, used under reducing conditions such as in deaerated acid. This alloy has low corrosion rates in air-free sulfuric acid up to 85% concentration at 30°C (85°F) and up to 60% at 95°C (205°F).[2] In acid exposed to the air, the corrosion rates for chromium-free grades (N04400 and N10665) are unacceptable, Table 8.13.[71]

Table 8.13 Corrosion Rates of Alloy 400 in Boiling Solutions of Various H_2SO_4 Concentrations

% H_2SO_4 Concentration	Boiling Temp (°C)	Corrosion Rate mpy (mm/y)
75	182	2,300 (58.4)
96	293	3,300 (83.8)

Nickel-Molybdenum Alloys

The high-molybdenum grades in either wrought form (alloy B-2 [N10665] and its variants, alloys B-3 [N10675] and B-4 [N10629]) or cast form (N12MV [ASTM A 494][86]) are resistant to 120°C (248°F) in the absence of oxidizing contaminants,[51,67] Figure 8.31.[88,92] The high-molybdenum grades are particularly useful where halide contamination (e.g., chlorides, fluorides) are present. Corrosion rates in 99% acid at 130°C (266°F) were only 9 mpy (0.23 mm/y) in 96-hour laboratory tests, Table 8.14.[67] Alloy 33, with its higher chromium content of 33% and low molybdenum content, has good resistance in 99% acid, as shown in Table 8.14.

Traces of nitric acid in sulfuric acid can increase corrosion of molybdenum-containing alloys such as B-2 (N10665). Table 8.15 shows an order of magnitude higher corrosion in a field test in an absorption tower.[17] In some of the other alloys and stainless steels, the corrosion rate is reduced by the presence of oxidizing agents such as nitric acid.

Because of the oxidizing nature of oleum, the chromium-free grades such as alloy B-2 (N10665) and cast N-12MV are not suitable.

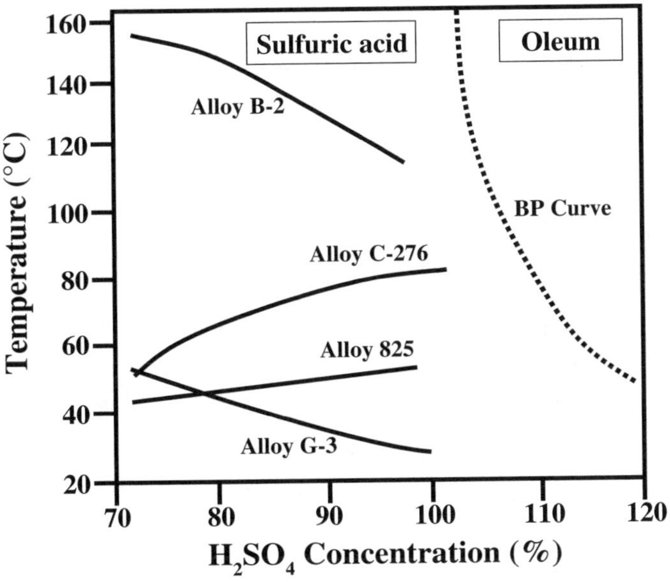

Figure 8.31 Isocorrosion Curves at 5 mpy (0.13 mm/y) for Nickel-Molybdenum Alloys in Concentrated Sulfuric Acid with 45 ppm Iron

Table 8.14 Corrosion Rate of Iron- and Nickel-Based Alloys in 99% Acid at 130°C (266°F)

Material	Corrosion Rate mpy (mm/y)
Type 316L	52 (1.3)
Alloy 2205	230 (5.8)
Alloy 255	7.5 (0.19)
Alloy C-276	50 (1.3)
Alloy G-3	73 (1.8)
Alloy G-30	42 (1.1)
Alloy B-2	8.5 (0.22)
Alloy 8204	1.7 (0.04)
Alloy 31	10 (0.25)*,[76]
Alloy 33	1.6 (0.04)**,[76]

* Tested in 98.5% concentration at 140° (284°F)

** Tested in 99.1% concentration at 150° (302°F)

Table 8.15 Corrosion Rate of Iron- and Nickel-Based Alloys in 99% Sulfuric Acid Absorption Tower at 100–120°C (212–248°F)

Material	Corrosion Rate mpy (mm/y)
Carbon steel	96 (2.4)
Cast iron	5 (0.1)
Ductile iron	10 (0.3)
Type 304L	<1 (0.03)
Type 316lL	2 (0.1)
Alloy 904L	8 (0.2)
Alloy 20 Cb-3	3 (0.1)
Alloy C-276	13 (0.3)
Alloy B-2	92 (2.3)
Alloy 26-1	<1 (0.03)
Alloy A-611	2 (0.1)

Nickel-Chromium-Molybdenum Alloys

These grades, typified in wrought form by alloys 625 (N06625), C-276 (N10276), C-4 (N06455), C-22 (N06022), 59 (N06059), C-2000 (N06200), and alloy 686 (N06686) and in castings by ASTM A 494[86] grades CW-12MW, CW-6M, CX-2M, CW-2M, and CW-6MC, tend to be more resistant in the presence of oxidizing contaminants because of their high chromium content.

Alloy 686 (N06686) shows slightly better resistance than alloy C-276 (N10276), alloy 22 (N06022), and alloy 59 (N06059) and is considerably better than alloy 625 (N06625), Figure 8.32.[74,93,94] Alloy C-276 (N10276) is resistant to 90–98.5% acid up to 90°C (194°F). However, as little as 17 ppm chloride increases the corrosion rate from 4 mpy (0.1 mm/y) to 32 mpy (0.8 mm/y). Also, in 98% acid at this temperature, rates are as high as 50 mpy (1.3 mm/y) with chloride contamination of 40 ppm or more.

The analogous casting alloy to alloy C-276, namely CW-12MV, is useful in throttling valves in 70–85% acid to about 74°C (165°F) and above 85% concentration to about 93°C (199°F). This alloy has excellent resistance to velocity effects, for example, in 95% acid at 50°C (122°F) and 70°C (158°F), Figure 8.33.[11]

The corrosion behavior of some wrought nickel-chromium-molybdenum alloys is compared with their cast equivalent in hot, concentrated acid, Table 8.16.[95]

When nickel-chromium-molybdenum castings are required in concentrated acid, alloys CW-6M (N30107) and CW-2M (N26455) are generally preferred. However, better erosion-corrosion resistance is often obtained with several less expensive, proprietary competitive alloys.[96]

The corrosion of some nickel-chromium-molybdenum alloys is compared with nickel-molybdenum alloys in various concentrations of reagent-grade sulfuric acid at a range of temperatures, Table 8.17.[97] This shows a complex behavior with chromium being beneficial under some conditions and detrimental under others. It also shows that all of these modern, nickel-based alloys, including the high-molybdenum alloy B-3 (N10675), have a good resistance to this range of oxidizing acid strengths up to at least 100°C (212°F).

Of the high-nickel materials, cast Illium® G was developed primarily to resist mixed nitric-sulfuric acids. In strong acid, the limiting temperature rises from 65°C (149°F) at 70% to 95°C (203°F) at 100% concentration. Because the alloy is relatively high in carbon and is not stabilized, it must be used in the solution-annealed condition. Illium® B

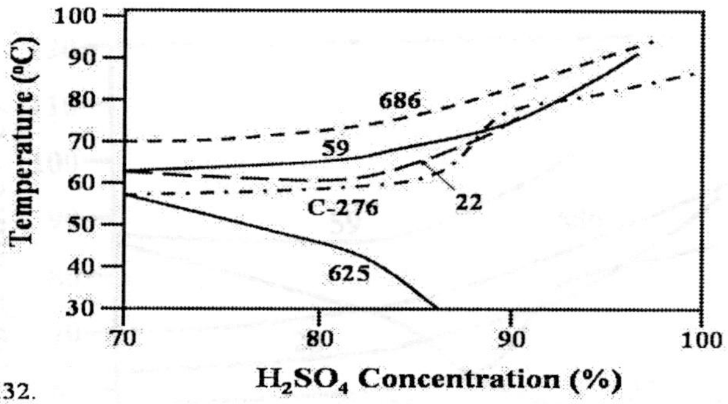

8.32.

Figure 8.32 Isocorrosion Curves at 5 mpy (0.13 mm/y) for Nickel-Chromium-Molybdenum Alloys in Concentrated Sulfuric Acid is more wear- and galling-resistant than Illium® 98 and will tolerate about 20°C higher temperatures in concentrated acid. Many pump impellers were upgraded from Illium® G to Illium® B during the period between 1958 and 1968.[98]

High-nickel alloys such as Illium® G and alloy C-276 (N10276) are suitable for oleum up to about 80°C (176°F), but are not usually competitive with stainless-type alloys.

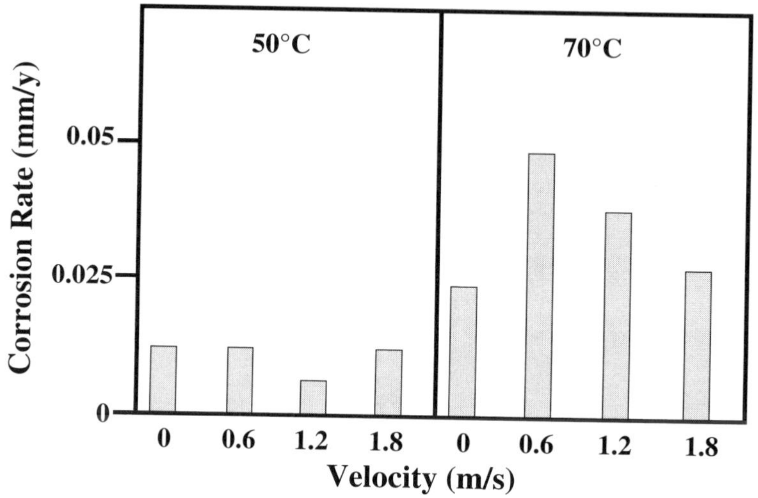

Figure 8.33 Effect of Velocity on Cast CW-12MV in 95% Sulfuric Acid at 50°C (122°F) and 70°C (158°F)

Table 8.16 Comparison of Corrosion of Cast (C) and Wrought (W) Nickel Alloys in Concentrated H_2SO_4 at 230°F (110°C)

UNS No.	Alloy	Corrosion Rate mpy (mm/y)
N06455	C-4 (W)	56 (1.4)
N26455	CW-2M (C)	42 (1.1)
N06022	C-22 (W)	62 (1.6)
N26022	CX-2MW (C)	77 (2)
N10276	C-276 (W)	13 (0.33)
N30107	CW-6M (C)	11 (0.28)
N06200	C-2000*	17 (0.43)

Table 8.17 Corrosion Rates mpy of Various Nickel Alloys in Reagent-Grade Sulfuric Acid

Temperature °C (°F)	Alloy	80%	90%	96%
66 (150)	B-3 (N10675)	—	—	—
	C-276 (N10276)	0.041	0.025	—
	C-2000 (N06200)	0.056	0.015	—
79 (175)	B-3	—	—	—
	C-276	0.137	0.048	0.043
	C-2000	0.279	0.074	0.051
93 (200)	B-3	0.013	0.015	0.023
	C-276	0.597	0.460	0.178
	C-2000	0.988	0.371	0.185
107 (225)	C-276	2.731	1.641	0.947
	C-2000	1.615	1.170	0.630
121 (250)	B-3	0.028	0.048	0.091
	C-276	5.664	4.788	—
	C-2000	2.370	2.243	—
Boiling	B-3	4.763	—	—
	C-276	—	—	—
	C-2000	—	—	—

High-Silicon Nickel Alloys

In hot, concentrated acid, the corrosion resistance of binary cast nickel-silicon alloys is improved by increasing the silicon content up to at least 15%. In more dilute solutions, the resistance passes through a maximum at about 7–8% silicon. These alloys become increasingly less ductile with increasing silicon content, although heat treatment can improve ductility. Nickel-based casting alloys with 10–10.5% silicon and 2.5–3.0% titanium have good resistance to boiling 80% and 94% acid while still having acceptable ductility.[99]

Cast nickel-silicon alloys typically containing 8 to 10% silicon were developed for handling hot or boiling sulfuric acid of most concentrations. A few wrought nickel-silicon alloys with 4 to 5% silicon have also been developed to handle high-temperature, high-concentration sulfuric acid. One of their weaknesses is the increased corrosion rates when the concentration of sulfuric acid falls below 95 percent. Also, their resistance to localized attack is poor in chloride-containing cooling waters.[100]

A silicon-rich alloy, Hastelloy® D-205 (20% Cr, 6% Fe, 5% Si, 2.5% Mo, 2% Cu, 0.03% C, balance Ni) has good resistance (<5 mpy [<0.13 mm/y]) in 20–60% acid to about 52°C (125°F). In boiling 99% acid it had a corrosion rate of 0.7 mpy (0.02 mm/y), while type 316 (S31600) corroded at 41 mpy (1.04 mm/y) and an iron alloy with 17% Cr, 20% Ni, and 5% Si had a corrosion rate of 1.1 mpy (0.03 mm/y). In concentrated acid it has generally better resistance than alloy C-276 (N10276), Figure 8.34.[101] The data for C-276 was determined using 98.5% commercial acid.[102] The data for D-205 in this figure was also done in similar-strength acid from the same supplier, but some ten years later, so there may have been compositional differences. C-276 measured in the same acid, however, gave almost identical results. Rotating-disc tests at up to 4,700 rpm were run in 96% acid at 130°C (266°F) with chloride additions without any measurable increase in corrosion rate of alloy D-205. The effect of acid concentration was also examined using the rotating disc and the 0.1 mm/y (3.9 mpy) corrosion rate was found to occur at combinations of acid strength and temperature shown in Table 8.18.[103]

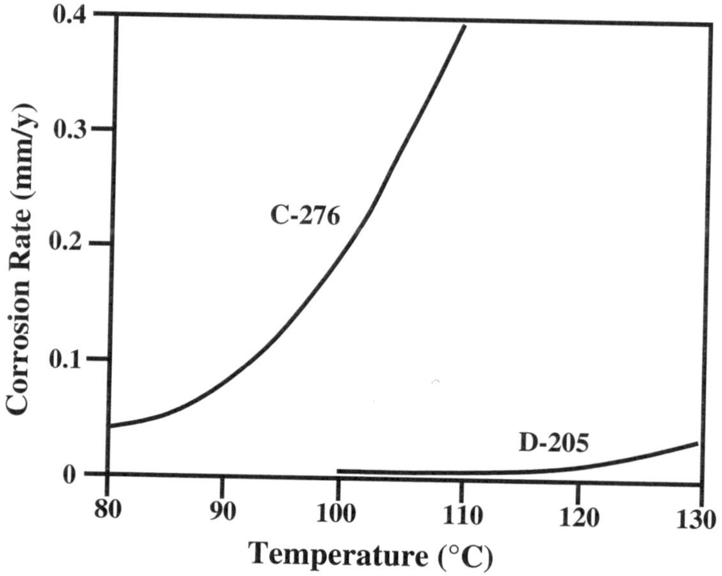

Figure 8.34 Effect of Temperature on Corrosion Rates of Alloys D-205 and C-276 in Concentrated H_2SO_4

Table 8.18 Acid Concentration and Temperature at Which the Corrosion Rate of Alloy D-205 is 0.1 mm/y (3.9 mpy)

Acid Concentration %	Temperature °C (°F)
98.5	170 (340)
95.0	110 (230)
90.0	70 (158)
85.0	60 (140)
80.0	58 (136)
70.0	65 (149)
60.0	80 (176)

The as cast weld material has poor ductility due to the presence of a brittle, silicon-rich eutectic phase. Heat treatment can improve ductility, but not enough to make it a viable engineering material for most applications. It is, however, being used in plate frame heat exchangers for strong acid cooling.[104]

Lead

Lead, that is, "chemical lead," has long been used for sulfuric acid service in dilute and intermediate concentrations, but has only limited usefulness in concentrated acids and is strongly attacked by oleum.

The corrosion resistance of lead is contingent upon the formation and retention of an insoluble film of lead sulfate products. Adherence of this protective film varies with mechanical effects, e.g., vibration, erosion or abrasion, and chemical conditions affecting solubility (i.e., acid concentration, contaminants, and temperature). Corrosion rates for lead in concentrated acid are given in Table 8.19.[105]

A maximum flow rate of 3 ft/s (1 m/s) is sometimes recommended for lead.

Diethylsulfate, used to absorb ethylene in the production of ethanol via the strong acid process, has caused accelerated general attack in 93–98% acid. Lead will pit in some process acids due to organic compounds that affect the lead sulfate film, e.g., during polymerization of butenes in 72% acid at 80°C (176°F). On the other hand, only 3 mpy (0.08 mm/y) is reported for lead used in the sulfonation of phenol in 93% acid at 120°C (248°F).

Contaminants such as nitrates and chlorides have an adverse effect on the resistance of lead because they increase solubility. Sulfur dioxide has no such effect (lead withstands all concentrations of sulfurous acid), but sulfur *trioxide* is very harmful, behaving like very strong sulfuric acid. Dissolved oxygen (DO) has no discernible effect. Lead alloys, e.g., tellurium lead and antimonial ("hard") lead, offer a slight advantage in corrosion resistance. There is also a slight strength advantage that favors lead alloys, although this benefit disappears at about 88°C (190°F).[106]

The behavior of lead under stress, or even under its own weight, may lead to potential creep or static fatigue problems. Currently, chemical lead is most often used as a membrane behind brick linings. Lead can be involved in galvanic corrosion effects if in direct contact with some materials. If in contact with stainless steels, lead initially corrodes more rapidly, but once a lead sulfate film is established, the stainless steel can suffer accelerated corrosion. When in contact with carbon or graphite,

Table 8.19 Corrosion Rate of Lead in Sulfuric Acid

Acid Concentration %	Temperature °C (°F)	Corrosion Rate mpy (mm/y)
98	24 (75)	>200 (>5)
93	90 (194)	50 (1.3)
90	90 (194)	20 (0.5)
80	140 (284)	20 (0.5)
80	110 (230)	5 (0.1)
70	125 (257)	5 (0.1)

the corrosion rate of lead is substantially increased and lead membranes behind brick linings must be isolated from graphite bricks or tiles, for example.

The mechanical and health/environmental problems associated with joining and fabrication (lead-burning) have decreased the use of lead and lead-lined equipment in favor of high-performance alloys and fluorinated plastics in sulfuric acid services.

Reactive and Refractory Metals

The reactive and refractory metals include titanium, zirconium, tantalum, and niobium. Of these, metals titanium and zirconium are unsatisfactory in strong sulfuric acid.

Tantalum

Tantalum is a premium material that resists up to 100% acid, subject to certain temperature and contamination limits. Early corrosion literature cited resistance of tantalum up to the atmospheric boiling point. However, this information ignored a relatively high initial corrosion rate in some circumstances and the ability of tantalum to absorb up to 740 times its own volume of hydrogen, with an attendant embrittlement. In pure sulfuric acid, significant attack begins at about 175°C (347°F). An older source had indicated a rate of 1.5 mpy (0.6 mm/y) in 95% acid at 200°C (392°F).[107]

Another source states that tantalum resists 98% acid at the same temperature.[108] However, other reports give tantalum a poor rating under such conditions, possibly due to traces of oxidizing contaminants. In 72% acid plus 3% chromic oxide at 125°C (257°F), tantalum corrodes at 5 mpy (0.13 mm/y). The two most aggressive contaminants are sulfur trioxide and fluorides. The former virtually precludes the use of tantalum in fuming acid or oleum.[17]

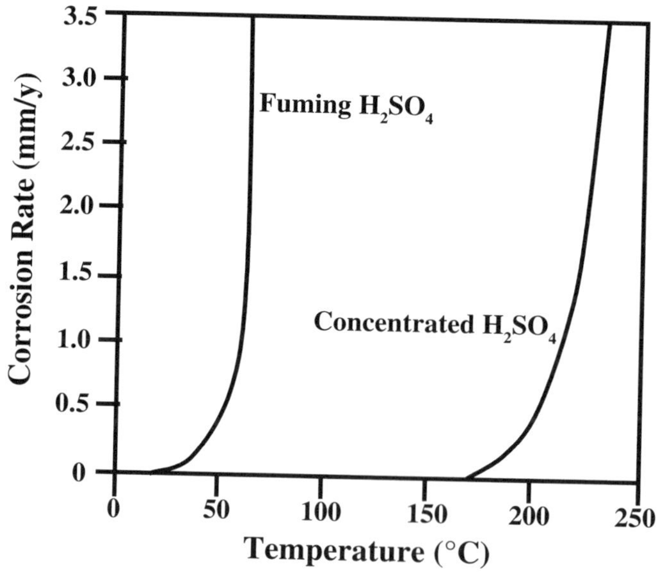

Figure 8.35 Effect of Temperature on the Corrosion of Tantalum in 98% and Fuming H_2SO_4

Tantalum shows a rate of 0.3 mpy (0.075 mm/y) in 15% oleum at 20°C (39°F), but the rate increases rapidly above room temperature, e.g., 92 mpy (2.3 mm/y) at 70°C (158°F).[107] Temperature has a strong effect on the corrosion rate of tantalum in 98% H_2SO_4 and oleum, Figure 8.35.[109] Fluorides greatly exacerbate the hydrogen embrittlement effect and are a possible contaminant in strong acid.

Because of the high cost of tantalum, it is used primarily in thin-walled items (e.g., heat exchangers) and as a lining and a pinhole-free coating applied by electrodeposition from a molten salt bath. It is also used as a mechanical patch in glass-lined vessels. As a patch, however, it must be electrically isolated from more anodic metals to prevent accelerated hydrogen absorption by tantalum, as the cathode in the galvanic couple. Sulfur-trioxide contamination is also a potential problem in glass-lined equipment operating in 98% or higher acid strength. If such vessels contain tantalum plugs from factory installation or maintenance operations then corrosion and embrittlement can occur. The user can specify alternative plug materials, or vessels may be purchased for a premium as plug-free.

Long-term corrosion tests with welded coupons were carried out to determine the application limits of tantalum and tantalum with 2.5% tungsten in sulfuric acid. The program included the following tests:

- Determination of corrosion behavior in 90–100% H_2SO_4 at temperatures up to 200°C (392°F)
- Determination of corrosion behavior in 96wt % H_2SO_4 comparing recovered nitration-spent acid and technical grade between 150 and 230°C (302 and 446°F)

Mainly immersion tests were performed. A comparison of tantalum and the alloyed tantalum showed that in technical H_2SO_4 the alloy performed better than the pure metal. Regardless of which material was considered, the higher the H_2SO_4 concentration, the lower the temperature necessary to achieve acceptable corrosion behavior. In technical H_2SO_4, the following application limits were determined:

Ta: 96wt % 200°C (392°F), 97wt % 150°C (302°F)
Ta-2.5% W: 96wt % 210°C (410°F), 97.5wt % 175°C (347°F)

Above 97.5 wt %, the corrosion resistance decreased rapidly. Testing in recovered nitration-spent acid showed much lower corrosion rates due to the presence of small amounts of nitric acid. In this type of acid containing oxidizing compounds, 230°C (446°F) was considered to be satisfactory, provided the wall of the heat exchanger is sufficiently thick. In highly concentrated H_2SO_4 (>96%) there are most likely no hydrogen ions present; therefore, only free H_2SO_4 molecules or HSO_4^- ions can be reduced in cathodic reactions. The hydrogen reduction is inhibited by the high overvoltage of the reduction reaction of H^+. There is therefore no risk of a hydrogen embrittlement for the application of tantalum or tantalum, 2.5% tungsten at elevated temperatures in the recovery operation for spent sulfuric acid.[110]

Niobium

Niobium is resistant to concentrated acid under oxidizing conditions, e.g., in concentrated acid containing ferric or cupric ions. In concentrated sulfuric acid, at 100°C (212°F) it has a corrosion rate of 0.25 mm/y (10 mpy), while in 70% sulfuric acid at 167°C (333°F) it corrodes at 5 mm/y (200 mpy). It is resistant at all concentrations up to 95% at room temperature.[111]

Precious/Noble Metals

Platinum is resistant at all concentrations up to 200°C (392°F). At higher temperatures platinum can corrode in >90% acid, especially if air is present.[112]

Gold and gold platinum alloys are resistant at all concentrations up to boiling point.

Silver is resistant to <60% acid at up to boiling point and at room temperature in up to 96% acid, corrosion rate 0.14 mm/y (5.5 mpy). It is not resistant to hot acid at >60% strength (see Figure 9.10).

Summary of Corrosion of Metals and Alloys in Strong Acid (>70%) and Oleum

Materials used in this range of acid strength have been summarized graphically, Figure 8.36.[113,114] This figure defines temperatures and concentrations of sulfuric acid and oleum in which certain metals and alloys have been successfully used under static conditions. The various areas represent corrosion rates of <0.5 mm/y (19.7 mpy). This figure is a composite from a number of sources and has been modified to include the general area of application of the high-silicon austenitic stainless steels.

The more resistant materials (i.e., not steel, cast iron, or low-alloy stainless steels) can also be used at lower concentrations and temperatures than indicated. The data that were used to generate these curves mostly come from alloys that have been modified and improved over the years, and the acid purity may well have changed, affecting corrosion behavior as discussed. The curves can, however, still give general guidance to areas of use in which different materials can be considered.

Suggested ranges of use for some stainless steels and nickel-based alloys in storage applications, i.e., at generally modest temperatures, in valves, instruments, etc., are shown below:[115]

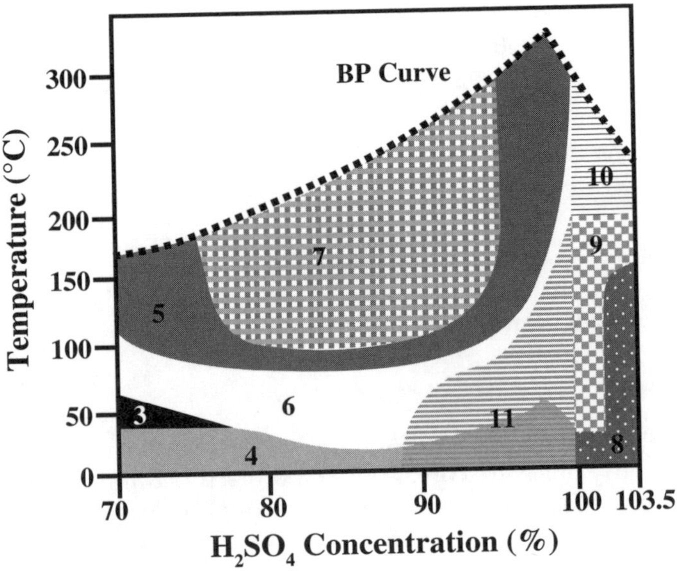

Legend

Zone 3

Impervious graphite
Tantalum
Gold
Platinum
Zirconium
Molybdenum
Monel if air-free
Glass
High-silicon cast iron (14.5% Si)
Hastelloy® B and D, not up to BP
Durimet 20, Worthite® up to 66°C (151°F)
Lead
Chlorimet

Zone 4

Impervious graphite up to 96%
Tantalum
Gold
Platinum
Zirconium
Ni-Resist®
Carbon steel
Type 316 (>80%)
Glass
High-silicon cast iron (14.5% Si)
Hastelloy® B, B-2, and D
Durimet 20
Worthite®
Lead up to 96%

Zone 5

Tantalum
Gold
Platinum
Glass
High-silicon cast iron (14.5% Si)
Hastelloy® B, B-2, and D, not up to BP

Zone 6

Tantalum
Gold
Platinum
Impervious graphite and lead up to 80°C (176°F) and 96% sulfuric
Glass
High-silicon cast iron (14.5% Si)
Hastelloy® B, B-2, and D, not up to BP
Durimet 20, alloy 20-Cb3, and Worthite® up to 66°C (150°F)

Zone 7

Gold
Platinum
Glass
High-silicon cast iron (14.5% Si)
Tantalum

Zone 8

Alloy 20Cb-3
Illium® G
Gold
Platinum
Glass
304
Durimet 20, alloy C-276
Carbon steel

Zone 9

Worthite®
Gold
Platinum
304
Glass
Durimet 20, alloy 20Cb-3

Zone 10

Platinum
Glass
Gold

Zone 11

High-silicon austenitic stainless steels

Figure 8.36 Area of Usefulness of Various Materials in Sulfuric Acid

70 to 100.5% H_2SO_4	S31254 (<25C [<77F]), N10276, N08367, N08024, N08026, N08825, N08028, N08904, N06022, N06030, N06455, N08926, N08031, N06625, N06985, N08020, N09925, N06059
90 to 100.5% H_2SO_4	S31600, S31603
93 to 100.5% H_2SO_4	S30400, S30403

Aluminum

Aluminum has good resistance to ambient 98%+ acid and is occasionally used in applications where dilution is unlikely.

Iron and Steel

Carbon steel should not be used in the acid strength ranges of 80–88% and 99.5–100.5% because of severe corrosion. The protective film on carbon steel can be easily removed or damaged, so turbulent or high-velocity conditions must be avoided. Cast iron is generally more resistant than carbon steel in concentrated acid, but gray iron must not be exposed to free sulfur trioxide or oleum.

Alloyed cast irons, particularly high-silicon ones, have good resistance to the range of acid strengths up to 100% and at elevated temperatures. The silicon irons have poor resistance in oleum and should not be used.

Stainless Steels and High-Performance Alloys

The stainless steels rely for their corrosion resistance on the presence of a passive film. A schematic polarization curve for a typical stainless steel is shown in Figure 7.3, illustrating the passive and transpassive regions. Should the passive film be removed, chemically or mechanically, the polarization curve will be similar to one for carbon steel, which does not have a passive region in hot, concentrated sulfuric acid, and corrosion rates are high. However, an activated stainless steel can corrode more rapidly than carbon or low-alloy steel. The passive film is enhanced by silicon content particularly, while molybdenum enhances the resistance under reducing conditions or if halides are present.

Type 304 (S30400) and standard duplex stainless steels should normally only be used above about 93% acid. The duplex grades are more tolerant of occasional dilution, such as under shutdown conditions.

The high-performance alloys include those stainless steels with high molybdenum such as alloy 254SMO (S31254) and 654SMO (S32654), and nickel-rich, chromium-bearing alloys such as alloy 20Cb-3 (N08020), alloy 926 (N08926), alloy 31 (N08031), and alloy 825 (N08825). Most of these have limited resistance in concentrated acid (see Figure 8.21). Actual performance of these high-performance alloys is discussed in relevant chapters.

The high-silicon austenitic stainless steels, such as SX®, SARAMET®, and ZeCor®, and the high-chromium alloy 33 (R20033), have good resistance to concentrated acid even at elevated temperatures and under velocity conditions.

Lead

Lead has good resistance in up to 70% acid at moderate temperatures. Its corrosion resistance falls dramatically at above about 85% acid (see Figure 9.12). It is also used in higher-strength acid, particularly where the protective film can itself be protected from erosion, e.g., behind brick. Lead is strongly attacked in oleum and should not be used.

Nickel Alloys

Pure nickel (N02200) can displace hydrogen from acid solutions and shows an active-passive transition in sulfuric acid solutions.[2] Without anodic protection or inhibitor additions, nickel can be used only for unaerated acid under essentially static conditions at ambient temperature.

The 16% chromium nickel-based alloy 600 (N06600) is somewhat less resistant than the copper-bearing alloy 400 (N04400) in air-free acid. The chromium-nickel-molybdenum alloys (e.g., N06625, N10276) are inherently less resistant than the nickel-molybdenum alloy N10665, but far superior when oxidizing contaminants are available to reinforce passivity. The high-molybdenum alloys are generally preferred when halides are present.

Alloys such as C-276 (N10276) and 625 (N06625) have good resistance to acid strengths up to at least 95% and have low corrosion rates in both the active and passive condition.

Reactive and Refractory Metals

Titanium and zirconium are not resistant in concentrated acid and should be avoided.

Tantalum is resistant to concentrated sulfuric acid but is attacked by free sulfur trioxide or oleum.

Niobium is resistant to concentrated acid under oxidizing conditions, e.g., if oxidizing ions are present. It is also resistant at all concentrations up to 95% at room temperature. In concentrated acid, niobium can embrittle.

Precious and Noble Metals

Platinum and gold have excellent resistance to concentrated acid even at elevated temperatures. Silver is attacked by >96% acid, and at all concentrations >60%, at elevated temperature.

References

1. C. P. Dillon, ed., "Concentrated Sulfuric Acid and Oleum," vol. MS-1, Materials Selector for Hazardous Chemicals (St. Louis, MO: MTI Inc., 1997): 212 pp.
2. B. D. Craig, D. B. Anderson, eds., "Handbook of Corrosion Data" (Materials Park, OH: ASM International, 1997): pp. 847–937.

3. D. Fyfe, R. Vanderland, J. Rodda, "Corrosion in Sulfuric Acid Storage Tanks," Chemical Engineering Progress, 73, March (1977): pp. 65–69.
4. T. N. Andersen, N. Vanorden, W. J. Schlitt, "Effects of Nitrogen Oxides, Sulfur Dioxide and Ferric Ions on the Corrosion of Mild Steel in Concentrated Sulfuric Acid," Metallurgical Transactions 11A (August 1980): p. 1421.
5. S. W. Dean, G. D. Grab, "Corrosion of Carbon Steel Tanks in Concentrated Sulfuric Acid Service," Materials Performance 25, 7 (1986): p. 48.
6. B. T. Ellison, W. R. Schmeal, "Corrosion of Steel in Concentrated Sulfuric Acid," J. Electrochemical Society 125, 4 (1978): p. 524.
7. F. Todt (1961) in Sulfuric Acid section, CD, "Dechema Corrosion Handbook" (Frankfurt, Germany: Dechema eV 2001).
8. J. R. Rodda, M. B. Ives, "The Corrosion Science of Sulfuric Acid Storage Tank Failures" (2004), unpublished draft paper.
9. Private communication, Union Carbide (1952 data) in C. P. Dillon, ed., "Concentrated Sulfuric Acid and Oleum," vol. MS-1, Materials Selector for Hazardous Chemicals (St. Louis, MO: MTI Inc., 1997): 212 pp.
10. M. G. Fontana, "Corrosion," Industrial and Engineering Chemistry 43, August (1951): p. 65A.
11. R. M. Kain, "Velocity Effects on the Corrosion Resistance of Several Cast and Wrought Alloys in Concentrated Sulfuric Acid at Elevated Temperatures," Corrosion/83, Research in Progress (Houston, TX: NACE International, 1983).
12. M. Tiivel, F. J. McGlynn, A. A. Trickett, "Carbon Steel Sulfuric Acid Storage Tank: Inspection Guidelines" (North York, ON, Canada: MARSULEX Inc., 1986): 46 pp.
13. API 653, "Tank Inspection, Repair, Alteration and Reconstruction" (Washington, DC: American Petroleum Institute).
14. M. Tiivel, F. McGlynn, "Design, Fabrication and Inspection of Sulfuric Acid Storage Tanks," Proc. Sulphur 87 (London: The British Sulphur Corp., 1987): pp. 265–308.
15. S. K. Brubaker, R. A. Tatnall, "Materials Selection and Design to Minimize Hydrogen Grooving," MP 34, 5 (1995): p. 60.
16. R. J. Landrum, "Designing for Corrosion Control: A Corrosion Aid for the Designer" (Houston, TX: NACE International, 1989): p. 158.
17. S. K. Brubaker, "Sulfuric Acid, Process Industries Corrosion—The Theory and Practice," eds., B. J. Moniz, W. I. Pollock (Houston, TX: NACE International, 1986), p. 243.
18. API 650, "Welded Steel Tanks for Oil Storage" (Washington, DC: American Petroleum Institute).
19. J. B. Rinckhoff, "Controlling Corrosion in Wet-Gas Sulfuric Acid Plants," Chemical Engineering 20, Nov. (1967).
20. ASTM A 285, "Specification for Pressure Vessel Plates, Carbon Steel, Low- and Intermediate Tensile Strength," Annual Book of ASTM Standards, Vol. 01.04 (West Conshohocken, PA: ASTM).
21. ASTM A 293, discontinued specification, Annual Book of ASTM Standards (West Conshohocken, PA: ASTM).
22. ASTM A516/A516M-90, "Standard Specification for Pressure Vessel Plates, Carbon Steel, for Moderate- and Lower-Temperature Service" (West Conshohocken, PA: ASTM, 2001).

23. Anon, "Corrosion in Sulfuric Acid," Proceedings of 1985 NACE Symposium (Houston, TX: NACE International, 1985).
24. ASTM A 515/A 515M-92, "Standard Specification for Pressure Vessel Plates, Carbon Steel, for Intermediate- and Higher-Temperature Service" (West Conshohocken, PA: ASTM, 1997).
25. Anon, "Guide on Methods for Assessing the Acceptability of Flaws in Welded Structures," BS 7910:1999 (London: BSI, 1999).
26. Anon, "Guidance on Some Methods for Assessing the Acceptability of Flaws in Metallic Structures," PD 6493:1991 (London: BSI, 1999).
27. C. P. Dillon, ed., "Forms of Corrosion—Recognition and Prevention," NACE Handbook No. 1 (Houston, TX: NACE International, 1982).
28. C. P. Dillon, private communication (1952), in "Concentrated Sulfuric Acid and Oleum," ChemCor 1 (St. Louis, MO: MTI, undated).
29. Private communication in C. P. Dillon, "Concentrated Sulfuric Acid and Oleum," vol. MS-1, Materials Selector for Hazardous Chemicals (St. Louis, MO: MTI Inc., 1997): 212 pp.
30. M. Rahmani, J. E. Strutt, Hydrodynamic Modeling of Corrosion of Carbon Steels and Cast Irons in Sulfuric Acid, MTI Publication No. T-2 (Houston, TX: NACE International, 1992).
31. Anon, "Materials of Construction—Mondi®," DKL Engineering (2003), http://members.rogers.com/acidmanual/materials_mondi.htm.
32. Anon, "Duriron and Durichlor 51M," Bulletin A/2j (Dayton, OH: Flowserve Corporation, 1998): 8 pp.
33. ASTM A 518, "Specification for Corrosion-Resistant High-Silicon Iron Castings," Annual Book of ASTM Standards, Vol. 01.02 (West Conshohocken, PA: ASTM).
34. A. Reynaud, "High-Chromium and High-Silicon Cast Irons," Materials Performance 35, 2 (1996): p. 93.
35. T. F. Banigan, "Failure of Cast and High Silicon Iron in Fuming Sulfuric Acid," Journal of Industrial and Engineering Chemistry, April (1922): p. 323.
36. T. C. Spence, private communication (Dayton, OH: Flowserve Corp., 2003).
37. D. J. Chronister, T. C. Spence, "Influence of higher silicon levels on the corrosion resistance of modified CF-type cast stainless steels," Corrosion/85, paper no. 305 (Houston, TX: NACE International, 1985): 10 pp.
38. Anon, "The Corrosion Resistance of Nickel-Containing Alloys in Sulfuric Acid and Related Compounds," CEB-1 (Suffern, NY: The International Nickel Co. Inc., 1983): 90 pp.
39. Anon, "Corrosion Handbook—Stainless Steels" (Sandviken, Sweden: AB Sandvik Steel, 1999): pp. I:I–II:88.
40. Anon, "Carpenter Stainless Steels," 4-99/12.5M (Reading, PA: Carpenter Technology Corp., 1999): 356 pp.
41. Anon, "General Corrosion" (Manchester, UK: Weir Material Services, 2003), http://www.weirmaterials.com/general_corrosion.htm.
42. G. Kobrin, J. Lilly, J. Mac Diarmid, B. Moniz, "Stainless Steels for Chemical Process Equipment," NiDI Reprint series no.14 038 (Toronto, ON, Canada: NiDI, 1998), pp. 1–9.
43. Private communication in C. P. Dillon, "Concentrated Sulfuric Acid and Oleum," ChemCor 1 (St. Louis, MO: MTI, undated).

44. D. R. McAlister, S. A. Ziebold, "Heat Recovery from Concentrated Sulfuric Acid," United States Patent No. 4,670,242 (June 2, 1987).
45. R. Matsuhashi, E. Sato, S. Abe, H. Abo, "The Effect of Ni Content on the Corrosion Behavior of Stainless Steels in Highly Concentrated Sulfuric Acid," Corrosion Engineering 39 (1990): pp. 89–100.
46. P. E. Morris, R. M. Kain, "Alloy Performance and Protection in the Chemical and Process Industries: An Electrochemical Approach," Chemical Engineering Progress 73, 6 (1977): p. 103.
47. J. E. Strutt, "Corrosion of Stainless Steels in 93% and 98.5% Sulfuric Acid," MTI Publication No. T-1 (Houston, TX: NACE International, 1992).
48. H. S. Tong, "Corrosion and Electrochemical Behaviour of Iron-Chromium-Nickel Alloys in Concentrated Sulfuric Acid Solutions," Electrochemical Corrosion Testing, ASTM STP 727, F. Mansfeld, U. Bertocci eds. (West Conshohocken, PA: ASTM, 1981): pp. 96–109.
49. M. Renner et al. (1986) in Metals Handbook—Corrosion, vol. 13, 9th ed., J. R. Davis, ed. (Metals Park, OH: ASM International, 1987): p. 1151.
50. Anon, "Corrosion Resistant Stainless Steels" (St Louis, MO: Enviro-Chem Systems), http://www.enviro-chem.com/plant-tech.
51. Anon, "Materials of Construction for Handling Sulfuric Acid," NACE Publication 5A151 (Houston, TX: NACE International, 1985): 6 pp.
52. R. M. Kain, P. E. Morris, "Anodic Protection of Fe-Cr-Ni-Mo Alloys in Concentrated Sulfuric Acid," Corrosion/76 paper no 149 (Houston, TX: NACE International, 1976): 10 pp.
53. Anon, "Protecting Equipment in Sulfuric Acid Plants," Sulfur, July–August (1985).
54. A. Desestret, J. Ferriol, G. Vallier, "Discussion on Two Special Stainless Steels Used in Nuclear Fuel Processing Plants," Materiaux et Techniques, Sept/Oct (1977), pp. 621–636.
55. J. D. Fritz, J. F. Grubb, R. E. Polinski, "Stainless Steels for Corrosion Resistance," Advanced Materials & Processes 159, 6 (2001): pp. 36–38.
56. M. Kohler, R. Kirchheiner, U. Heubner, "Nicrofer® 2509 Si 7, a New Corrosion Resistant Material for Handling of Hot, Highly Concentrated Mineral Acids," Werkstoffe u Korrosion (in German) 46 (1995): pp.18–26.
57. J. R. Rodda, M. B. Ives, "Development of Corrosion Resistant Stainless Steel Alloys for Hot Concentrated Sulfuric Acid Service" (2004), unpublished draft paper.
58. M. Davies, D. S. Hodgson, J. Rodda, "Application of SARAMET in H2SO4 Plants," 13th International British Sulphur Conf., Vienna (1988): 10 pp.
59. Anon, "SARAMET® Austenitic Stainless Steel" (Vancouver, BC, Canada: Kvaerner Chemetics, 2002): 4 pp.
60. G. Harding, "Developments in Sulphuric Acid Resistant Metal, 'SARAMET®' Technology" (Vancouver, BC, Canada: Aker Kvaerner Chemetics, 2003): 18 pp.
61. Anon, "Sandvik SX—the Sulphuric Acid Steel," data sheet (Goteborg, Sweden: Edmeston AB, undated): 2 pp.
62. Anon, "Nicrofer 3033—Alloy 33: A New Corrosion Resistant Austenitic Material for Many Applications," VDM Report No. 24 (ThyssenKrupp VDM, Werdohl, Germany, June 1998).

63. C. H. Clayton, T. E. Johnson, "Engineering Materials for Pumps and Valves," CEP Sept. (1978): pp. 54–57.
64. Anon, "Iso Corrosion SX® ≤ 1 mils year^{-1} in Sulphuric Acid" (Goteborg, Sweden: Edmeston AB, undated): 1 pp.
65. Anon, "ZeCor® Alloy Beats Sulfuric Acid Corrosion" (St Louis, MO: Enviro-Chem Systems, 2002): 4 pp.
66. Anon, "Incoloy® alloy 020," publication no. SMC-018 (Hereford, U.K.: Special Metals, undated): 4 pp.
67. N. Sridhar, "Behavior of High-Performance Alloys in Sulfuric Acid Streams," Corrosion/87, paper no. 19 (Houston, TX: NACE International, 1987): 16 pp.
68. H. Thielsch, "Copper in Stainless Steels," Welding Research Council Bulletin Series 9 (1951): pp. 1–5.
69. C. W. Kovack, "High-Performance Stainless Steels," NiDI Reference Book Series No. 11 021 (Toronto, ON, Canada: NiDI, 2000): 96 pp.
70. J. F. Grubb, "AL-6XN®," Edition no. 2 (Pittsburgh, PA: Allegheny Ludlum Steel Corp., 1995): 38 pp.
71. Anon, "Solutions to Materials Problems," CD (Huntington, WV: Inco Alloys International, 1997).
72. Anon, "Carpenter Alloys for Controlling Severe Corrosives" (Reading, PA: Carpenter Technology Corp., 1989).
73. Anon, "Nicrofer® 3127 hMo—Alloy 31," material data sheet no 4031 (Werdohl, Germany: Krupp VDM GmbH, 1997): 14 pp.
74. J. R. Crum, "Comparison of 5 mpy Isocorrosion Lines for Several Iron Base Alloys in Sulfuric Acid" (Huntington, WV: Special Metals Corporation, 2004): 1 p.
75. Anon, "Incoloy® alloy 27-7MO," publication no. SMC-092 (Huntington, WV: Special Metals Corporation, 2002): 4 pp.
76. L. Paul, Thyssen Krupp VDM, input during review (2004).
77. J. E. Niesse, D. R. McAlister, "Stainless Steels for Heat Recovery from High Temperature Sulfuric Acid," Corrosion/87, paper no. 22 (Houston, TX: NACE International, 1987): 11 pp.
78. U. Heubner, J. Klower, et al., "Nickel Alloys and High Alloy Special Stainless Steels. Properties, Manufacturing, Application," 3rd ed. (Germany: Expert-Verlag, 2003): pp. 73–74.
79. M. Kohler, M. Heubner, K.-W. Eichenhofer, M. Renner, "Alloy 33, A New Corrosion Resistant Austenitic Material for the Refinery Industry and Related Applications," Corrosion '95, paper no. 338 (Houston TX: NACE International, 1995): 14 pp.
80. A. Gildenpfennig, U. Gramberg, G. Hohlneicher, "Passivation and Corrosion of the Metallic High Performance Materials Alloy 33 and MC-Alloy in Different Environments," Corrosion Science 45 (2003): pp. 575–595.
81. C. P. Dillon, "Corrosion of Stainless Steel by Nitric Acid," MP 31, 7 (1992): pp. 51–53.
82. R. M. Forbes-Jones, R. M. Kain, "The Effect of Microstructure on the Corrosion Resistance of Several Cast Alloys," paper no. 67, Corrosion '75 (Houston, TX: NACE International, 1975): 26 pp.
83. Anon, "Durcomet 100," bulletin A/7k (Dayton, OH: Flowserve Corp., 1999): 8 pp.
84. Anon, "Durimet 100," bulletin A/1f (Dayton, OH: Duriron Co., 1981): 6 pp.

85. Anon, "Chlorimet 2, Chlorimet 3," bulletin A/3k (Dayton, OH: Duriron Co., 1981): 6 pp.
86. ASTM A 494/A 494M-00, "Standard Specification for Castings, Nickel and Nickel Alloy" (West Conshohocken, PA: ASTM, 2000).
87. R. W. Monroe, S. J. Pawel, "Corrosion of Cast Steels," in Metals Handbook—Corrosion, vol. 13, 9th ed., J. R. Davis, ed. (Metals Park, OH: ASM International, 1987): pp. 573–582.
88. Anon, "Concentrated Sulfuric Acid and Oleum," ChemCor 1 (St. Louis, MO: MTI, undated).
89. R. J. Borges, "Nickel/Chrome Super Alloy Materials for Critical High Temperature and High Velocity Sulfuric Acid Service," Corrosion/87, paper no. 23 (Houston, TX: NACE International, 1987).
90. G. E. McClain, W. A. Mueller, "Corrosion Problems in Acid Flow Control," CEP (February 1982): pp. 48–50.
91. Anon, "Lewmet Alloys," brochure no. L8/93-33M (St Louis, MO: Chas. S. Lewis & Co. Inc., 1993): 4 pp.
92. J. Crum, Special Metals, corrections during review, 2004.
93. Anon, Special Metals publication on DKL Engineering (2003), http://members.rogers.com/acidmanual/materials_metals.htm.
94. Anon, "Isocorrosion Diagram for the alloy Nicrofer 5923hMo (alloy 59) in Pure Sulphuric Acid," Thyssen Krupp VDM web site, http://www.wdisweb.de/wdisweb/wdis.
95. T. C. Spence, D. R. Stickle, "Corrosion-Resistant Casting Alloys," Advanced Materials & Processes 160, 1 (2002): pp. 51–54.
96. C. Houska, "Castings—Stainless Steels and Nickel-Base," NiDI Ref book no. 11 022 (Toronto, ON, Canada: NiDI, 2001): 88 pp.
97. P. Crook, personal communication (Kokomo, IN: Haynes International, 2003).
98. C. H. Clayton, T. E. Johnson, "Engineering Materials for Pumps and Valves," Chem. Eng. Prog., Sept. (1978): p. 54.
99. W. Barker, T. E. Evans, K. J. Williams, "Effect of Alloying Additions on the Microstructure, Corrosion Resistance and Mechanical Properties of Nickel-Silicon Alloys," Br. Corros. J. 5, 3 (1970): pp. 76–86.
100. D. C. Agarwal, "Nickel Base Alloys and Newer 6Mo Stainless Steels Meet Corrosion Challenges of the Modern Day Chemical Process Industries," Anti-Corrosion Methods and Materials 48, 5 (2001): pp. 287–297.
101. Alfa Laval brochure in "Corrosion," DKL Engineering (2003), http://members.rogers.com/acidmanual/corrosion.
102. L. Novak, S. Sjogren, "Performance of Alloy C-276 Plate Heat Exchangers for Sulphuric Acid Duties," Corrosion '85, paper no. 300 (Houston, TX: NACE, 1985).
103. L. Novak, personal communication (Lund, Sweden: Alfa Laval Lund AB, 2004).
104. M. Caruso, e-mail communication (Kokomo, IN: Haynes International, March 6, 2003).
105. M. G. Fontana, "Corrosion," Industrial and Engineering Chemistry 43, August (1951): p. 105A.

106. M. G. Fontana, "Corrosion Engineering," 3rd Edition (New York: McGraw-Hill, 1986): 465 pp.
107. H. H. Uhlig, "Corrosion Handbook" (New York: John Wiley, 1948): 1,188 pp.
108. M. Schussler, C. Pokross, "Corrosion of Tantalum," in J. R. Davis, ed., "Corrosion," Metals Handbook, 9th Edition, vol. 13 (Materials Park, OH: ASM International, 1987): pp. 725–739.
109. D. F. Taylor (1956) in M. Schussler, C. Pokross, "Corrosion of Tantalum," in J. R. Davis, ed., "Corrosion," Metals Handbook, 9th Edition, vol. 13 (Materials Park, OH: ASM International, 1987): p. 730.
110. M. Renner, K. Andersson, D. Michalski-Vollmer, "Application Limits of Ta and Ta-2.5% W for Sulfur Acid Handling," Metals and Corrosion 49 (1998): pp. 877–887.
111. Anon, "Niobium," Technical Data Sheet NioNio-056 (Albany, OR: Wah Chang, 2003): 42 pp.
112. E. Rabald, "Corrosion Guide," (Amsterdam, Netherlands: Elsevier Scientific Publishing Co., 1968): pp. 466–474.
113. Various authors, in Sulfuric Acid section, CD, "Dechema Corrosion Handbook" (Frankfurt, Germany: Dechema eV, 2001).
114. DuPont (1986) in C. P. Dillon, ed., "Concentrated Sulfuric Acid and Oleum," vol. MS-1, Materials Selector for Hazardous Chemicals (St. Louis, MO: MTI Inc., 1997): 212 pp.
115. Anon, "Materials for the Handling and Storage of Concentrated (90 to 100%) Sulfuric Acid at Ambient Temperatures," RP0391 (Houston, TX: NACE International, latest edition).

9

Corrosion of Metals and Alloys in Weak and Intermediate-Strength Acid

The behavior of metals and alloys in weak and intermediate-strength sulfuric acid is strongly influenced by the acid concentration, temperature, and presence of impurities. Corrosion behavior of common metals and alloys will be discussed in this chapter in three ranges of acid strength: 0–5%, 5–25%, and 25–70%.

pH Range

When dealing with solutions of pure, strong acids in distilled water, the concentration of acid is directly related to pH. A convenient mnemonic device is that a pH of 4 is approximately 4 ppm of acid (i.e., 3.65 ppm HCl, 4.9 ppm sulfuric, etc.). Because the pH scale is logarithmic, the following relationships obtain:

pH	**% Sulfuric Acid**
4	0.00049
3	0.0049
2	0.049
1	0.49
0	4.9

Negative values are no longer linear because of diminished hydrogen ion activity, i.e., a pH of minus 1 is <49%. At pHs *higher* than 4–4.5, the corrosivity of water solutions of acids toward iron and steel is primarily governed by dissolved oxygen (DO).

The addition of other chemical species may buffer or otherwise modify this relationship. In appraising the possible corrosive nature of an acidic stream, the total acidity (TA) may be a more important consideration than pH.[1]

Sulfate salts other than sodium and potassium sulfate give low pHs because they are the product of a weak base and a strong acid. Ferric sulfate, cupric sulfate, and ammonium sulfate (in the absence of an excess of free ammonia) form acid solutions. The ferric salt is the strongest acid of the three, but less reliable as an oxidant because both ferrous and ferric ions are inherently stable. Cupric sulfate is a more powerful oxidant because the cuprous ion is unstable ($2\ Cu^+ \longrightarrow Cu^{++} + Cu^\circ$). Ammonium sulfate is a much weaker acid because ammonium hydroxide is a moderately strong base (less so than sodium or potassium hydroxide, but more so than ferrous hydroxide, for example).

Oxidizing and Reducing Conditions

Contaminants in dilute sulfuric acid may change its inherently reducing nature and profoundly affect its corrosion characteristics as to whether attack will be general or localized. Localized forms of attack are often ion-specific, e.g., the effect of chlorides.

Oxidizing Contaminants

The most common oxidizing species in dilute sulfuric acid is dissolved oxygen (DO), with a solubility of about 42 to 48 mg/L in this concentration range (0–5%).[2]

Corrosion test data in C.P. acid in up to 5% acid concentration show that DO greatly increases corrosion of copper, alloy 400 (N04400), and alloy 200 (N02200) as compared with rates under anaerobic conditions, whereas DO can maintain passivity of stainless-steel alloys at ambient temperatures. Although the effect of DO is often noted in laboratory test data, more powerful oxidants are encountered in industrial practice, notably ferric and cupric salts. A first approximation of overall oxidizing capacity of an acidic solution can be made by iodometric analysis (i.e., measuring the ability of a solution to liberate iodine from potassium iodide) and calculating the apparent equivalent ferric or cupric ion concentration.

Reducing Contaminants

Reducing contaminants in dilute sulfuric acid are encountered as both inorganic and organic species. Inorganic reductants may be cations (e.g., stannous, antimonious, arsenious) or anions (e.g., sulfides, fluorides, chlorides). When halides are present in a low-pH solution, they act substantially like the halogen acids themselves (e.g., H^+ and Cl^-). Chloride-contaminated sulfuric acid behaves like a mixture of sulfuric and hydrochloric acid (see Chloride Ion Contamination in Chapter 10).

Organic contaminants may be inherently reducing or may simply consume DO to produce anaerobic conditions.

Weak and Intermediate-Strength Acid (0–5%)

The lowest concentrations of "dilute" sulfuric acid to be considered are those in the pH range 4 to 0 (0–5% acid). Sulfuric acid streams of this order of concentration are found in waste waters, pH control, metal pickling, and various chemical and petrochemical processes. It is in this range of concentration that corrosion data and materials-selection advice are most difficult to find.

From a practical standpoint, corrosion rates in excess of about 20 mpy (0.5 mm/y) are usually intolerable in terms of reasonable equipment life and process contamination. In this and subsequent chapters, we attempt to confine our discussions to conditions in which the materials have a reasonable chance of practical application. Corrosion rates should be limited to a maximum of about 4–5 mpy (0.13 mm/y) for thin-wall construction (e.g., heat-exchanger tubing).

Furthermore, alloys that depend for their corrosion resistance upon a passive film (e.g., the stainless steels) are either resistant or not (i.e., they show very high rates upon activation, as discussed in Chapter 7). When rates of 10–20 mpy (0.25–0.5 mm/y) or higher are reported for this type of alloy, it usually indicates that the alloy has been cycling between active and passive conditions and may be inherently untrustworthy for the applicable conditions. Aqueous solutions of acid salts also give acidities of this order of magnitude.

Aluminum

Aluminum may be used to handle dilute sulfuric up to 1% concentration, depending upon the nature and extent of contamination, to about 25°C (77°F). One source reports useful resistance to 10% at room temperature. Although there is some attack, the attack is not sufficiently rapid to prevent its use in special applications.[3] Some use has been made of aluminum pipe to handle mine-water from soft coal, although acid mine-waters are generally corrosive. Aluminum has been used to process thin, aqueous slurry containing 1% sulfuric acid at room temperature.[4] An aluminum condenser has been used to cool sludge vapors from coking pots, giving an acceptable life despite some corrosion. Corrosion was <1 mpy (<0.025 mm/y) up to 70°C (158°F) in a 1.5% acid containing 45% ammonium sulfate.[5] Rates in sulfuric acid of pH 1 at room temperature should be less than 0.3 mm/y (12 mpy).[6] Because of the dubious response of aluminum, especially in the presence of contaminants, confirmatory corrosion tests are recommended whenever its use is contemplated. Furthermore, the use of aluminum equipment in plants is often jeopardized by potential external corrosion problems arising from insulation and related materials of an alkaline nature.

Chloride contamination can lead to pitting and crevice corrosion of aluminum. Contamination with heavy-metal salts (e.g., copper, mercury, and lead) can cause cementation attack.

Iron and Steel

Steel and low-alloy steel are unsuitable in dilute sulfuric acid, although organic corrosion inhibitors can diminish corrosivity somewhat, e.g., in chemical cleaning using inhibited dilute acid.

Unalloyed gray cast iron and ductile cast iron are unsuitable in dilute sulfuric acid. Inhibitors are ineffective, owing to the galvanic influence of the contained discrete graphite particles.

Alloy Cast Irons (Ni, Si) Although nickel additions improve resistance, corrosion rates for a nickel cast iron (F41000) at 30°C (86°F) are on the order of 17 mpy (0.4 mm/y).[7]

Silicon cast irons (e.g., F47003) are fully resistant to the atmospheric boiling point in the absence of fluoride contamination.

Stainless Steels

All stainless steels can be classified into three groups according to metallurgical structure and response to heat treatment. These are the ferritic, martensitic, and austenitic groups. Further subdivisions include duplex alloys with austenitic/ferritic microstructures and precipitation-hardening (PH) grades strengthened by an age-hardening treatment. Stainless steels are commonly used in weak sulfuric acid.

Surface contamination with chlorides prior to service (e.g., piping shipped or stored under marine atmospheres) will result in formation of HCl *in situ* upon exposure to sulfuric acid. A preliminary wash with potable or other low-chloride water is mandatory under these conditions.

Ferritic, Martensitic, and PH Stainless Steels The ferritic stainless steels (e.g., S43000) have no application in this concentration of acid except in the presence of strongly oxidizing contaminants (Cu^{++}, Fe^{+++}, Sn^{+++}, or nitric acid). Halide contamination encourages localized corrosion of these steels. The corrosion rate of grade 444 (S44400) in 2% sulfuric acid at 30°C (86°F) was 10.2 mm/y (400 mpy), while type 304 (S30400) austenitic stainless steel corroded at only 0.03 mm/y (1.3 mpy).[8] The super-ferritic grades (e.g., S44625) should have approximately the same or slightly better resistance than S31603. Data for some superferritic stainless steels compared with other alloys in this range of acid strength are presented below, Table 9.9.

Martensitic stainless steels should not be exposed to dilute sulfuric acid and, in the hardened condition, are subject to hydrogen-assisted cracking (HAC).

The precipitation-hardening grades (e.g., S17400, S15700) are somewhat less resistant than their 18-8 austenitic counterparts.

Duplex Stainless Steels The duplex grades are generally not suitable in this concentration of sulfuric acid in the absence of strongly oxidizing contaminants. There is an inherent tendency to develop galvanic corrosion between the ferrite and austenite phases under reducing conditions.

Although the lower duplex grades would not be suitable above about 70°C (158°F) in nominally pure acid, the superduplex grades do have useful resistance, as discussed in High-Performance Alloys, below.

Austenitic Stainless Steels The corrosion resistance of the conventional 18-8 grades—types 304 (S30400), 316 (S31600), and 317 (S31700)—and their low-carbon analogs depends upon aeration and temperature. Types 316 (S31600) and 317 (S31700) will tolerate somewhat more severe conditions in dilute sulfuric acid than will type 304 (S30400), Table 9.1.[9]

Various contaminants in the acid can have a significant influence on the corrosion of S30400, S31600, and S31700, Table 9.2.[9] These data illustrate the strong effect that oxidizing or reducing species can have on the corrosion rate of these stainless steels.

The standard stainless steels, S30400, S31600, and S31703, have reasonable resistance in this very weak acid, even at elevated temperatures, Figure 9.1.[10] These data also show the better resistance of the molybdenum-containing grades S31600 and S31700 over the standard 304 grade, S30400, in weak acid.

The passivity in a given test may be strongly influenced by the concentration of dissolved oxygen, while more powerful oxidizing or reducing contaminants can

Table 9.1 Corrosion Rates mpy (mm/y) of Types 304, 316, and 317 SS in Weak Acid

% H_2SO_4	Temperature °F (°C)	Type 304 (S30400)	Type 316 (S31600)	Type 317 (S31700)
0.25	140 (60)	36 (0.91)	0	—
	BP	2.7 (0.07)	—	—
0.50	100 (38)	37 (0.94)	0	0
	150 (66)	22.5 (0.57)	0	0
	200 (93)	790 (20)	3.8 (<0.10)	0
	BP	—	54 (1.37)	—
1.00	RT	25 (0.64)	0	0
	100 (38)	57.5 (1.46)	0	0
	150 (66)	220 (5.6)	0	0
	175 (79)	380 (9.7)	0	0
	200 (93)	790 (20)	0	0
	B.P.	316 (8.0)	49 (1.25)	—
2.50	RT	31 (0.79	0	—
	104 (40)	36 (0.91)	0	—
	140 (60)	160 (4.1)	3.6 (0.09)	—
	176 (80)	—	36 (0.91)	—
3.00	100 (38)	67 (1.7)	0	0
	150 (66)	388 (9.9)	8.2 (0.21)	0
	175 (79)	528 (13.4)	40 (1)	58 (1.5)
	200 (93)	1,300 (33)	94 (2.4)	120 (3.1)
5.00	RT	47 (1.2)	0	0
	100 (38)	222 (5.6)	0	0
	150 (66)	1,100 (28)	36 (0.91)	3.8 (0.1)
	175 (79)	1,630 (41)	740 (18.8)	90 (2.3)
	200 (93)	5,400 (137)	140 (3.6)	230 (5.8)

completely govern the corrosion mechanism. A comparison between these 18-8 grades and the acid-resistant, nickel-rich austenitic alloys is given below. Active-passive regions for types 304 (S30400) and 316 (S31600) in reagent-grade acid vary with temperature and acid strength, Figure 9.2.[11] The as-mixed line in this figure is taken from earlier work.[12] Note that the activation temperature for type 316 (S31600) is <40°C (104°F) in air-free 5% acid.

Under oxidizing conditions, e.g., with small percentages of nitric acid (e.g., 500 to 1,500 ppm), the limiting temperatures are much higher in this range of sulfuric acid.

Under reducing conditions, activation and rapid corrosion can ensue. However, despite anticipated organic contaminants, a 0.4% acid from the rubber industry showed similar characteristics as in C.P. solutions for S30400 and S31600 at 21–43°C (70–110°F). A 2% acid (approximately) from the pulp and paper industry was noncorrosive to both S30400 and S31600 (unlike C.P. acid in the 1–2.5% range), possibly due to chlorate inhibition.[9]

Intergranular corrosion would be a problem if the non-L, regular-carbon grades were exposed in the sensitized condition (i.e., as-welded or thermally stress-relieved).

Table 9.2 Corrosion Rate mpy (mm/y) of Types 304, 316, and 317 SS in Contaminated Weak Acid

% H_2SO_4	Contaminants	pH	Temperature °F (°C)	Type 304	Type 316	Type 317
0.03–0.05	15% glycol	—	130–150 (54–66)	<0.1[d] (0.002)	<0.1[d] (0.002)	<0.1 (0.002)
0.03–0.05	15% glycol	—	210 (99)	<0.1[a d] (0.002)	<0.1[a d] (0.002)	<0.1 (0.002)
0.1	W and Mo salts	2–3.5	170–185 (77–85)	147[c] (3.73)	0.4 (0.01)	—
0.1	0.04% SO_2	2	164–177 (73–80)	2.3[c] (0.06)	<0.1[d] (0.002)	<0.1 (0.002)
0.02–0.2	Clay; organics; iron-chloride vapors	—	125–305 (52–150)	10[c d e] (0.25)	5.5[c] (0.14)	—
0.38	Chlorides	—	70–110 (21–43)	8.8[d] (0.22) 4[d] (0.10)	<0.1[d] (0.002)	—
<1	Iron sulfates	1.8–4.5	430–470 (806–878)	16 (0.04)	43 (10.9)	—
0.027–1.87	<150 ppm Cl^-	1.1–8.1	120–180 (49–82)	<0.1 (0.002) 0.1 (0.002)	<0.1 (0.002) 0.2 (0.005)	—
2.0	Saturated NaCl	—	120–128 (49–53)	7.2[a d] (0.18)	5.9[a] (0.15)	—
2.5	$CuSO^4$ 0.1%; alcohols	—	200–250 (93–127)	73 (1.85)	20 (0.51)	18.8 (0.48)
4.0	Large amounts of NaCl	—	70–82 (21–28)	32[c d] (0.81)	25[a d] (0.63)	—
5.0	3–4% $ZrSO^4$	—	90 (32)	0	1 (0.254)	—
5.0	Copper 0.5–0.56 oz/gal	—	176–185 (80–85)	0.3[d] (0.07)	0.1[d] (0.002)	—
5.0	Sodium dichromate 1.11 lb/gal	—	70–85 (21–29)	0.3[d] (0.07)	<0.1[d] (0.002)	—

[a] Slight pitting (1–5 mils)
[b] Moderate pitting (5–10 mils)
[c] Severe pitting (>10 mils)
[d] Crevice corrosion
[e] scc

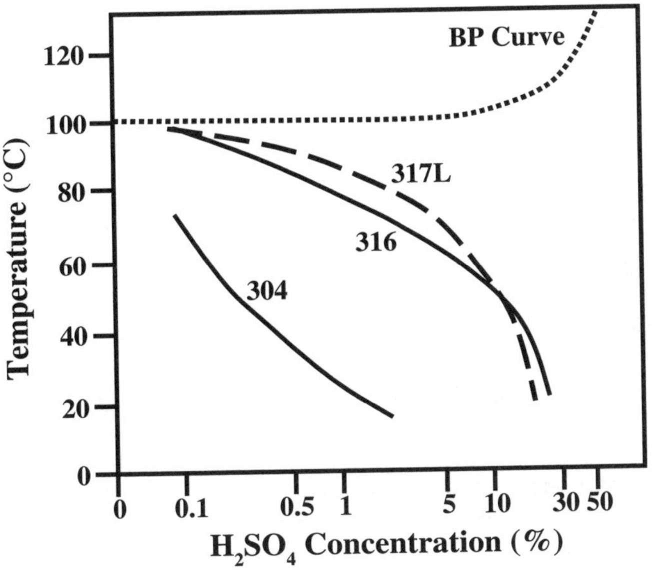

Figure 9.1 Isocorrosion Curves at 4.4 mpy (0.11 mm/y) for Types 304, 316, and 317L Stainless Steels

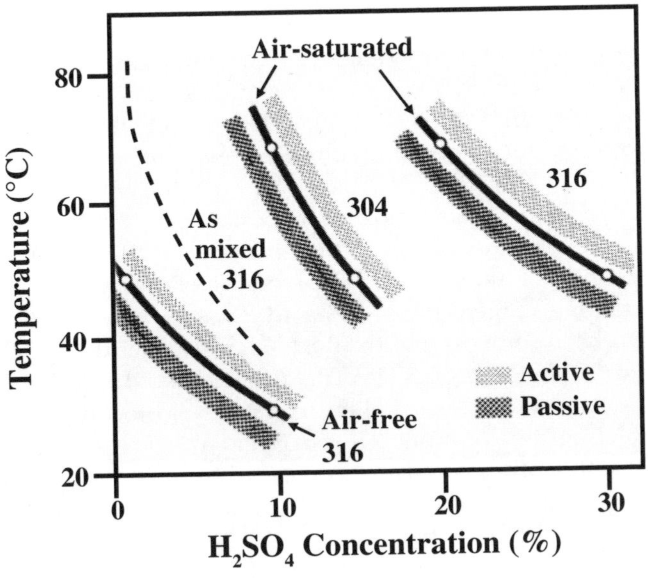

Figure 9.2 Effect of Air on the Active-Passive Behavior of Types 304 and 316 Stainless Steels in Weak and Intermediate-Strength Acid

High-Performance Alloys

The classic austenitic alloy for dilute sulfuric acid service was alloy 20 (20% Cr, 29% Ni, 3% Cu, 2% Mo) and the cast variant, CN7M (N08007). In modern, wrought niobium-stabilized products, the nickel range is 4.5–7.5% higher (N08020; 32–38% Ni) than the cast form (N08007), which conforms to the original nominal 29% nickel composition. The modern alloys are resistant to pure acid in the 0–5% range to the atmospheric boiling point, although the corrosion rate of N08020 is 16 mpy (0.41 mm/y) in boiling 5% acid. The original alloy 20 materials with lower nickel content can be attacked significantly at slightly higher concentrations above about 88°C (190°F). Unfortunately, the temperature limitations of such alloys have sometimes been overlooked in practice because of the excellent reputation of such alloys for this type of service. Higher-molybdenum versions of alloy 20 have also been developed to better resist chloride contamination, e.g., alloy 20Mo-6 (N08026).[13]

In 5% acid at 80°C (176°F), both N08904 and alloy 20 showed <1 mpy (<0.025 mm/y) in one test; in another, alloy 20 corroded at 90 mpy (2.3 mm/y), suggesting that these alloys are borderline at this temperature.[1]

New austenitic stainless steels, also known as superaustenitic stainless steels, have been developed primarily for improved resistance to chloride ion effects. These alloys have a composition of approximately 20% Cr, 25% Ni, 6% Mo and are exemplified by alloy 254SMO (S31254) and alloy 6XN (N08367). They have a resistance somewhat similar to a 20-type alloy in this acid range, Figure 9.3.[14–16] For more severe chloride contamination, even more highly alloyed grades such as high-chromium 6% molybdenum alloys (e.g., alloy 31 [N08031]) and alloys with 7–8% molybdenum (e.g., alloy 654 SMO [S32654]) are now available.[17] The corrosion behavior of some high-molybdenum alloys is compared with that of standard stainless steels, titanium, and other high-performance alloys in 0–70% sulfuric acid in Figure 9.8 below.

The superduplex stainless steels have useful resistance in low and intermediate acid; for example, alloy 7Mo-PLUS® (S32950) is said to have a rate of not more than 5 mpy (0.13 mm/y) at up to the atmospheric boiling point in 5% acid.[18] Even better resistance is shown by the more highly alloyed (subgroup D-4) grades, such as alloy 2507 (S32750).[17] Annealed alloy 255 (S39255), on the other hand, corrodes at 46 mpy (1.17 mm/y) in boiling reagent grade 5% acid.[7] The corrosion resistance of duplex stainless steels has been compared with 18-8 (304; S30400) and 17-12-2.5 (316; S31600) in 0–70% sulfuric acid, Figure 9.4.[19,20] The proprietary duplex alloy Zeron 100® (S32760) was developed to have excellent resistance to localized corrosion in seawater and high-chloride solutions. It also has good resistance to weak and intermediate-strength sulfuric acid, as shown in this figure.

Alloys with high concentrations of alloying elements such as chromium, nickel, molybdenum, and copper have been assigned UNS numbers in the N series (and specifications in the ASTM B series for nonferrous metals) when the total alloy content exceeds 50%. These alloys are included here as high-performance alloys; true nickel-based alloys with >50% nickel are discussed below.

Lacking molybdenum and copper, alloy 800 (N08800) is not usually considered for dilute sulfuric acid service. Its order of resistance should be very similar to type 304 (S30400) stainless steel. Laboratory tests in 5% acid at 50°C (122°F) measured rates of 20 mpy (0.5 mm/y), increasing to 50 mpy (1.3 mm/y) with aeration.[5] Once activation has occurred, DO aggravates corrosion by acting to enhance cathodic depolarization; that is, it is not a strong enough oxidant to effect repassivation.

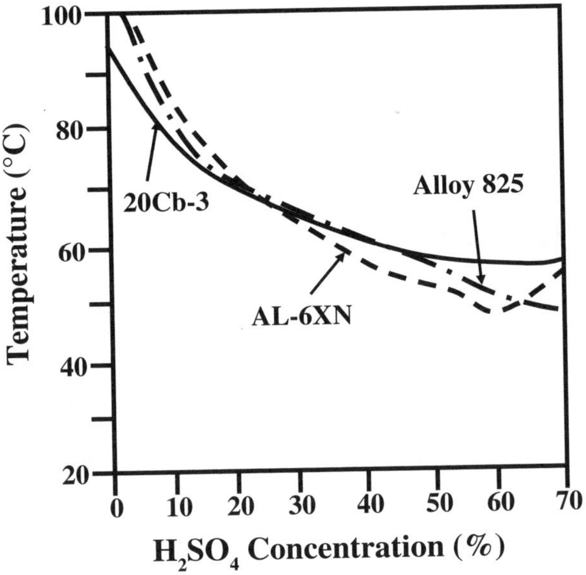

Figure 9.3 Isocorrosion Curves at 0.1 mm/y (3.9 mpy) of Alloys 825, 20Cb-3, and AL-6XN® in 0–70% Sulfuric Acid

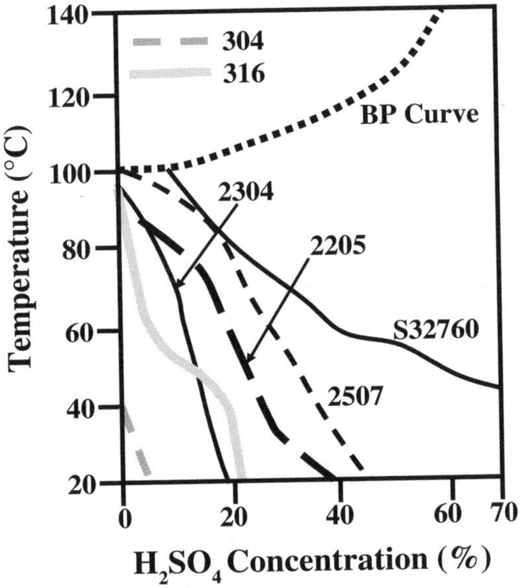

Figure 9.4 Isocorrosion Curves at 0.1 mm/y (3.9 mpy) of Duplex Stainless Steels Compared with 304 and 316 in 0–70% Sulfuric Acid

The titanium-stabilized alloy 825 (N08825) has resistance substantially equivalent to alloy 20Cb-3 (N08020) in this concentration range. Alloy G-30 (N06030) is also satisfactory, although its corrosion resistance is somewhat diminished in the presence of 200 ppm chloride ion contamination. On the other hand, its resistance is enhanced by oxidizing contaminants. In boiling 2% sulfuric acid its corrosion rate is 8 mpy (0.20 mm/y) compared with 6 mpy (0.15 mm/y) for the nickel-based alloy 625 (N06625).[21]

Nickel and Its Alloys

The alloys containing more than 50% nickel must be considered in two categories, chromium-free and chromium-bearing. The second category may be divided into those that also contain molybdenum and those that do not.

Chromium-Free Nickel 200 (N02200) will resist up to 2% sulfuric acid at room temperature in the absence of DO or oxidizing agents. In 5% acid at 30°C (86°F), the rate is about 9 mpy (0.23 mm/y) unaerated and 61 mpy (1.55 mm/y) when air-saturated.[7] Like the less expensive and more commonly used alloy 400 (N04400) (see below), nickel is very susceptible to increased corrosion by oxidizing ion contamination.

Laboratory data show that the nickel-copper alloy 400 (N04400) is able to resist boiling 5% sulfuric acid, with a rate of about 3.4 mpy (0.09 mm/y).[7] This may be due to the removal of dissolved oxygen (DO) in the boiling solution, because the alloy otherwise shows high corrosion at 5% concentration in the 60–90°C (140–194°F) temperature range, with the maximum rate at about 80°C (175°F). Contamination with DO or oxidizing ions causes a large increase in corrosion, as do even relatively low velocities (<1 ft/s). One might expect autocatalytic corrosion from alloy 400 (N04400) under some circumstances because the alloy contains about 35% copper, but this effect has not been reported. However, a weak acid solution contaminated with cupric ions reportedly caused rapid corrosion, particularly of welds, with simultaneous copper plating of a very adherent nature.

The molybdenum-containing nickel alloy B-2 (N10665) is reported to corrode at 0.5 mpy (<0.02 mm/y) in boiling 2% and at 3 mpy (0.08 mm/y) in boiling 5% acid.[7] Rates are as high as 10 mpy in the 50–90°C (122–194°F) temperature range. The presence of DO increases corrosion but should not be a major factor, although even trace amounts of stronger oxidants (e.g., Fe^{+++}, Cu^{++}) will cause greatly accelerated attack. The cast alloy grade N-7M (N30007) is similar in corrosion resistance, as are alloys B-2 (N10665), B-3 (N10675), and B-4 (N10629).

Chromium-Bearing The nickel-chromium alloy 600 (N06600) is of little practical use in dilute sulfuric acid. Although rates of <4 mpy (<0.1 mm/y) are reported in boiling 0.16% acid, higher rates are observed in aerated 1% acid at lower temperatures. DO accelerates corrosion, although stronger oxidants enhance passivity so this alloy can be used in acid mine waters and brass-pickling solutions. In 5% acid, rates were 9, 10, and 30 mpy (0.23, 0.25, and 0.76 mm/y) at 30, 60, and 80°C (86, 140, and 176°F), respectively, under anaerobic conditions but were increased 5–8-fold by aeration.[22]

The nickel-chromium-molybdenum alloys include N06625, N10276, N06022 and similar compositions. Alloy 625 (N06625) is attacked at >5 mpy in 0–5% acid above about 70°C (158°F). The more highly alloyed products show rates of <5 mpy (0.13 mm/y) in this acid range up to about 90°C (194°F) and are very tolerant of chloride contamination. Also, their resistance is enhanced rather than diminished by oxidizing contaminants. Alloy 686 (N06686) is particularly resistant to weak acid and has a corrosion rate of <5 mpy (<0.13 mm/y) in up to 5% acid at temperatures exceeding the boiling point, up to about 120°C (250°F), see Figure 9.9 below.

Alloy 59 (N06059) has good resistance to weak and intermediate-strength acid, even if chlorides are present. In 5% acid at 60°C (140°F) its corrosion rate was 0.004 mm/y (0.16 mpy), while at 100°C (250°F) the rate was 0.104 mm/y (4.1 mpy).[23]

Copper and Its Alloys

Copper and its alloys (other than the high-zinc yellow brasses, which can suffer dezincification) are inherently resistant to dilute sulfuric acid as long as DO or other oxidants are absent. Their use, however, is fraught with danger because the corrosion products (i.e., cupric ions, Cu^{++}) are themselves of an oxidizing nature and the corrosion is therefore autocatalytic (increasing exponentially as oxidizing ions accumulate in the solution).

Copper, phosphor (tin) bronzes, and aluminum bronzes are usefully resistant under controlled reducing (or at least non-oxidizing) conditions.

Lead

Below 5% acid, corrosion increases and 3% antimonial lead (L52901) is recommended.[24] The conventional chemical lead (L51120) showed rates of 6 mpy (0.15 mm/y) in nickel sulfate and zinc sulfate solutions at 100°C (212°F). Resistance is due to the formation of insoluble lead sulfate films, which are susceptible to mechanical damage, erosion, abrasion, etc., with attendant higher rates of attack.

Lead and its alloys are usefully resistant to 5% acid up to the atmospheric boiling point, with rates typically <5 mpy (0.13 mm/y). Rates of 1–5 mpy (0.03–0.13 mm/y) were reported at 47°C (117°F) in a 5% sulfuric acid solution saturated with ammonium sulfate.[25]

In many modern industrialized countries, lead has fallen into disuse because of poor mechanical properties (susceptibility to "static fatigue" in free-standing or loose-lined construction), the severe shortage of skilled craftsmen ("lead burners"), and health and toxicity considerations. Nevertheless, it is still offered as an automatically applied lining (e.g., in Germany). In other parts of the world it is apparently still being used where its corrosion resistance is deemed acceptable.

Reactive and Refractory Metals

All four reactive metals (titanium, zirconium, tantalum, and niobium) will suffer corrosion in the presence of fluoride contamination, due to the formation of soluble fluoride complexes with the possibility of hydriding.

Titanium and Its Alloys Because of the reducing nature of this sulfuric acid range, unalloyed titanium would not normally be considered. While titanium (e.g., R50400) shows less than 1 mpy (<0.025 mm/y) at 60°C (140°F) in aerated 1–3% acid, it will corrode at anywhere from <1 to >200 mpy (<0.025 to >5 mm/y) at 100°C (212°F).[4] At 5% acid, even with aeration, corrosion is severe at 60°C (140°F). On the other hand, oxidizing agents—e.g., cupric or ferric ions (at 1,000 ppm), nitric acid, and chlorine—can maintain passivity.[26]

The titanium-palladium grade 7 (R52400) and titanium-nickel-molybdenum grade 12 (R53400) alloys were developed to extend their resistance to reducing acids. In contrast with unalloyed titanium, grade 7 shows <1 mpy (<0.025 mm/y) in 1% at 204°C (399°F) with aeration, and only 5 mpy (0.127 mm/y) at 190°C (374°F) under nitrogen. In 5% acid at 70°C, under nitrogen, grade 7 gives a rate of only 6 mpy (0.15 mm/y), compared with active dissolution of R50400. The R53400 alloy is only slightly less resistant, showing borderline resistance from <1 to 36 mpy (<0.025 to 0.91 mm/y) in 0.75% acid at 100°C (212°F) and from <1 to 65 mpy (<0.025 to 1.65 mm/y) in 3% acid at 66°C (151°F).[4]

The titanium-ruthenium grades 26 (R52404) and 27 (R52254) have similar performance to the titanium-palladium grades in 0–5% sulfuric acid.[27]

Zirconium Zirconium alloys (e.g., R60702, R60705) will resist these concentrations of sulfuric acid up to and above the atmospheric boiling point, i.e., up to 250°C (482°F). There may be problems with pyrophoric corrosion products in the event of contamination with oxidizing species, chlorides, and elevated temperatures.

Tantalum Tantalum will resist 0–5% sulfuric acid to the atmospheric boiling point and above the atmospheric boiling point to 250°C (482°F).

Niobium Niobium will resist 0–5% sulfuric to the atmospheric boiling point. Corrosion rates of 5 mpy (0.13 mm/y) in pure sulfuric acid are reported. The presence of Fe^{+3} and Cu^{+2} may reduce the corrosion rate.[28]

Cobalt Alloys

Hard-facing alloys of the 27% Cr, 6% Mo variety (e.g., R30021) show rates of less than 2 mpy (<0.05 mm/y) in these concentrations to at least 66°C (150°F), as do the high-tungsten (tungsten >13%) variety, e.g., R30001, R30004.[4]

Precious/Noble Metals

Silver, gold, and platinum are resistant to dilute sulfuric acid to the atmospheric boiling point.

Weak and Intermediate-Strength Acid (5–25%)

Dilute sulfuric acid in the 5–25% concentration range is above the strength that can conveniently be described in terms of pH. It is inherently a reducing acid (in the absence of oxidizing contaminants), attacking base metals with the evolution of hydrogen as the cathodic reaction.

As noted above, corrosion rates in excess of about 20 mpy (0.5 mm/y) are usually intolerable, and rates should be limited to a maximum of about 4–5 mpy (0.10–0.13 mm/y) for thin-wall construction (e.g., heat-exchanger tubing). For passive alloys, rates of <5 mpy (<0.13 mm/y) are usually sought because of their active/passive fluctuation under more severe conditions.

The previous discussion of oxidizing and reducing contaminants also applies in this concentration range, although the solubility of DO diminishes from 42 ppm in 5% to about 27 ppm in 25% sulfuric acid.[2]

Aluminum

Aluminum may be used to handle dilute sulfuric up to 10% concentration, depending on the nature and extent of contamination, to about 25°C (77°F). Although there is some attack, the corrosion rate will be <0.11 mm/y (<4.3 mpy) and is acceptable in special applications.[29]

Iron and Steel

Steels and low-alloy steels are unsuitable in this range of concentrations.

Unalloyed gray or ductile cast iron is also unsuitable in 5–25% sulfuric acid.

Alloy Cast Irons (Ni, Si) Only at room temperature under anaerobic conditions are the corrosion rates for a nickel cast iron (F41000) below 20 mpy (0.51 mm/y) in this range of acid strengths.[3]

At low levels of silicon, corrosion is increased in this range of acid strengths, but above a threshold level of silicon, the corrosion rate decreases, Table 9.3.[30]

Silicon cast irons (e.g., F47003) are considered resistant to the atmospheric boiling point and to about 8% acid, in the absence of fluoride contamination. However, rates >5 mpy (>0.13 mm/y) but below 20 mpy (0.51 mm/y) are expected in 15 and 25% acid at temperatures between 75 and 70°C (167 and 158°F), respectively.[31]

Table 9.3 Corrosion Rate mpy (mm/y) of Iron-Silicon Alloys in Various Concentrations of Sulfuric Acid at 50°C (122°F)

	% H_2SO_4		
% Si	5	10	20
0	19 (0.48)	17 (0.43)	54 (1.4)
1.5	124 (3.1)	116 (2.9)	117 (3)
2.5	224 (5.7)	197 (5)	200 (5.1)
7.5	249 (6.3)	353 (9)	506 (12.9)
9.5	274 (7)	365 (9.3)	711 (18)
11.5	—	767 (19.5)	—
14.5	0.088 (<0.01)	0.075 (<0.01)	—

Stainless Steels

All stainless steels can be classified into three groups according to metallurgical structure and response to heat treatment. These are the martensitic, ferritic, and austenitic groups. Further subdivisions include duplex alloys with austenitic/ferritic microstructures and precipitation-hardening (PH) grades strengthened by an age-hardening treatment. Stainless steels are commonly used in weak sulfuric acid.

Surface contamination with chlorides prior to service (e.g., piping shipped or stored under marine atmospheres) will result in formation of HCl *in situ* upon exposure to sulfuric acid. A preliminary wash with potable or other low-chloride water is mandatory under these conditions.

Martensitic, Ferritic, and PH Grades of Stainless Steels The martensitic and ferritic grades have no useful application in this concentration range. The superferritic grades do find some applications in this range of acid strength; these are discussed in a later section under High-Performance Alloys (see Table 9.9).

The precipitation-hardening grades (e.g., S17400, S15700) are somewhat less resistant than their 18-8 counterparts.

Duplex Stainless Steels There is an inherent tendency to develop galvanic corrosion between the ferrite and austenite phases under reducing conditions. The lower duplex grades, such as S32900, are not suitable above about 7.5% concentration of sulfuric acid above about 70°C (158°F) in the absence of strongly oxidizing contaminants. The superduplex grades are discussed in a later section under High-Performance Alloys.

Austenitic Stainless Steels In C.P. acid, the molybdenum-free grade, type 304 (S30400), and its low-carbon and stabilized variants, is, at best, resistant only to about 15% concentration and about 40°C (104°F) maximum under completely aerated conditions.[7] The active/passive boundaries for types 304 (S30400) and 316 (S31600), aerated and deaerated, are shown in Figure 9.2. These data show that type 316 (S31600) has very limited application in this range of acid strengths in air-free conditions and is only passive up to about 20% at 70°C (158°F) and 30% at 50°C (122°F). This figure also shows that type 304 (S30400) stainless steel is only passive in air-saturated 15% acid up to around 45°C (113°F).

The standard stainless steels, types 304 (S30400), 316 (S31600) and 317L (S31703) have limited resistance to this range of acid strengths (see Figure 9.1).

Other tests using somewhat different techniques of sample preparation showed 0.113 mm/y (4.46 mpy) corrosion rates at 20°C (68°F) for 304 (S30400) in 5% acid, at 40°C (104°F) for 316 (S31600) in 20% acid, and at 50°C (122°F) for 317 (S31700) in 20% acid.[32] Under anaerobic conditions the limiting concentration is more likely to be less than 10% at 30°C (86°F) for type 316 (S31600) and its low-carbon analog (S31603). This stainless steel will tolerate aerated 25% acid to about 50°C (122°F) and as high as 65–70°C (149–158°F) in aerated acid in the lower strengths in this range.

5% acid solutions from the textile industry are reportedly noncorrosive to both S30400 and S31600 at room temperature, contrary to laboratory tests in C.P. acid, which typically corrode the molybdenum-free grades at 10–50 mpy (0.25–1.5 mm/y) (probably varying with dissolved oxygen content). Presumably, the textile solutions contained unreported oxidizing species.[9]

An 8% metal-pickling acid at 54°C (130°F) was moderately corrosive to S30400 but only attacked S31600 at about 1 mpy (0.025 mm/y) (although alloy 20 suffered 6 mpy [0.15 mm/y]), undoubtedly due to oxidizing metal cations. A 10% metal-pickling acid in the 50–70°C (120–160°F) range gave rates of less than 1 mpy (0.025 mm/y) for S31600, S31700, and alloy 20. Inexplicably, N08825 was attacked at 13 mpy (0.33 mm/y) in this same environment. Similarly, a 10% acid from the sugar industry was noncorrosive to all three grades of stainless steel in the 20–67°C (68–152°F) range.

An acid of approximately 12% concentration from a metal-pickling operation was moderately corrosive to S30400 at about 88°C (190°F), some 13 mpy (0.31 mm/y) compared with inches per year in C.P. acid in the 10–20% range. S31600 had a rate of about 3 mpy (0.08 mm/y). Contrary to the data for C.P. acid, a field exposure in plant acid at approximately 20% concentration and 40°C (104°F) gave rates of <1 mpy (<0.025 mm/y) for S30400 and S31600, undoubtedly due to oxidizing contaminants.[9]

The corrosion rate for type 304 (S30400) in 5N acid (24.5%) at room temperature is about 1,000 mpy (25.4 mm/y).[33] Chloride contamination induces localized forms of attack. The general corrosion rate diminishes from 1,000 mpy (25.4 mm/y) to about 10 mpy (0.25 mm/y), but intergranular cracking occurs at 0.1 N NaCl (0.585%) and transcrystalline SCC at 0.5 N NaCl (2.93%) in this acid strength.[34] Of course, as mentioned previously, chlorides promote activation and accelerated general attack at elevated temperatures. Intergranular corrosion occurs if the regular-carbon grades are exposed in the sensitized condition, i.e., as-welded or thermally stress-relieved, at slightly elevated temperatures.

Under oxidizing conditions, the limiting temperatures are much higher in this range of sulfuric acid. However, aside from the dangers of crevice effects (in which oxidants cannot be replenished), there is a danger of intergranular attack (IGA). The Strauss Test (ASTM A393[35]), now obsolete, used a boiling 25% sulfuric acid containing cupric sulfate to detect susceptibility to IGA in sensitized 18-8 stainless steels, whereas the annealed material was passivated by the cupric ions. (The modern test for IGA is the *copper-accelerated* cupric sulfate test; ASTM A-262,[36] Practice E).

A byproduct 15% sulfuric acid from coal treatment in the temperature range 15–30°C (59–86°F) gave a corrosion rate of 24 mpy (0.61 mm/y) for S30400 (with severe pitting), although S31600 and N08020 were only attacked at slightly over 1 mpy (0.025 mm/y). Presumably, sulfides and chlorides leached from the coal cause reducing conditions toward the molybdenum-free grade.[9]

Mixtures of sulfuric acid with extraneous chemicals can profoundly alter corrosion characteristics of stainless steels, such as type 304 (S30400), 316 (S31600), and 317 (S31700), Table 9.4.[9] These data illustrate the effects of oxidizing species present in the acid—for example, the corrosion rate in boiling 10–15% acid containing copper salts shows negligible corrosion.

Laboratory tests in dilute C.P acid for three periods of 48 hours each at 80°C (176°F) show the high-manganese austenitic stainless steel with 22% Cr, 13% Ni, 5% Mn (S20910) to be superior to type 316 (S31600), Table 9.5.[13]

Table 9.4 Corrosion Rate mpy (mm/y) of Various Alloys in Contaminated Weak Acid

% H_2SO_4	Mix	Temperature °F (°C)	Type 304 (S30400)	Type 316 (S31600)	Type 317 (S31700)
5.5	8.9% selenious acid, 0.08% tellurous acid	150–160 (66–71)	80 (2)[c, d]	25 (0.64)[c, d]	—
6	11% alum	78 (26)	<0.1 (<0.002)	0.7 (0.02)	—
7–8	0.8–0.9 oz/gal $NaNO_3$	155–165 (68–74)	36 (0.91), 53 (1.3)[d]	0.1 (0.002), 3.1 (0.08)	0, 2.2 (0.06)
7.5–8.0	3% aluminum sulfate, 1% ferric sulfate, small amounts calcium and magnesium sulfates	200–210 (93–99)	25 (0.64)	6 (0.15)[a]	—
9	3% sodium sulfate, 1% HF, 1% orthosilic acid, 0.5% sodium fluorosilicate	80–120 (27–49)	15 (0.38)	7 (0.18)	—
5–10	25% Na_2SO_4, trace H_2S	100–140 (38–60)	53 (1.4), 62 (1.6)[c, d]	18 (0.46), 25 (0.64)[c, d]	3.3 (0.08), 4 (0.10)[c, d]
5–10	0.25% $CuSO_4$	100–200 (38–93)	<0.1 (<0.002)[e]	<0.1 (<0.002)[e]	—
10	MnO_2 sludge	95–122 (35–50)	78 (2)	8.3 (0.21)	8.7 (0.22)
10	1% MnO_2, some permanganic acid	131–208 (55–98)	4.5 (0.11)	4.7 (0.12)	5.3 (0.13)
13	$MnSO_4$, MnO, water	80–210 (27–99)	34 (0.86)	31 (0.79)	—
10–15	sulfuric acid sludge, large amounts of sludge oil	150–200 (66–93)	95 (2.4), 77 (2)	72 (1.8)	77 (2)
10–15	$CuSO_4$	BP	0.1 (0.002)[e]	0.1 (0.002)	—
8.5–15.6	0.4–1.0 selenious acid, small amounts H_2SO_3	70–80 (21–27)	32.4 (0.82)[c]	0.4 (0.01)[d]	—
17.5	3.5% H_2CrO_4	160–200 (71–93)	3 (0.08),[a] 2 (0.05)	1.5 (0.04)	—
18	3% NaCl	170–190 (77–88)	530 (13)	210 (5.3)	170 (4.3)
18.5	3% H_2CrO_4	180–185 (82–85)	9.1 (0.23), 3.1 (0.08),[a, e] 5.1 (0.13), 24 (0.61)	16 (0.41), 40 (1),[e] 7 (0.18), 14 (0.36)[e]	21 (0.53)[a]

continued

Table 9.4 Corrosion Rate mpy (mm/y) of Various Alloys in Contaminated Weak Acid *(continued)*

% H_2SO_4	Mix	Temperature °F (°C)	Type 304 (S30400)	Type 316 (S31600)	Type 317 (S31700)
20	<5% $CuSO_4$	140 (60)	188 (4.8)	14 (0.36)	15 (0.38)
20	3% HF	79 (26)	—	>1,000 (>25)	11 (0.28)
20	4% $Na_2Cr_2O_7$	150–160 (66–71)	14 (0.36)	23 (0.58)	11 (0.28)
19–22	3.5–4.5% $CuSO_4$, 2–3% $NiSO_4$ and 500 ppm Cl-	130–150 (54–66)	>0.1 (>0.002)	<0.1 (>0.002)	—
22–25	5–6% $Na_2Cr_2O_7$	150–160 (66–71)	21 (0.53)[c, f]	11 (0.28)	13 (0.33)

[a] Slight pitting (1–5 mils)
[b] Moderate pitting (5–10 mils)
[c] Severe pitting (>10 mils)
[d] Crevice corrosion
[e] Not resistant when sensitized
[f] Stress corrosion cracking

Table 9.5 Corrosion Rate mpy (mm/y) of High-Manganese (S20910) and Type 316 Stainless Steels in Weak Acid at 80°C (176°F)

% H_2SO_4	S20910	S31600
5	0.2 (<0.01)	33 (0.84)
10	15 (0.38)	112 (2.85)

Silicon-Rich Stainless Steels The 5% silicon stainless steels (e.g., alloy SX®; S32615) are not usually applied in this range of acid concentration but would be reasonably resistant in the absence of aggressive reducing contaminants. The modified SARAMET® alloy is also reasonably resistant to this range of acid strengths (see Figure 8.16).

High-Performance Alloys

This type of alloy includes the more highly alloyed austenitic, duplex, and ferritic stainless steels and some nickel-rich austenitic alloys.

Alloy 904L (N08904) is reportedly resistant to about 66°C (150°F).[37] Another source shows a limit of about 90°C (195°F) at 5% and 70°C (158°F) at around 25% acid.[19] Yet another source shows a temperature limit of between 90 and 100°C (194 and 212°F) for this range of acid strength and also shows that this copper-bearing alloy is more resistant than the copper-free N08320, which is otherwise similar in composition.[7]

The classic austenitic alloy for dilute sulfuric acid service is the original alloy 20 (20% Cr, 29% Ni, 3% Cu, 2% Mo [N08020]) and its cast analog, CN7M (N08007). In modern practice, alloy 20 has been replaced with alloy 20Cb-3 (20% Cr, 35% Ni, 3% Cu, 2% Mo [N08020]), resulting in better resistance, Table 9.6.[9,38]

Table 9.6 Corrosion Rate mpy (mm/y) of Alloy 20 Type Alloys Compared with Types 304 and 316 Stainless Steels in a Range of Acid Strengths

% H_2SO_4	Temperature °C (°F)	Type 304	Type 316	Alloy 20	Alloy 20Cb-3
8.7	80 (176)	460 (11.7)	<1 (<0.025)	25 (0.64)	5 (0.13)
10	15 (59)	20 (0.51)	<1 (<0.025)	<1 (<0.025)	<1 (<0.025)
	25 (77)	70 (1.8)	<1 (<0.025)	<1 (<0.025)	<1 (<0.025)
	40 (104)	292 (7.4)	4 (0.1)	<5 (<0.13)	<5 (<0.13)
	60 (140)	—	90 (2.3)	<5 (<0.13)	<5 (<0.13)
	66 (151)	>1,500 (>38)	80 (2)	80 (2)	<5 (<0.13)
	80 (176)	—	>150 (>3.8)	>150 (>3.8)	5 (0.13)
20	15 (59)	440 (11.2)	<1 (<0.025)	<1 (<0.025)	<1 (<0.025)
	40 (104)	440 (11.2)	110 (2.8)	25 (0.64)	<5 (<0.13)
	66 (151)	>1,500 (>38)	>150 (>3.8)	>150 (>3.8)	<5 (<0.13)
25	15 (59)	53 (1.3)	<1 (<0.025)	<1 (<0.025)	—
	40 (104)	>400 (>10)	>400 (>10)	—	<5 (<0.13)

Alloy 20Cb-3 (N08020) corrodes at 16 mpy (0.4 mm/y) in boiling 5% acid. Higher-molybdenum variants of alloy 20Cb-3 (N08020) have been developed specifically to resist chloride contamination (e.g., N08024 and N08026 with 4% and 6% molybdenum, respectively). The modified alloy 20Mo-6 (N08026) has about the same or lower corrosion rate as N08020 in up to 7% boiling acid but is much less resistant as acid strength increases. For example, N08020 corrodes at about 28 mpy (0.71 mm/y) in boiling 15% while N08026 corrodes at about 60 mpy (1.52 mm/y) in the same boiling reagent-grade acid.[13] The alloy 20Mo-6 (N08026) proves to be better than alloy 20Cb-3 (N08020) in 10–25% acid at 80°C (175°F).[13]

Other alloys with lower nickel content can be as good or better in some applications, depending on the specific chemistry of the sulfuric acid solution (see Figure 9.4). The beneficial effect of nickel content of stainless steels and nickel-based alloys has been demonstrated in 15% sulfuric acid at 80°C (176°F), Figure 9.5.[39]

Corrosion rates of various iron-based and nickel-based alloys in boiling 10% and 20% sulfuric acid are shown in Table 9.7. The data have been presented in descending order from least to most resistant in the boiling 10% acid. This table shows that none of the common iron-based alloys is suitable for these conditions. It also shows that data derived from different sources testing under nominally the same conditions are not always in agreement. This is largely due to variations in acid purity and testing techniques used.

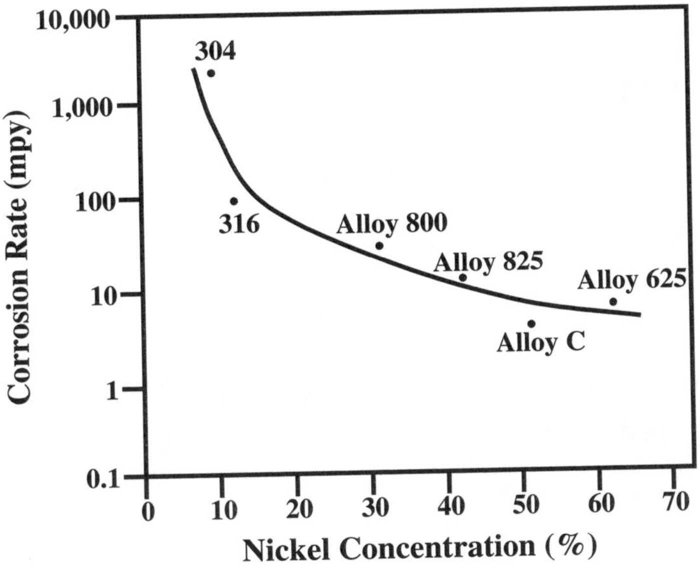

Figure 9.5 Effect of Nickel Content of Stainless Steels and Nickel-Based Alloys in 10% Sulfuric Acid at 80°C (176°F)

Other alloys have been developed (primarily for improved resistance to chloride ion effects) having a composition of approximately 20% Cr, 25% Ni, 6% Mo. These materials, exemplified by alloy 254SMO (S31254) and alloy 6XN (N08367), have resistance somewhat similar to a 20-type alloy in this acid range (see Figure 9.8 below). compiled with data from various sources. Figure 9.8 also includes data for alloy 31 (N08031), which has a similar molybdenum content but higher nickel and chromium content. For more severe chloride contamination, even more highly alloyed grades with 7–8% molybdenum (e.g., alloy 654 SMO [S32654]) are now available.[17]

Lacking molybdenum and copper, the nickel-iron-chromium alloy 800 (N08800) is not usually considered for dilute sulfuric acid service. Its order of resistance should be very similar to type 304 (S30400) stainless steel. Laboratory tests in 5% acid at 50°C (122°F) show rates of 20 mpy (0.5 mm/y), increasing to 50 mpy (1.3 mm/y) with aeration.[7] (Once activation has occurred, DO aggravates corrosion by acting to enhance cathodic depolarization; it is not a strong enough oxidant to effect repassivation.)

Corrosion rates for alloy G-30 (N06030) in 20% acid at 79°C (174°F) increased from <1 to >20 mpy (<0.025 to >0.51 mm/y) by the addition of 1–3% HF. The temperatures for the 5 mpy (0.13 mm/y) isocorrosion line diminish from about 93°C (199°F) at 5% to about 75°C (167°F) at 25%; 200 ppm chloride ion contamination depressed the isocorrosion line roughly another 10°C (50°F).[37]

Table 9.7 Corrosion Rates mpy (mm/y) for Various Alloys in Boiling 10% and 20% Sulfuric Acid

Alloy (UNS No.)	10%	20%
E-Brite® (S44627)	3,500 (89)	—
316L (S31603)	636 (16.2), 372 (9.4), 1,869 (47.5)	—
317L (S31703)	298 (7.6), 490 (12)	—
600 (N06600)	270 (9)	—
200 (N02200)	130 (3.3), 80 (2)	—
904L (N08904)	101 (2.6), 140 (3.6)	—
7Mo-PLUS® (S32950)	100 (2.5)	—
ULTIMET® (R31233)	99 (2.5)	—
2507 (S32750)	96 (2.4)	—
28 (N08028)	83 (2.1)	—
AL-6XN (N08367)	72 (1.8)	0.24 (<0.01)
2205 (S31803)	64 (1.6)	—
25-6MO (N08926)	59 (1.5)	—
27-7MO (S31277)	58 (1.5)	—
Ferralium 255 (S32550)	40 (1.0)	—
31 (N08031)	38 (0.96)	—
625 (N06625)	37 (0.9), 46 (1.2), 25 (0.6), 21 (0.5)	124 (3.15), 91 (2.3)
G-30 (N06030)	31 (0.8), 31 (0.8)	54 (1.37)
C-4 (N06455)	22 (0.6)	36 (0.9)
825 (N08825)	20 (0.5)	—
G-3 (N06985)	19 (0.5)	30 (0.76)
C-276 (N10276)	14 (0.4), 20–30 (0.5–0.8)	42 (1.1)
C-22 (N06022)	13 (0.3)	33 (0.84)
20 Cb-3 (N08020)	13 (0.3), 51 (1.3), 25 (0.64), 16 (0.4)	8.4 (0.2), 38 (0.97)
59 (N06059)	5.9 (0.15)	—
400 (N04400)	2.4 (0.6)	7.5 (0.19)
Allcorr (N06110)	2 (<0.1)	—
B-2 (N10665)	1.5 (<0.1), 2.0 (0.05)	0.7 (<0.02)
B-3 (N10675)	0.8 (<0.1)	1.2 (0.03)
29-4-2 (S44800)	0.8 (<0.1)	—
Tantalum (R05210)	<1 (<0.1)	—

Table 9.8 Corrosion Rate mpy (mm/y) of Iron-Based and Nickel-Based Alloys in Sulfuric Acid at Various Strengths and Temperatures

	60°C (140°F)			80°C (176°F)			100°C (212°F)		
% H_2SO_4	Alloy 20	Alloy C-276	Alloy 31	Alloy 20	Alloy C-276	Alloy 31	Alloy 20	Alloy C-276	Alloy 31
20	<5 (<0.13)	<1 (<0.025)	<0.1 (<0.002)	10 (0.25)	4 (0.10)	<0.1 (<0.002)	>25 (>0.64)	>1 (>0.025)	0.3 (0.008)
40	<5 (<0.13)	<2 (<0.051)	<0.1 (<0.002)	10 (0.25)	3 (0.08)	<0.2 (<0.005)	>25 (>0.64)	10 (0.25)	0.6 (0.015)
60	>5 (>0.13)	<2 (<0.051)	<0.1 (<0.002)	11 (0.28)	4 (0.10)	0.4 (0.01)	>50 (>1.27)	11 (0.28)	1 (0.025)
80	5 (0.13)	<1 (<0.025)	0.2 (0.005)	18 (0.46)	15 (0.38)	0.8 (0.02)	>50 (>1.27)	240 (6.1)	240 (6.1)

The nickel-rich alloy 31 (N08031; 31% Ni, 27% Cr, 32% Fe, 6.5% Mo) was tested against similar alloys with lower molybdenum content in various intermediate strengths of sulfuric acid, Table 9.8.[40] These data show the benefit of molybdenum under these conditions.

The nickel-rich alloys can be molybdenum-free, but the molybdenum-bearing grades containing anywhere from 2 to 7% molybdenum are more commonly used in sulfuric acid service.

Superferritic Stainless Steels The first superferritic steels were based on 26% Cr, 1% Mo (S44627) and the niobium-stabilized XM-27®; E-Brite® (S44627). These were developed to provide better resistance to chloride SCC than the austenitic 300 series. These ferritic steels have low-interstitial content with high chromium and very low carbon levels. Most modern superferritic stainless steels are based on a 29% Cr, 4% Mo alloy and they need low C + N levels, i.e., less than 0.025%, to avoid intergranular corrosion caused by chromium depletion from precipitation of carbides and nitrides. Some of the current ferritics contain higher levels of C + N and have additions of titanium or niobium as carbon/nitrogen stabilizers. Superferritic steels include AL 29-4C® (S44735), AL 29-4-2® (S44800), Sea-Cure® (S44660), and Monit® (S44635).

Sea-Cure® (S44660) has a corrosion rate of 4.3 mpy (0.11 mm/y) in boiling 10% sulfuric acid.[41] The superferritic grade E-Brite® (S44627) has generally inferior resistance to S31603 in weak acid while the newer alloys, e.g. S44800, are considerably better and more like the nickel alloy 625 (N06625), Table 9.9.[42] Most of these tests were carried out for five 48-hour periods, but where E-Brite® had very high corrosion rates it was only tested for one period.

Table 9.9 Corrosion Rate mm/y (mpy) of Ferritic and Other Alloys in Boiling Dilute Sulfuric Acid Solutions

Alloy	Sample Condition	1%	5%	10%
29-4-2	NA	0.005 (0.2)	0.03 (1.3)	0.02 (0.8)
(S44800)	A	0.07 (2.6)	0.27 (10.7)	0.46 (18.2)
625	NA	0.07 (2.6)	0.32 (12.7)	0.52 (20.6)
(N06625)	A	0.06 (2.2)	0.23 (8.9)	0.64 (25.3)
E-Brite®	NA	0.02 (0.7)	0.36 (14)	900 (35,000)
(S44627)	A	13.7 (541)	77 (3,020)	2,600 (>100,000)
316	NA	0.55 (21.7)	2.49 (98.2)	8.61 (339)
(S30600)	A	0.66 (25.8)	2.71 (107)	8.73 (334)

NA: Not activated
A: Activated at the start of each test period

Superduplex Stainless Steels The superduplex grades typically contain nitrogen that permits the use of this type of steel, often without the need for heat treatment.

Alloy 2507 (S32750) resists 10% H_2SO_4 to the atmospheric boiling point, but its limiting temperature drops to about 45°C (113°F) at 25%, see Figure 9.4.

The proprietary grade, alloy 7-Mo PLUS® (S32950) is reported to have a rate of not more than 5 mpy (0.13 mm/y) up to the atmospheric boiling point at 5%.[16] At 70°C (158°F), the limiting concentration is 25% for a 5 mpy (0.13 mm/y) corrosion rate for the wrought form. A wider range of applicability was obtained for commercial grade acid, but increased temperature or concentration excursions resulted in rapidly increasing corrosion rates. The influence of impurities, velocity, and material conditions was not investigated. Corrosion rates of this alloy are compared with other duplex stainless steels in boiling solutions of various acid concentrations, Table 9.10.

Another proprietary duplex grade, Ferrallium 255® (S32550), also has good resistance to concentrated sulfuric acid, e.g., in 10% acid up to about 90°C (194°F).[43] In 20%

Table 9.10 Corrosion Rates mpy (mm/y) of Duplex Stainless Steels in Boiling Sulfuric Acid of Various Concentrations

% H_2SO_4	B. Pt. °C	Alloy 2205	Alloy 2507	Alloy 7-Mo PLUS®
10	102	64 (1.63)	96 (2.44)	100 (2.54)
20	104	40 (1.02)	62 (1.57)	83 (2.11)
30	108	25 (0.64)	45 (1.14)	73 (1.85)
40	114	—	—	65 (1.65)
50	123	—	—	59 (1.50)
60	140	—	—	53 (1.35)
70	165	—	—	47 (1.19)

acid at 79°C (174°F), the corrosion rate for this alloy was <0.01 mm/y (<0.4 mpy), but this increased several orders of magnitude to >10 mm/y (>400 mpy) by the addition of 1% HF and to >100 mm/y (>4,000 mpy) when 3–5% HF was added.[44]

Nickel and Its Alloys

The alloys containing more than 50% nickel must be considered in two categories, chromium-free and chromium-bearing. The second category may be divided into those that also contain molybdenum and those that do not.

Chromium-Free Nickel 200 (N02200) will resist 5% acid at room temperature, with a rate of about 9 mpy (0.2 mm/y) unaerated, and >60 mpy (>1.5 mm/y) when air-saturated. Like the less expensive and more commonly used alloy 400 (see below), nickel is very susceptible to increased corrosion by oxidizing ion contamination.[7] Sulfuric acid starts to become inherently oxidizing in nature at about 5N (25%) concentration.

The copper-containing nickel alloy 400 (N04400) will resist boiling 5% sulfuric acid, with a rate of about 3–4 mpy (0.086 mm/y). This may be due to the removal of dissolved oxygen (DO) from the boiling solution, because the alloy otherwise shows high corrosion at 5% concentration in the 60–90°C (140–194°F) temperature range, with the maximum at about 80°C (175°F). Contamination with DO or oxidizing ions causes a large increase in corrosion, as do even relatively low velocities (<1 ft/s). An agitated 6% acid solution contaminated with 0.5% copper sulfate corroded at 9.14 mm/y (9,360 mpy) at 82°C (180°F).[7] Corrosion rates in various strengths of boiling acid are shown in Table 9.11.[15] These data show that this alloy has good resistance to intermediate-strength acid as long as strong oxidizers are absent.

The nickel-molybdenum alloy B-2 (N10665) is reported to corrode at 0.5 mpy (0.013 mm/y) in boiling 2% acid and at 3 mpy (0.076 mm/y) in boiling 5% acid. Rates are as high as 10 mpy (0.25 mm/y) in the 50–90°C (122–194°F) temperature range for up to 20% acid.[45] Although DO increases corrosion somewhat, it should not be a major factor, compared with even trace amounts of stronger oxidants (e.g., Fe^{+++}, Cu^{++}), which will cause greatly accelerated attack. The cast alloy N-7M (N30007) is similar in corrosion resistance, as is alloy B-3 (N10675).

Table 9.11 Corrosion Rates of Alloy 400 in Boiling Solutions of Various H_2SO_4 Concentrations

% H_2SO_4	Boiling temp (°C)	Corrosion rate mpy (mm/y)
5	101	3.4 (0.086)
10	102	2.4 (0.061)
19	104	7.5 (0.19)
50	123	650 (16.5)

Chromium-Bearing The nickel-chromium alloy 600 (N06600) is of little practical use in dilute sulfuric acid. DO accelerates corrosion, although stronger oxidants enhance passivity at acceptable temperatures. In 5% acid, corrosion rates were 9, 10, and 30 mpy (0.23, 0.25, and 0.76 mm/y) at 30, 60, and 80°C (86, 140, and 176°F), respectively, under anaerobic conditions, but these were increased 5–8 fold by aeration.[7]

The nickel-chromium-molybdenum alloys include N06625, N10276, N06022, and alloys of similar composition. They show rates of <5 mpy (0.13 mm/y) in this acid range up to about 90°C (194°F) and are fairly tolerant of chloride contamination, suffering only a 10–15°C (50–59°F) diminution in temperature limit. However, in boiling 10% acid, contamination with 10,000 ppm chloride increased the corrosion rate for N06022 from <5 to >130 mpy (<0.13 to >3.3 mm/y).[34] In tests in 20% acid with 3% HF at 79°C (174°F), a rate of 14 mpy (0.36 mm/y) was observed.[37] The resistance of nickel-chromium-molybdenum alloys is enhanced rather than diminished by oxidizing contaminants.

Alloy 686 (N06686) is particularly resistant to weak acid and has a corrosion rate of <5 mpy (<0.13 mm/y) in up to 15% acid at temperatures up to the boiling point and up to about 90°C (250°F) in 25% acid, see Figure 9.9 below.

Alloy 59 (N06059) has good resistance to weak acid, especially in the presence of chlorides. In 10% acid at 100°C (250°F), the corrosion rate for this alloy was 0.174 mm/y (6.9 mpy), while in 20% acid at the same temperature the rate rose to an unacceptable 0.343 mm/y (13.5 mpy).[23]

Once the gases from power-generating plants have been scrubbed to remove sulfur dioxide, they are often mixed with around 10% of the unscrubbed gas to raise the gas temperature from 60° to 70–80°C (from 135°F to 160–180°F) to increase stack efficiency. This practice has produced serious corrosion in the stack breeching section and stack liner. Under normal operating conditions the acid concentration in this region is on the order of 25–55%. Alloy 625 (N06625) and alloy C-276 (N10276) have been successfully used in this application, often as a cladding or thin metal liner.[33]

Copper and Its Alloys

Copper-based alloys may be used with 5–25% sulfuric acid only under anaerobic and strictly reducing conditions. Yellow brasses should not be used under any circumstances, because of susceptibility to dezincification. Use of copper-based materials is fraught with danger because the corrosion products (i.e., cupric ions, Cu^{++}) are themselves of an oxidizing nature and the corrosion is autocatalytic.

Copper, phosphor (tin) bronzes, and aluminum bronzes are usefully resistant under controlled reducing conditions. Silicon bronze (C65500) can be used in this range of acid strengths, while aluminum bronze is used in 10–20% acid, for example, in steel-pickling applications.[4]

Lead

Lead and its alloys are usefully resistant in this acid range up to the atmospheric boiling point, with rates typically <5 mpy (<0.13 mm/y). Rates of 1–5 mpy (0.03–0.13 mm/y) were reported in a 5% sulfuric acid solution, saturated with ammonium sulfate, at 47°C (117°F).[25] Resistance is due to the formation of insoluble lead-sulfate films, which are susceptible to mechanical damage, erosion, abrasion, etc., with attendant higher rates of attack.

In modern industrialized countries, lead has fallen into disuse because of poor mechanical properties and the shortage of skilled craftsmen ("lead burners") and due to health and toxicity considerations.

Reactive and Refractory Metals

All four reactive metals (titanium, zirconium, tantalum, and niobium) will suffer corrosion in the presence of fluoride contamination, due to the formation of soluble fluoride complexes with the possibility of hydriding.

Titanium and Its Alloys Because of the reducing nature of this sulfuric acid range, unalloyed titanium would not normally be considered. In 5% acid, even with aeration, corrosion is severe at 60°C (140°F). On the other hand, oxidizing agents (e.g., cupric or ferric ions, nitric acid, and chlorine) can maintain passivity. About 1,000 ppm Fe^{+++} or Cu^{++} is required to maintain passivity in boiling 15% acid.

The titanium-nickel-molybdenum grade 12 (R53400) and titanium-palladium grade 7 (R52400) alloys were developed to extend resistance to reducing acids. In 5% acid at 70°C (158°F), under nitrogen, grade 7 has a rate of only 6 mpy (0.15 mm/y), compared with active dissolution of the unalloyed R50400. The 0.13 mm/y (5 mpy) isocorrosion line for grade 12 (R53400) drops from about 52°C (125°F) to below 25°C (77°F) at 10% concentration, whereas that for grade 7 (R52400) diminishes from 100°C (212°F) only to 52°C (126°F). In boiling 15% sulfuric acid, corrosion of unalloyed titanium diminishes from >2,000 mpy (>50 mm/y) to <10 mpy (<0.25 mm/y) with the addition of about 1,000 ppm Fe^{+++}. (Cupric ions diminish the rate only to some 30 mpy [0.76 mm/y]). However, in 10% acid at 160–170°C (320–340°F), with 30 ppm fluoride ion contamination, rates were 0.38 mm/y (15 mpy) with the possibility of hydriding.

Zirconium Zirconium (e.g., R60701, R60702) will resist these concentrations of sulfuric acid to the atmospheric boiling point. Above the atmospheric boiling point, zirconium can be used to 250°C (482°F) at 5% and lower temperatures as the concentration increases. At 25% the temperature limit is 200°C (392°F). There may be problems with pyrophoric-corrosion products in the event of contamination with oxidizing species, chlorides, and elevated temperatures.

Tantalum Tantalum will resist 5–25% sulfuric acid up to and above the atmospheric boiling point to 250°C (482°F).[46,47]

Niobium Niobium should not generally be used above about 70°C (158°F) in this range of acid strengths.[28] It is, however, said to be completely resistant to 20% sulfuric acid at 100°C (212°F).[48]

The corrosion rates of niobium, niobium-tantalum, and tantalum in boiling 20, 40, 60, and 80% sulfuric acid was found to increase with increasing acid concentration and decrease with time due to oxide formation. Additions of tantalum to niobium improved its corrosion resistance.[49] This investigation did not include a study of mechanical effects and embrittlement.

Cobalt Alloys The 30% Cr, 5% W, 1.5% Mo hard-facing alloys are resistant at room temperature but are attacked in excess of 200 mpy (>5 mm/y) in boiling 10% acid, unless oxidizing contaminants are present.[50] The cobalt-based ULTIMET® alloy corrodes at 99 mpy (2.5 mm/y) in boiling 10% sulfuric acid.[51]

Hard-facing alloys of the 27% Cr, 6% Mo variety (e.g., R30021) show rates of less than 2 mpy (<0.05 mm/y) in these concentrations to at least 66°C (150°F), as do the high-tungsten (>13% tungsten) grades, e.g., R30001 and R30004.

Precious/Noble Metals

Silver, gold, and platinum are resistant to 5–25% sulfuric acid to the atmospheric boiling point.

Weak and Intermediate-Strength Acid (25–70%)

Sulfuric acid starts to become inherently oxidizing at about 5N concentration (24.5%), as evidenced by the reduction of the anion during some corrosion processes. As noted previously, finely divided nickel (Raney Nickel) will evolve hydrogen sulfide from a 25% sulfuric acid solution at room temperature, while corrosion of alloy 400 (N04400) in the boiling acid also evolves hydrogen sulfide and sulfur dioxide (which interact in the vapors to produce elemental sulfur and water). A similar effect was noted with boiling 53–57% acid in contact with the now obsolete silicon-nickel casting alloy, Hastelloy® D.[1]

This concentration of sulfuric acid is *reducing* toward some metals and alloys, dissolving some ferrous alloys and conventional 18-8-type stainless steels with the evolution of hydrogen as the cathodic reaction.

As previously described, specific oxidizing or reducing contaminants profoundly affect the corrosion characteristics and may introduce specific ion effects.

At the upper end of the range (about 67% plus), sulfuric acid solutions approach the character of concentrated acid.

Aluminum

Aluminum and its alloys are inherently nonresistant to sulfuric acid in these concentrations, and rates increase with temperature and concentration to a maximum in 70% acid.[52]

Iron and Steel

Steels are definitely unsuitable below about 65% concentration, but some companies have used steel tanks and piping in the 60–67% range, accepting rates of 25 mpy (0.6 mm/y) for tanks and 50 mpy (1.2 mm/y) for piping even at low velocities.[53] In this acid concentration, ringworm attack occurs in the heat-affected zones (HAZ) of welds, due to spheroidal carbides, whereas lamellar carbides elsewhere provide more effective anchoring for a protective ferrous sulfate film. Normalizing at approximately 845°C (1,550°F) after welding prevents the localized attack. Where this is impractical, a lead overlay was sometimes used to protect the steel substrate weldment. Steel is now rarely used except for storage of cold concentrated acid.

Both corrosion characteristics and brittleness militate against the use of gray cast iron in this concentration range. Ductile iron, with its lesser susceptibility to mechanical breakage, could conceivably be used as low as 65% acid; based on the experience with steel, however, significant corrosion would be expected and it is not recommended.

Alloy Cast Irons (Cr, Si) Chromium-bearing cast irons do not resist these concentrations of acid and may show higher rates than unalloyed cast iron. The austenitic cast irons (e.g., Ni-Resist® I [F41000]) corrode at <5 mpy (<0.13 mm/y) only up to about 5% at 20°C (68°F). If a corrosion rate of 20 mpy (0.51 mm/y) is acceptable, the austenitic-nickel cast irons can be used to about 40°C (104°F) in 25–70% acid.[7] A corrosion rate of 20 mpy (0.51 mm/y) was found on Ni-Resist® I in 5% and 10% acid at 30°C (86°F), but at a rate of only 8 mpy (0.2 mm/y) was found in 25% sulfuric acid containing acid sludge at 60°C (140°F). Presumably contaminants in the acid sludge inhibited the corrosion in this environment.[54]

Silicon as an alloying element has a strong influence on the corrosion of alloys in strongly oxidizing environments. At low levels of silicon, corrosion is increased, but above a threshold level corrosion rate decreases, especially for the higher concentrations in this range, Table 9.12.[30]

Provided there is no fluoride contamination, the 14.5% silicon iron (F47003) will resist up to 5% acid to the atmospheric boiling point, as well as acid in the 55–70% range. Relatively high rates of 15–20 mpy (0.38–0.51 mm/y) can be found at about 33% acid concentration and in an ill-defined zone between about 10 and 45% acid at 80°C (175°F).[50] In most practical applications, this minor aberration can be ignored. Silicon cast irons are also very brittle and almost impossible to weld. On initial exposure there is a high corrosion rate while the silica-rich film is developed, but as the film becomes protective the corrosion rate decreases to low levels in, for example, boiling 30% acid, Figure 9.6.[31]

Table 9.12 Corrosion Rates mpy (mm/y) of Iron Silicon Alloys in Various Concentrations of Sulfuric Acid at 50°C (122°F)

	% Sulfuric Acid		
% Si	35	50	70
0	144 (3.7)	126 (3.2)	0.95 (0.02)
1.5	145 (3.7)	200 (5.1)	0.99 (0.025)
2.5	250 (6.4)	278 (7.1)	1.00 (0.025)
7.5	545 (13.8)	182 (4.6)	2.17 (0.06)
9.5	374 (9.5)	133 (3.4)	0.87 (0.02)
11.5	—	—	0.50 (0.01)
14.5	—	—	0.00

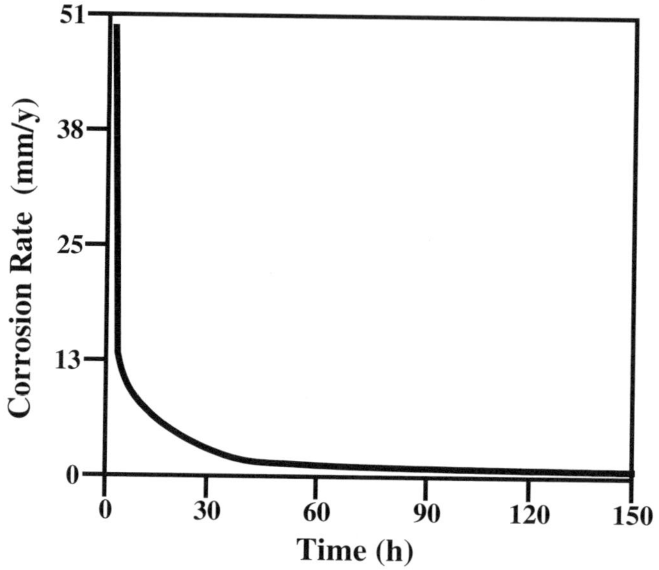

Figure 9.6 Effect of Exposure Time on the Corrosion of High-Silicon Cast Iron in Boiling 30% Sulfuric Acid

Stainless Steels

All stainless steels can be classified into three groups according to metallurgical structure and response to heat treatment. These are the martensitic, ferritic, and austenitic groups. Further subdivisions include duplex alloys with austenitic/ferritic microstructures and precipitation-hardening (PH) grades strengthened by an age-hardening treatment. Stainless steels are commonly used in weak sulfuric acid.

Surface contamination with chlorides prior to service (e.g., piping shipped or stored under marine atmospheres) will result in the formation of HCl *in situ* upon exposure to sulfuric acid. A preliminary wash with potable or other low-chloride water is mandatory under these conditions.

Martensitic, Ferritic, and PH Grades of Stainless Steels The martensitic and ferritic grades have no known applications in this concentration range.

The PH grades should be analogous to their 18-8 counterparts (S17700 comparable with S30400; S15700 comparable with S31600) in this range of acid and are generally unsatisfactory. The corrosion resistance will vary with the specific heat treatment and contaminants.

Duplex Stainless Steels Duplex alloys such as type 329 (S32900) may be less resistant than type 316 (S31600) under some circumstances because of galvanic corrosion of the ferrite phase under reducing conditions.

Austenitic Stainless Steels None of the conventional 18-8 stainless steels (types 304 through 317) are satisfactory in pure as-mixed or even fully aerated acid in this concentration range.[7]

Table 9.13 Corrosion Rates mpy (mm/y) of Types 304 and 316 Stainless Steels in a Range of Acid Strengths and Temperatures

% H_2SO_4	Temp. °C (°F)	Type 304 (S30400)	Type 316 (S31600)
30	15 (59)	35 (0.89)	65 (1.7)
35	25 (77)	600 (15.2)	70 (1.8)
40	15 (59)	50 (1.3)	200 (>5)
	40 (104)	>800 (>20)	>800 (>20)
50	15 (59)	>150 (>3.8)	>250 (>6.4)
	32 (90)	>200 (>5)	>200 (>5)
60	15 (59)	>200 (>5)	>100 (>2.5)
	40 (104)	>300 (>7.6)	>1,000 (>25)
66	32 (90)	>200 (>5)	>100 (>2.5)
70	15 (59)	40 (1)	50 (1.3)
	40 (104)	90 (2.3)	>500 (>12.7)

High corrosion rates can be expected in this range of acid strengths for 304 (S30400) and 316 (S31600), even at moderate temperatures, Table 9.13.[9]

The 18-8 type of stainless steels, even without molybdenum (e.g., 304L and 347), can be rendered passive in 60–65% sulfuric acid at 80–85°C (176–185°F) by additions of about 2,000 ppm Cu^{++}. In 30% acid at 93°C (200°F), cupric sulfate is effective at lower concentrations than chromic oxide or nitric acid (0.5% vs. 1 and 2%, respectively).[7] At 16°C (60°F), 1% nitric acid will passivate type 304 (S30400) stainless steel over this entire concentration range.

Inhibition by deliberate addition of oxidizing species is not attempted in normal industrial practice because it is impossible to maintain the inhibitor concentration within the crevices inherent in equipment. However, it is sometimes effective in commercial practices where the inhibitor ion is supplied continuously as part of the process. The addition of about 2% copper as an alloying element to types 304 and 316 (S30400 and S31600) is also beneficial, although not a commercial practice.[55]

High-Performance Alloys

This group comprises alloys with higher chromium and nickel contents than the conventional 18-8 grades, usually together with additions of other alloying elements, such as molybdenum and copper.

Among these more highly alloyed stainless steels, type 310 (S31000) can ostensibly resist aerated acid to about 40% and 70°C (158°F) and resist SCC in 50% acid, contaminated with 3% NaCl, at 30°C (86°F).[7]

The basic high-performance austenitic alloy for intermediate strengths of sulfuric acid has traditionally been alloy 20Cb-3 (N08020), Figure 9.7.[38] The cast version, CN-7M (N08007), is also widely used in this range of acid strengths. Most problems in the application of these alloys arise either from failure to recognize their inherent temperature limitation or from the adverse effects of halides or inherently reducing contaminants (e.g., stannous ions, Sn^{++}).

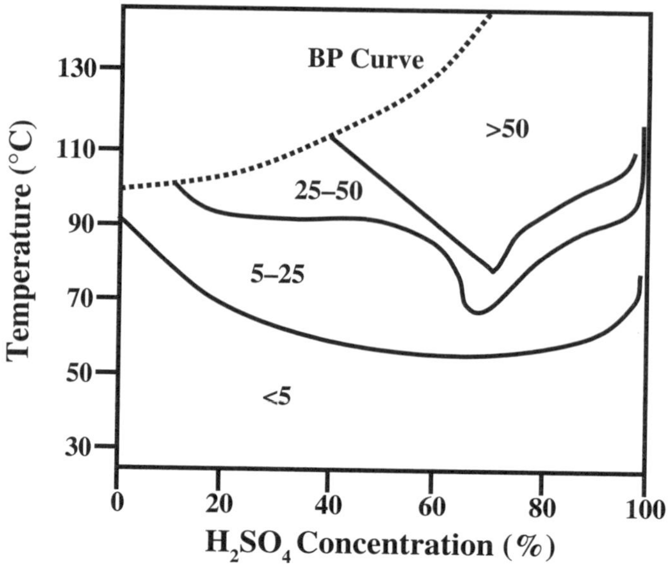

Figure 9.7 Isocorrosion Curves (mpy) for Alloy 20Cb-3 in Sulfuric Acid

Table 9.14 Average Corrosion Rate of Alloy 20 (N08020) in Different Strengths of Acid at Different Temperatures

% H_2SO_4	Temp. °C (°F)	Corrosion Rate mpy (mm/y)
10	Boiling	13.2 (0.3)
20	60 (140)	0
	85 (185)	18 (0.46)
	Boiling	17 (0.43)
		8.4 (0.2)
40	Boiling	6 (0.2)
70	50 (122)	2 (0.05)
	60 (140)	3 (0.075)
	75 (167)	5 (0.127)
	85 (185)	7 (0.178)

Alloy 20 strip has good resistance to the higher-strength acid even at elevated temperatures, Table 9.14.[56,57]

Alloy 20Cb-3 (N08020) is distinctly superior to type 316 (S31600) in some solutions of sulfuric acid in the 25–70% range. The superior resistance of alloy 825 (N08825) in 67% acid containing alkyl sulfates may be an anomaly. However, alloy 20Cb-3 (N08020) shows maximum corrosion between 65 and 74% acid at about 80°C (175°F) and the lower cast alloy CN-7M (N08007) shows maximum corrosion at somewhat lower temperatures in this range.

In general, copper-bearing alloys such as alloy 904L (N08904) are very similar to alloy 20Cb-3 in their corrosion behavior in this range of acid strengths, their exact relationship depending more on specific conditions and contaminants than on the acid concentration.

The 6% molybdenum austenitic stainless steels, such as alloys 254SMO (S31254), AL-6XN (N08367), and 25-6MO (N08926), have somewhat better resistance over the lower end of this acid strength range by virtue of their 6% molybdenum content. As the strength increases toward 70%, the high-molybdenum content reduces corrosion resistance in the more oxidizing conditions, and alloys such as 904L (N08904) and 20Cb-3 (N08020) become more resistant than the higher-molybdenum steels, Figure 9.8.[13,19] Alloy 6XN (N08367) resists corrosion by 60% sulfuric acid at 40°C (105°F), with a corrosion rate of about 5 mpy (0.13 mm/y). Corrosion rates tend to increase slightly for all such alloys as the concentration increases from 25 to 70%. Alloy 20Mo-6 (N08026) is more resistant than alloy 20Cb-3 (N08020) in 25–70% acid at 80°C (175°F), both corroding at <10 mpy (<0.25 mm/y) in this range of acid strengths.[13]

There are now also austenitic stainless steels and nickel-based alloys, such as 654SMO (S32654) and 27-7MO (S31277), with 7% molybdenum. The corrosion resistance of 654SMO in mid-strength acid is shown in Figure 9.8. This shows that 654SMO has better resistance than the alloys with lower molybdenum until the acid strength becomes high enough to become oxidizing. Above about 50% acid, alloy 20Cb-3 (N08020) is more resistant than the high-molybdenum stainless steels. At higher acid strength, the additional molybdenum decreases corrosion resistance. This same effect is shown in Figure 8.22, in which the high-molybdenum alloys 25-6MO (N08926) and 27-7MO (S31277) are both better than alloy 825 (N08825) in low- and intermediate-strength acid but are less resistant in more concentrated acid (above about 80–90%).

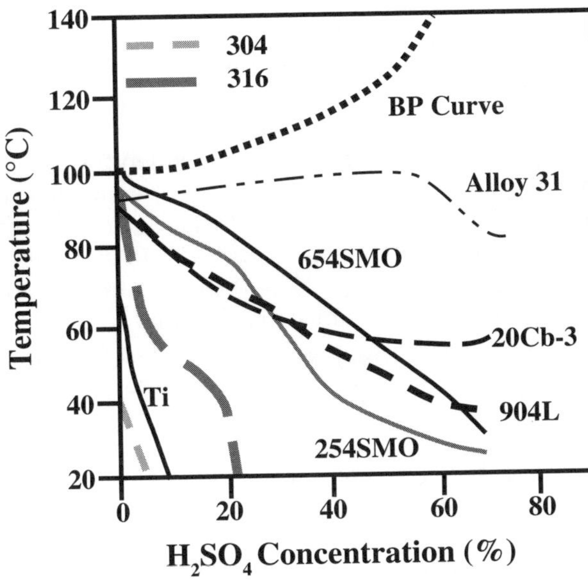

Figure 9.8 Isocorrosion Curves at 0.1 mm/y (3.9 mpy) of High-Performance Stainless Steels Compared with Titanium, 304, and 316 in 0–70% Sulfuric Acid

Table 9.15 Approximate Maximum Temperature °C (°F) of Use in Various Strengths of Sulfuric Acid

Alloy (UNS No.)	Total Alloy % (Cr+Ni+Mo+Cu)	20% H_2SO_4	50% H_2SO_4	80% H_2SO_4
316 (S31600)	31.5	30 (86)	—	—
904L (N08904)	52	60 (140)	40–50 (104–122)	40 (104)
AL-6XN (N08367)	52.5	70–80 (158–176)	40–50 (104–122)	40 (104)
20Cb-3 (N08020)	61	70–80 (158–176)	60 (140)	55 (131)
Sanicro 28 (N08028)	62.5	70–80 (158–176)	60 (140)	55 (131)
654SMO (S32654)	55	90 (194)	60 (140)	—
Alloy 31 (N08031)	65.7	104 (220)	107 (224)	75 (167)
C-276 (N10276)	87	75 (167)	50–60 (122–140)	100 (212)

The approximate maximum recommended temperature of use of some of these and other alloys in various strengths of sulfuric acid can be based on manufacturer's data for a maximum corrosion rate of 0.1 mm/y (3.9 mpy), Table 9.15.[58]

The temperature rating of the best alloy in this series, alloy 31 (N08031), is better than that of the nickel alloy N10276 (added to this table for comparison) in the lower-strength acids. On the basis of these comparisons it appears that the resistance of the superaustenitic alloys to low- and mid-strength acid improves with the increase in total alloy content in terms of chromium, nickel, molybdenum, and copper. The higher molybdenum content (7.5%) in alloy 654SMO (S32654) seems to be particularly beneficial. The nickel alloy C-276 (N10276), with the highest total alloy content, has a total chromium, molybdenum, and copper content similar to that of type 316 (S31600).

Alloy 800 (20% Cr, 35% Ni, 43% Fe [N08800]) lacks molybdenum and copper and is therefore not usually considered for this concentration range of sulfuric acid. It is resistant to the lower acid concentration in this range, but only at low temperatures. It was, however, resistant to SCC in 50% sulfuric acid, contaminated with 3% NaCl, at 30°C (86°F) in laboratory tests of 30 days' duration.[7]

Alloy 825 (N08825) has useful resistance in concentrations up to 40% at boiling, up to 60% at 175°F (80°C), and at all concentrations at temperatures <150°F (<65°C), Table 9.16.[15] The data in this table was derived from tests run for 48 hours at boiling temperature and 168 hours at the lower temperatures. The only data that give a reasonable comparison between C.P. acid and plant acid (at approximately 50%) show that the plant acid is much more aggressive to this alloy. The plant acid presumably contained either reducing contaminants and/or chlorides to produce this higher corrosion rate.

Table 9.16 Corrosion Rates mpy (mm/y) of Alloy 825 under Various Conditions

Acid Concentration %	50°C (122°F)	100°C (212°F)	Boiling
40 (C.P. acid)	0.5 (0.013)	14 (0.36)	11 (0.28)
50 (C.P. acid)	1.0 (0.025)	14 (0.36)	20 (0.51)
60 (C.P. acid)	4.0 (0.10)	20 (0.51)	120 (3.05)
80 (C.P. acid)	5.0 (0.13)	20 (0.51)	1,360 (31.5)
25.3 (plant acid)	0.5 (0.013)	—	16 (0.41)
50.3 (plant acid)	5.0 (0.13)	50 (1.3)	1,920 (48.8)

Alloy 825 (N08825) is quite similar to alloy 20Cb-3 but is somewhat less resistant in pure acid and may occasionally demonstrate better resistance in contaminated acid, Table 9.17.[7] This table compares data from the original alloy 20 (not the modern 20Cb-3 [N08020]) and also the NI-O-NEL® alloy that has similar analysis to alloy 825 (N08825).

Table 9.17 Corrosion Rates mpy (mm/y) of Alloy 20, NI-O-NEL, and Types 304, 316, and 317 in Contaminated Intermediate-Strength Acid Solutions

% H_2SO_4	Contaminant	Temp. °F (°C)	Type 304 (S0400)	Type 316 (S31600)	Type 317 (S31700)	Alloy 20[a]	NI-O-NEL®[b] (~Alloy 825)
25	15% H_3PO_4, 3% Na_2SO_4, 0.3% HF	RT-212 (RT-100)	Corroded	Corroded	—	10 (0.25)[d]	—
30	Na_2SO_4	175–202 (80–95)	340 (8.6)	390 (9.9)	460 (11.7)	—	—
15–35	H_2SO_4 sludge, tar, oil	120–200 (49–93)	121 (3.1)	59 (1.5)[d]	45 (1.1), 482 (12.2)	17 (0.43)[d], 157 (4)[d]	—
45	8% $Na_2Cr_2O_7$, 8% Na_2SO_4, 3% chromic sulfate	105 (41)	50 (1.3)	41 (1)	—	—	—
67	Minor hydrocarbons	77–122 (25–50)	15 (0.38)	1.5 (0.04)	—	1.3 (0.03)[d]	1.4 (0.04)
67	C_3H_8, C_3H_6	104–167 (40–75)	75 (1.9)[d]	80 (2)[d]	—	1.3 (0.03)[c]	1.4 (0.04)
69	0.4–0.5% $NaSO_4$, 0.3–0.55 Ni, 0.1–0.15% As	100 (38)	140 (3.6)	0.5 (0.01)	—	0.9 (0.02)	—

[a] 20% Cr, 29% Ni, 3% Cu, 2% Mo
[b] 22% Cr, 42% Ni, 3% Mo, 1.8% Cu, 1% Ti
[c] Slight pitting (3–5 mils)
[d] Severe pitting (>10 mils)

Alloy 31 (Cronifer® 3127hMo[N08031]) has excellent resistance in sulfuric acid in the concentration range from 20 to 60% up to 100°C (212°F) and in 80% acid up to 80°C (175°F). In 72% sulfuric with 8% nitric and 4% HF at 54°C (130°F), it has a corrosion rate of around 0.01 mm/y (0.4 mpy) compared with about 0.025 mm/y (1 mpy) for alloy G-30 (N06030).[59]

Alloy G-30 (N06030) was originally developed for phosphoric and sulfuric acid mixtures but finds some applications in mid-strength sulfuric. A major application of alloy G-30 (N06030) is in the digestion in sulfuric acid of phosphate rock containing halide contaminants.

Superferritic Stainless Steels Superferritic and ELI grades (e.g., S44625 and S44626) have no practical applications in this range of acid. However, they might be selected for specific purposes (e.g., to resist water side attack in coolers) should there be sufficient oxidizing contaminants in the acid.

Superduplex Stainless Steels Superduplex stainless steels such as alloy 7-Mo PLUS® (S32950), 3RE60 (S31500), and others may be less resistant than type 316 (S31600) under some circumstances because of galvanic corrosion of the ferrite phase under reducing conditions. Generally the resistance of superduplex stainless steels is not good in this range of acid concentrations, Table 9.18.[16]

Commercial acid, presumably containing iron, is less corrosive to S32950 than C.P. acid, Table 9.19.[25]

Corrosion rates can increase very rapidly if the temperature and concentration limits in Table 9.19 are exceeded. The effects of other impurities, velocity, and material condition (i.e., welded, cold-worked, etc.) were not investigated.

In boiling 50% sulfuric acid containing 6,000 ppm Fe^{+++} (i.e., ASTM A262,[36] Practice B), alloy 7-Mo PLUS® shows <19 mpy (<0.48 mm/y) for as-welded strip, with slight IGA of the weld and HAZ, compared with almost 400 mpy (10.2 mm/y) for type 329 (S32900). For autogenously welded strip, alloy 7-Mo PLUS® is substantially equivalent to type 316L (S31603), and alloy 255 (S32550), slightly better than alloy 2205 (S31803), and far superior to type 329 (S32900).[25]

Table 9.18 Corrosion Rates mpy (mm/y) of Superduplex Stainless Steels in a Range of Boiling Acid Strengths

% H_2SO_4	Alloy 2205 (S31803)	Alloy 2507 (S32750)	Alloy 7-Mo PLUS® (S32950)
10	64 (1.6)	96 (2.4)	100 (2.5)
20	40 (1)	62 (1.6)	83 (2.1)
30	25 (0.6)	45 (1.1)	73 (1.9)
40	—	—	65 (1.7)
50	—	—	59 (1.5)
60	—	—	53 (1.3)
70	—	—	47 (1.2)

Table 9.19 Temperature °C (°F) and Acid Strength Limits for 5 mpy (0.13 mm/y) Corrosion Rate for S32950 in C.P. and Commercial Acid

% H_2SO_4	C.P. Acid	Commercial Acid*
10	B.Pt.	B.Pt.
20	85 (185)	B.Pt.
30	70 (158)	B.Pt.
40	65 (149)	95 (203)
50	60 (140)	50 (122)
60	55 (131)	55 (131)
70	50 (122)	60 (140)

* Presumably containing Fe+++ ions

The more highly alloyed duplex grades such as alloy 255 (S32550) corrode at about 4 mpy (0.1 mm/y) at around 70°C (158°F) in 25% acid. In 65–70% acid, however, the isocorrosion line drops to about 50°C (122°F) for the wrought and 30°C (86°F) for the cast form.[25] In 25–40% acid at 79°C (174°F), the corrosion rate of alloy 255 (S32550) was increased from <0.02 mm/y (<1 mpy) to >20 mm/y (>800 mpy) by additions of 1–5% HF. High rates of attack were experienced in 60–70% acid, and the rate was largely unaffected by HF contamination.[44]

It has been shown that small additions of ruthenium (up to 0.3%) to duplex stainless steels are beneficial in resisting corrosion in intermediate-strength sulfuric acid solutions. The effect was particularly marked on the high-chromium (29%) duplex steels. For example, the 29% Cr, 14% Ni, 3% Mo steel corroded at 0.6854 mm/y (27 mpy) in 55% sulfuric at 40°C (104°F), while the same alloy with an addition of 0.28% ruthenium corroded at only 0.0276 mm/y (1.2 mpy) under the same conditions. Similar benefits were seen for acid strengths ranging from 10 to 90%. Ruthenium was found to work by increasing the efficiency of hydrogen evolution, thereby changing the cathodic reaction kinetics, but also inhibiting anodic dissolution of the alloys.[60]

Nickel and Its Alloys

Chromium-Free Nickel 200 (N02200) is not usually employed in this range of acid because it is inferior to and more expensive than the nickel-copper alloys. It is employed only under anaerobic conditions and in the absence of other, stronger oxidizing contaminants. Dissolved oxygen greatly exacerbates corrosion. Finely divided (i.e., Raney) nickel reduces the sulfate ion at ambient temperature, releasing H_2S and SO_2.

In 25–70% sulfuric acid, alloy 400 (N04400) is resistant at 30°C (86°F) even with aeration. At 60°C (140°F), corrosion is excessive (>0.1 mm/y [>3.9 mpy]) only with aeration, but higher rates are observed at 95°C (203°F). In boiling acid, the rate is >0.2 mm/y (>7.5 mpy) at 19% but increases dramatically at higher acid concentrations. For example, in boiling 50% acid the corrosion rate is 650 mpy (16.5 mm/y).[15] Corrosion is further aggravated by HF contamination.

The nickel-molybdenum alloy B-2 (N10665) has corrosion rates of only 1–2 mpy (0.025–0.050 mm/y) in 60% up to boiling, and the corrosion rate is 9 mpy (0.23 mm/y) in boiling 70% acid. The cast alloy N-7M (N30007) is similar in corrosion resistance, but the older N-12MV (N30012) has rates of over 0.13 mm/y (5 mpy) above about 50°C for 25% acid, permissible temperature rising to about 90°C (195°F) at 70%. Contamination by 2% NaCl lowers the isocorrosion line to about 110°C (230°F) above about 50%. Dissolved oxygen has no significant adverse effect, but even traces (>5–10 ppm) of stronger oxidants (e.g., Fe^{+++}, Cu^{++}) would greatly accelerate corrosion. Traces of HF can also accelerate attack.

Alloy B-3 (N10675) has similar rates to N10665 (see Table 9.20 for data on alloy B-3).

Chromium-Bearing In 25–70% acid, corrosion rates for alloy 600 (N06600) are excessive even at ambient temperatures, and the alloy has no known application in this service.

The addition of chromium in lieu of about half the molybdenum content in a nickel-molybdenum alloy greatly increases resistance to 25–70% acid when it is contaminated with oxidants. Typical of these nickel-chromium-molybdenum alloys are alloy 825 (N08825), alloy 20 (N08020), alloy 625 (N06625), and alloy 276 (N10276). The 5-mpy (0.13-mm/y) isocorrosion line for alloy C-276 (N10276) drops from about 75°C (170°F) at 25% acid to 50°C (122°F) at 70%. For alloy C-276 (N10276) and the more highly alloyed derivatives, such as C-22 (N06022), alloy 59 (N06059), and alloy 686 (N06686), the isocorrosion lines are substantially the same in this range of acid, but additions of 1% ferric chloride have a pronounced beneficial effect for the latter alloys.[61]

The presence of 200 ppm chloride as NaCl lowers the isocorrosion lines, markedly in the 25–40% range and somewhat less so at 50–70% concentration. Similarly, HF contamination increases corrosion for the straight nickel-chromium-molybdenum alloys like N06625, N10276, and N06022. Ultimately the relative behavior of alloy C-276, as compared with the more recent and more highly alloyed variants, is entirely contingent on the presence or absence of contaminants of one kind or another.

The corrosion of some of the nickel-chromium-molybdenum alloys is compared with nickel-molybdenum alloys in various concentrations of reagent-grade sulfuric acid at a range of temperatures, Table 9.20.[62] This shows a complex behavior with the chromium being beneficial under some conditions and detrimental under others. It also shows that the high-molybdenum B-3 alloy (N10675) has good resistance to the whole range of intermediate acid strengths, including the hot 70% acid that is so aggressive to many alloys.

Because of necessary compositional adjustments and formation of intermetallic compounds, castings are often less resistant than the nominally equivalent wrought materials. The representative castings and wrought equivalents (where applicable) are compared under various conditions in Table 9.21.[63] In these tests the cast alloys generally showed higher corrosion rates than their wrought equivalents.

Generally, the alloy CW-6M is preferred in the 20–70% range (as well as in concentrated acid) when nickel-chromium-molybdenum castings are required.

Alloy 686 (N06686) is particularly resistant to weak acid and has a corrosion rate of <5 mpy (<0.13 mm/y) in 25% acid up to about 90°C (250°F). This alloy is compared with alloy 59 (N06059) and other nickel-based alloys in Figure 9.9.[64–66]

Table 9.20 Corrosion Rates in ~~mpy~~ mm/y of Various Nickel Alloys in Reagent-Grade Sulfuric Acid

Temperature °C (°F)	Alloy	% H_2SO_4							
		5	10	20	30	40	50	60	70
38 (100)	B-3	0.069	0.041	0.031	0.018	—	—	—	—
66 (150)	B-3	0.150	0.109	0.081	0.061	0.030	0.025	0.018	—
	C-0276	—	—	—	—	—	0.020	0.018	0.046
	C-2000	—	—	—	—	—	<0.01	0.003	0.005
79 (175)	B-3	0.130	0.112	—	—	—	—	—	—
	C-276	—	0.033	0.048	0.061	0.191	0.262	0.300	0.157
	C-2000	—	<0.01	0.005	0.005	0.013	0.018	0.023	0.071
93 (200)	B-3	0.107	0.114	0.112	0.094	0.064	0.043	0.028	0.013
	C-276	—	0.140	0.396	0.417	0.483	0.617	0.665	0.503
	C-2000	—	0.020	0.025	0.038	0.051	0.163	0.366	0.417
107 (225)	C-276	—	—	—	—	1.016	1.133	1.026	1.064
	C-2000		—	—	—	0.721	0.676	0.841	1.405
121 (250)	B-3	—	—	—	—	—	—	—	0.031
	C-276	—	—	—	—	—	2.329	2.870	13.68
	C-2000	—	—	—	—	—	1.715	2.807	4.318
Boiling	B-3	0.010	0.013	0.015	0.018	0.020	0.028	0.048	0.152
	C-276	—	0.180	0.493	0.826	1.875	3.645	13.08	—
	C-2000	—	0.089	0.178	0.419	1.130	3.353	9.271	—

Table 9.21 Comparison of Corrosion Rates mpy (mm/y) of Cast and Wrought Nickel Alloys

UNS No.	Alloy	20%, 225°F (107°C)	50%, 202°F (94°C)
N06455	C-4	62 (1.6)	13 (0.33)
N26455	CW-2M	82 (2.1)	17 (0.43)
N06022	C-22	54 (1.4)	16 (0.41)
N26022	CX-2MW	116 (3)	52 (1.3)
N10276	C-276	54 (1.4)	13 (0.33)
N30107	CW-6M	31 (0.79)	16 (0.41)
N06200	C-2000*	5 (0.13)	6.4 (1.6)

* No cast equivalent

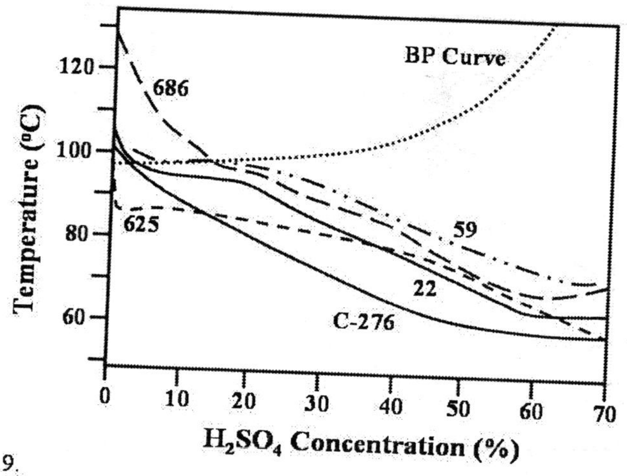

9.9.

Figure 9.9 Isocorrosion Curves at 5 mpy (0.13 mm/y) of Nickel-Chromium-Molybdenum Alloys in 0–70% Sulfuric Acid

Copper and Its Alloys

In static sulfuric acid at room temperature, about 21°C (70°F) under an air atmosphere, corrosion is minimal (e.g., <1 mpy; < 0.025 mm/y) in the 60–70% range, corresponding to the minimum oxygen solubility. Cupronickel has been successfully employed in 40–50% acid at temperatures approaching 100°C (212°F). At elevated temperatures of 80–90°C (175–195°F), corrosion is minimal if oxygen is rigorously excluded, but it may accelerate in practice with the accumulation of cupric ions. An initial corrosion step apparently involves cuprous ions, which seem to be unstable, depositing colloidal copper and releasing cupric ions:

$$2\ Cu^{+} \longrightarrow Cu^{\circ} + Cu^{++}$$

The cupric ion is itself an *oxidizing* contaminant.

Brasses are not employed in dilute or mid-strength sulfuric acid because of their susceptibility to dezincification. From a practical standpoint, bronzes of all types—i.e., tin, silicon, aluminum, and nickel bronzes—show essentially the same corrosion resistance as copper.

Lead

There are a number of lead alloys used for increased corrosion resistance, endurance limit, and hardness (or ability to work-harden) and to prevent excessive grain growth. Corroding lead is a high-purity (99.94%) product used where there would be undesirable effects from impurities otherwise introduced by corrosion. Lead has largely fallen into disuse in industrialized countries because of health and environmental considerations.

Chemical lead (L51120) has improved corrosion resistance and mechanical strength and is commonly used in chemical processes. The solubility of the protective lead sulfate is at a minimum in this range of acid, with decreasing solubility from 0°C (32°F) to 50°C (122°F).[24] Unless the film is disturbed by mechanical effects, corrosion rates are below 0.13 mm/y (5 mpy) up to 50% acid at the atmospheric boiling point and to about 120°C (250°F) in the 50–70% range. The addition of 2–6% antimony results in hard lead, or antimonial lead (L52901), which has improved mechanical properties but about half the resistance of chemical lead in pure acid and in acid contaminated with species that solubilize the otherwise protective film. As little as 5% hydrochloric acid will increase corrosion by at least an order of magnitude (>10-fold) due to the increased solubility of lead chlorides. Nitric acid has a similar solubilizing effect, and 5% is probably the limiting permissible level of contamination.[25]

Organic contaminants can also solubilize the protective film. This effect has been noted in sulfonation of a variety of oils (e.g., tallow, neatsfoot, fish, vegetable, and peanut) with 25% acid at 60°C (140°F). Alkyl (i.e., ethyl and isopropyl) sulfates solubilize the lead sulfates in 60–65% acid at 80–90°C (175–195°F). Lead has in the past been used to protect the weld heat-affected zone (HAZ) in steel in 60–67% acid.[25]

Reactive and Refractory Metals

All of these reactive metals will suffer corrosion in the presence of fluoride contamination due to the formation of soluble fluoride complexes with the possibility of hydriding.

Titanium and Its Alloys Titanium is attacked rapidly in this range of sulfuric acid unless there are powerful oxidizing contaminants present. Titanium alloys are used in the 25–50% range when the acid is saturated with chlorine or other strong oxidizing agents.[67] In a copper extraction leach system operating at 93–99°C (199–210°F), titanium was resistant (<0.04 mm/y [<1.5 mpy]) in 23–29% acid containing about 17% copper salts.[7]

Titanium-nickel-molybdenum grade 12 (R53400) is unsuitable in 25–70% acid even in the presence of oxidants, but the titanium-palladium grade 7 (R52400) will resist 25% acid up to about 50°C (122°F) and 50% at ambient temperatures.[26]

Zirconium Unalloyed zirconium (R60702) has excellent corrosion resistance to pure 25–70% sulfuric acid to 200°C (392°F)—i.e., above atmospheric boiling point—at 25% and to just below the boiling point at 70%. Above 70% the corrosion is very temperature-dependent due to breakdown of the ZrO_2 film, and very high corrosion rates are possible.

In 55% acid at 130°C (266°F) the rate of attack was <1 mpy (<0.025 mm/y), whereas rates for alloy 20Cb-3 (N08020), alloy 825 (N08825), and alloy C-276 (N10276) were 100, 120, and 300 mpy (2.54, 3.05, and 7.62 mm/y), respectively.[67]

It has recently been found that the tin content of Zr 702 (R60702) is an important factor in its corrosion resistance in intermediate-strength sulfuric acid, particularly at higher concentrations and temperatures. The upper limit for this alloy in 60% sulfuric is around 170°C (338°F) in samples with either typical (2,499 ppm Sn) or low (1,400 ppm Sn) tin levels. At 65%, welded R60702 with typical tin content has a limit of around 165°C (329°F), while for the low-tin version it is around 175°C (347°F), welded or unwelded. At 75%, welded and unwelded material with typical tin content is limited to around 160°C (320°F), while for low-tin material, welded or unwelded, the limit is 175°C (347°F).[68]

It should be noted that the niobium-bearing alloy R60705 is slightly less resistant than R60702.

A Zr 702 (R60702) lined mist eliminator experienced cracking in 67–70% acid at around 100°C (210–215°F) in about two months. The cracking occurred parallel to the rolling direction of the 0.083-in (2-mm) liner, initiating at the welds. The plant acid apparently contained traces of ferric ions.

Welds of zirconium (R60702 and R60705) have similar corrosion rates as the base metal in 25–50% acid. Above 50% acid, the weld corrosion rate is high enough that heat treatment may be warranted. This is related to intermetallic precipitates in the grain boundaries of the weld and heat-affected zone. Weld corrosion rate can be made the same as that of the parent material by heat treatment at 725–775°C (1,330–1,420°F) for 60 minutes per inch of thickness, which is sufficient to dissolve and redistribute the intermetallic particles.[69]

U-bend specimens failed in 65% plant acid in a 72-hour test. Cracking occurred in both 65 and 67% acid in the presence of as little as 50 ppm Fe^{+++} ions. No cracking occurred when the bends were shot-peened, even in the presence of the ferric ion contaminant. The shot-peening reduces the stress in the metal surface to protect against SCC. Cracking tendency is mitigated by hydrocarbon contaminants, which change the redox potential toward more reducing conditions, and by stress relief (thermal or mechanical).[70]

Stress corrosion cracking can occur in the 64–69% range with high stresses (e.g., U-bends). Stress-relieving to 530–590°C (990–1,100°F) for 60 minutes per inch of thickness will eliminate stress corrosion cracking.[69,71–73] The lower-temperature heat treatment will not dissolve and redistribute the intermetallic particles in the welds, so the weld will have a higher corrosion rate if the temperature is above the weld limit line. Fabrication problems have been reported with contaminated welds and forming that leave severe localized plastically deformed areas (possibly with a press brake). Heat-treating to improve the weld did not eliminate the SCC problem.[71] There have been several applications in which the zirconium was heat-treated to 725–775°C (1,330–1,420°F) and SCC problems were not seen. Special attention to fabrication procedures and correct selection of heat-treatment conditions are critical in the 64–69% range.

In clad vessels, the combination of zirconium cladding and base material may not allow the heat treatment of welds. For applications in 64–69% acid in which stress relief is used to avoid SCC, the welds will still have lower corrosion rates, limiting temperature and concentration use.[69]

Pyrophoric films have been reported on zirconium in 77.5% acid below the boiling point.[74] The presence of ferric or cupric ions may allow pyrophore formation in 25–75% acid. Oxidizing contaminants mimic the redox potential of >70% acid, resulting in corrosion and possible pyrophoric corrosion products in weaker acid.[75] It has been reported that under plant conditions with oxidizing ions, breakaway corrosion has been observed in as low as 60% acid at about 85°C (185°F), with the simultaneous formation of pyrophoric corrosion products.[76]

If pyrophoric film is suspected to be present, the equipment should be treated with steam before opening for maintenance or inspection.[69] One example is to treat with steam at 250°C (482°F) for 30 minutes.[74]

Tantalum Tantalum will resist these concentrations of sulfuric acid to at least the atmospheric boiling point, unless contaminated with fluorides. In fact, reboilers for 70% acid in concentrators are commonly fabricated from tantalum.

Niobium Niobium has limited resistance to this range of acid strengths, especially if oxidizing species are not present. It may embrittle and is not generally used in this environment.

The corrosion rates of niobium, niobium-tantalum, and tantalum in boiling 20, 40, 60, and 80% sulfuric acid were found to increase with increasing acid concentration and to decrease with time due to oxide formation. Additions of tantalum to niobium improved its corrosion resistance.[49] This investigation did not include a study of mechanical effects and embrittlement.

Cobalt Alloys The 30% Cr, 5% W, 1.5% Mo hard-facing alloys are attacked at about 13 mpy (0.33 mm/y) by boiling 50% acid containing about 4% ferric sulfate. The low-carbon variety (1.2% C) only corroded at 5 mpy (0.13 mm/y).

Hard-facing alloys of the 27% Cr, 6% Mo variety (e.g., R30021) show rates of less than 2 mpy (0.05 mm/y) in these concentrations to at least 66°C (150°F), as do the high-tungsten (>13% tungsten) types, e.g., R30001 and R30004.

Precious/Noble Metals

Fine silver will resist 25 to 50% acid to the atmospheric boiling point, but rates are excessive in hot 60% acid.[77] Sterling silver contains a distinct copper-rich phase in a duplex structure and might be less resistant.

Gold will resist this range of sulfuric acid to the atmospheric boiling point.

Platinum has excellent resistance to all concentrations of sulfuric acid within this range to at least 100°C (212°F). It has been used for stills and condensers in the concentration of sulfuric acid from the intermediate strengths.

Summary of Corrosion of Metals in Alloys in Weak and Intermediate-Strength Acid (0–70%)

Various metals have been tested to assess the effect of acid strength on corrosion in boiling sulfuric acid, Figure 9.10.[78] These data show that while many metals can resist weak sulfuric acid, very few are resistant in the intermediate strength.

Materials that can be used in this range of acid strength have been summarized graphically, Figure 9.11.[79,80] This figure shows areas where different types of materials have been successfully used and provides an initial guide to materials selection. The areas shown represent corrosion rates of <0.5 mm/y (<19.7 mpy).

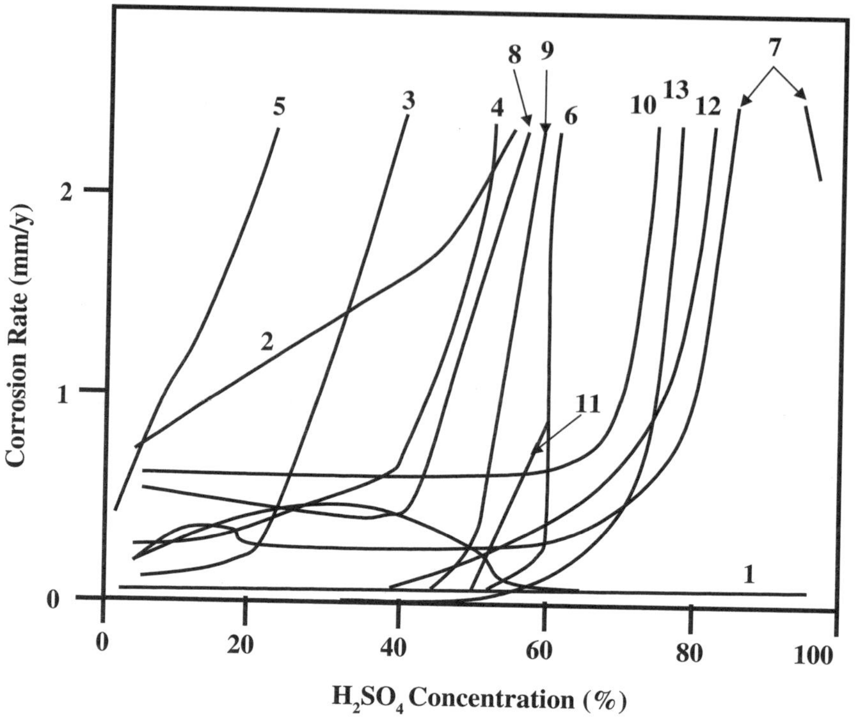

Legend

1. High-silicon cast iron (14.5% Si)
2. Alloy 20
3. Alloy 825
4. Illium® G
5. Hastelloy C
6. Hastelloy B
7. Hastelloy D
8. Alloy 400
9. 10% Al-Bronze
10. Lead (with tellurium)
11. Silver
12. Molybdenum
13. Zirconium

Figure 9.10 Corrosion Rates of Various Metals and Alloys in Boiling Sulfuric Acid

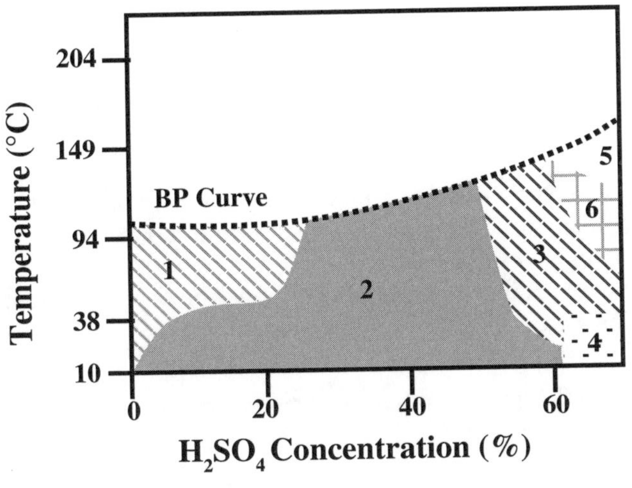

Legend

Zone 1

Impervious graphite
Tantalum
Gold
Platinum
Silver
Zirconium
Ni-o-nel ~ alloy 825
Tungsten
Molybdenum
Type 316 up to 10% aerated
10% Al Bronze, copper, alloy 400 if air-free
Illium® G
Glass
Hastelloy® B and D
Durimet 20
Worthite®
Lead
Haveg 43
Rubber up to 77°C (170°F)
Chlorimet 2 up to 70°C (158°F)
Cast alloy C

Zone 2

Ni-Resist® up to 20%
Impervious graphite
Tantalum
Gold
Platinum
Silver
Zirconium
Nionel ~ alloy 825
Tungsten
Molybdenum
Type 316 up to 25% at 24°C (75°F) aerated
10% Al Bronze, copper, Monel if air-free
Glass
High-silicon cast iron (14.5%)
Hastelloy® B and D
Durimet 20 up to 66°C (150°F)
Worthite up to 66°C (150°F)
Lead
Haveg 43
Rubber up to 77°C (171°F)
Chlorimet

Zone 3

Impervious graphite
Tantalum
Gold
Platinum
Zirconium
Molybdenum
Monel if air-free
Glass
High-silicon cast iron (14.5%)
Hastelloy® B, B-2, and D
Cast alloy C
Durimet 20, Worthite® up to 66°C (150°F)
Lead
Chlorimet

Zone 4

Impervious graphite up to 96%
Tantalum
Gold
Platinum
Zirconium
Ni-Resist®
Carbon steel
Type 316 (>80%)
Glass
High-silicon cast iron (14.5%)
Hastelloy® B, B-2, and D
Durimet 20
Worthite®
Lead up to 96%

Zone 5

Tantalum
Gold
Platinum
Glass
High-silicon cast iron (14.5% Si)
Hastelloy® B, B-2, and D, not up to BP

Zone 6

Tantalum
Gold
Platinum
Impervious graphite and lead up to 80°C (176°F) and 96% sulfuric
Glass
High-silicon cast iron (14.5% Si)
Hastelloy® B, B-2, and D, not up to BP
Durimet 20 and Worthite® up to 66°C (150°F)

Figure 9.11 Summary of Alloy Use in 0–70% Sulfuric Acid

Iron and Steel

Cast-iron and carbon steel are useful in aqueous sulfuric acid solutions only in the pH range of 4.5–6.5 (<5 ppm acid), where corrosion rate is determined primarily by dissolved oxygen, or in acid concentrations very closely approaching 70%. At any concentration in between, iron or steel is rapidly attacked, with attendant hydrogen gas evolution.

Dissolved oxygen in dilute acid (or an oxidizing ion capable of reduction at the cathode, e.g., Fe^{+++}) aggravates corrosion by facilitating depolarization of the cathode and increasing the corrosion current.

Alloyed cast irons, particularly high-silicon ones, have good resistance to the range of acid strengths and at elevated temperatures.

Stainless Steels and High-Performance Alloys

The stainless steels rely for their corrosion resistance on a passive film, which may be damaged in reducing acids. A schematic polarization curve for a typical stainless steel is shown in Figure 7.3, illustrating the passive region and the transpassive region. Should the passive film be removed, chemically or mechanically, the curves will be similar to those for conventional steel, with attendant high corrosion rates. An activated stainless steel can corrode more rapidly than carbon or low-alloy steel.

The passive film is enhanced by silicon content particularly, while molybdenum enhances the resistance under reducing conditions. The passivity in a given test may be strongly influenced by the concentration of dissolved oxygen, while more powerful oxidizing or reducing contaminants can completely govern the corrosion mechanism in dilute to intermediate concentrations.

The high-performance alloys include stainless steels with high molybdenum, such as alloy 254SMO (S31254), alloy 926 (N08926), and alloy 31 (N08031), and nickel-rich chromium-bearing alloys, such as alloy 20Cb-3 (N08020) and alloy 825 (N08825), whose austenitic structure and metallurgical/corrosion characteristics lead to electrochemical behavior somewhat similar to the 18-8 variety of stainless steels. Most of these are corrosion resistant in dilute sulfuric within specific limitations of temperature and contamination and are superior to conventional grades of stainless steel. Actual performance of these high-performance alloys is discussed in specific chapters. Isocorrosion curves for N08904, 20Cb-3 (N08020), and the high-molybdenum 254 (S31254) and 654 (S32654) alloys in 0–70% acid is shown in Figure 9.8.

Zinc, Tin, and Lead

Zinc undergoes rapid attack by dilute sulfuric acid with the evolution of hydrogen.

Tin is fairly resistant to dilute sulfuric acid in the absence of dissolved oxygen (DO) because of a high overvoltage of hydrogen. In 6% acid at 20°C (68°F), the corrosion rate was only 7 mpy (0.18 mm/y) under hydrogen, but this increased to 1,100 mpy (28 mm/y) under an oxygen atmosphere.[81]

In this group of materials, lead is resistant to the broad range of dilute sulfuric acid from 5–40% to the atmospheric boiling point by virtue of an insoluble film of lead sulfate formed by initial attack. This protective layer is removed by mechanical action (i.e., abrasion, erosion). A rate of about 5 mpy (0.13 mm/y) in 20% acid doubled at 5 ft/s (1.5 m/s). The sulfate film is increasingly soluble at elevated temperatures, especially above about 30% concentration. Below 5% acid, there is a marked increase in corrosion rate and the 3% antimonial grade is preferred. Lead is resistant to acid concentrations up to 70% at moderate temperatures. Because of health and environmental considerations, lead is not in common use in industrialized countries today.

Copper Alloys

Because copper cannot displace hydrogen, copper and its alloys, excepting the high-zinc brasses that are susceptible to dezincification, are inherently resistant to dilute sulfuric acid *in the absence of oxygen or oxidizing agents*. Unfortunately, the cupric ions formed as corrosion products (cuprous ions are unstable) are themselves oxidizing and the corrosion is therefore autocatalytic.

Some of the bronze alloys can be used in acid strengths from 0–70% at temperatures below about 70°C (160°F).

Nickel Alloys

Pure nickel (N02200) can displace hydrogen from acid solutions and shows an active-passive transition in sulfuric acid solutions.[4] Without anodic protection or inhibitor additions, nickel can be used only for unaerated acid under essentially static conditions at ambient temperature.

The 16% Cr alloy 600 (N06600) is somewhat less resistant than the copper-bearing alloy 400 (N04400) in air-free acid. The chromium-nickel-molybdenum alloys (e.g., N06625 and N10276) are inherently less resistant than the nickel-molybdenum alloy N10665, but far superior when oxidizing contaminants are available to reinforce passivity. The high-molybdenum alloys are generally preferred when halides are present.

Reactive and Refractory Metals

The corrosion resistance of all of these metals is adversely affected by the presence of fluoride ions.

Titanium is inherently nonresistant to dilute sulfuric acid, suffering active corrosion with possible hydriding from formation of atomic hydrogen at cathode sites. It can be maintained in a passive condition by the presence of strong oxidizing contaminants. The titanium-palladium and titanium-ruthenium alloys have somewhat greater, although still limited, resistance.[26,27]

Zirconium is inherently resistant to pure dilute sulfuric acid up to 70%. It is resistant up to 250°C (482°F), well above the atmospheric boiling point, at low concentrations. The corrosion of zirconium in weak and intermediate acid concentrations is compared with that of other alloys in Table 9.22.[82,83]

As concentration increases, the upper limit is the atmospheric boiling point at 70%.[69] In practice, the critical conditions are affected by temperature and by specific oxidizing contaminants. "Breakaway" corrosion may occur at temperatures and concentrations above the resistant region and with sufficient oxidizing ions. This might lead to pyrophoric corrosion products being formed.

Tantalum is resistant to all concentrations of sulfuric acid in the 0–70% range up to 250°C (482°F) well above the atmospheric boiling point, in the absence of fluoride contamination.[46,47] The corrosion behavior of tantalum and other alloys has been tested in a range of acid strengths and temperatures, Figure 9.12.[84]

Niobium has very limited use: only at less than 10% and atmospheric boiling point. At higher temperatures and concentrations it may embrittle.[28]

Table 9.22 Corrosion Rates in mpy (mm/y) of Zirconium and Other Alloys in Sulfuric Acid

Concentration %	Temperature °C (°F)	Zr 702	310L	316L	B-2	C-276
2	225 (437)	<0.1 (<0.002)	—	—	14.9 (3.8)	39.7 (1)
5	232 (450)	0.1 (0.002)	—	—	110 (2.8)	153 (3.9)
10	102 (216)	<0.1 (<0.002)	45 (1.1)	574 (14.6)	<1 (<0.025)	7.0 (0.18)
10	225 (437)	0.1 (0.002)	—	—	1,023 (26)	661 (16.8)
20 + 8% Fe^{+3}	80 (176)	<0.1 (<0.002)	—	—	—	>20 (>0.5)
30	108 (226)	<0.1 (<0.002)	1,137 (28.9)	>5,000 (>127)	2 (0.05)	55 (1.4)
40	80 (176)	<0.1 (<0.002)	>28,000 (>710)	—	—	—
55	132 (270)	0.1 (0.002)	>100,000 (>2,500)	>10,000 (>250)	1.89 (0.05)	295 (7.5)
55	168 (334)	19.6 (0.5)	—	—	37 (0.94)	212 (5.4)

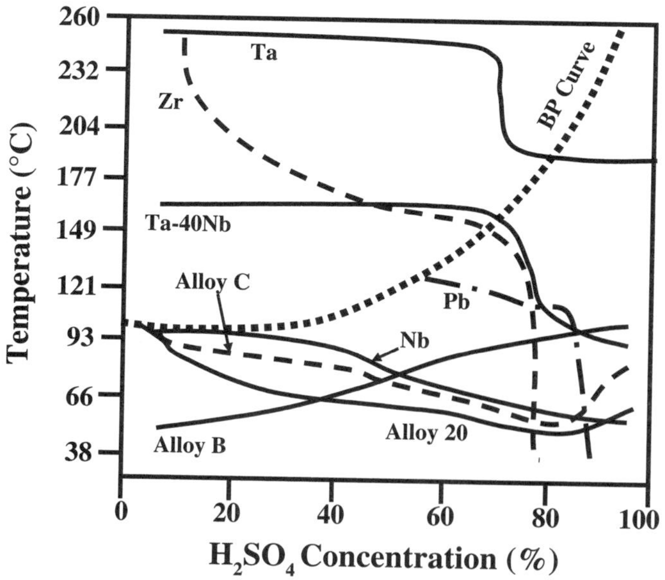

Figure 9.12 Isocorrosion Curves at 5 mpy (0.13 mm/y) Comparing Various Reactive and Refractory Metals and Alloys with Other Alloys in Sulfuric Acid

References

1. C. P. Dillon, "Corrosion Control in the Chemical Process Industries," 2nd Edition, publication No. 45 (St Louis, MO: MTI Inc., 1994): 420 pp.
2. Anon, "Resistance to Corrosion," publication no. 3M 8-88 S-37, 4th edition (Huntington, VA: Inco Alloys International, 1985): 44 pp.
3. H. H. Uhlig, "Corrosion Handbook" (New York: John Wiley, 1948): 1,188 pp.
4. B. D. Craig, D. B. Anderson, eds., "Handbook of Corrosion Data" (Metals Park, OH: ASM International, 1989): pp. 847–937.
5. Anon, "Corrosion Handbook," Vol. 8, Table 91 (Frankfurt, Germany: Dechema eV, 1991): p. 100.
6. E. H. Hollingsworth, H. Y. Hunsicker, "Corrosion of Aluminum and Aluminum Alloys," in J. R. Davis, ed., "Corrosion," Metals Handbook, 9th Edition, vol. 13 (Materials Park, OH: ASM International, 1987): pp. 583–609.
7. Anon, "The Corrosion Resistance of Nickel-Containing Alloys in Sulfuric Acid and Related Compounds," CEB-1 (Suffern, NY: The International Nickel Co. Inc., 1983): 90 pp.
8. A. Sabata, W. J. Schumacher, "Martensitic and Ferritic Stainless Steels," in CASTI Handbook of Stainless Steels and Nickel Alloys, S. Lamb, ed. (Edmonton, AB, Canada: CASTI Publishing Inc. 2000): p. 144.
9. J. P. Polar, "A Guide to Corrosion Resistance" (New York: Climax Molybdenum, 1961): pp. 229–245.
10. H. Abo, M. Ueda, S. Noguchi in Anon, "The Corrosion Resistance of Nickel-Containing Alloys in Sulfuric Acid and Related Compounds," CEB-1 (Suffern, NY: The International Nickel Co. Inc., 1983): 90 pp.
11. E. H. Phelps, D. C. Vreeland, "Corrosion of Austenitic Stainless Steels in Sulfuric Acid," Corrosion 13 (1957): pp. 21–25.
12. G. C. Kiefer, W. G. Renshaw, "The Behavior of the Chromium Nickel Stainless Steels in Sulfuric Acid," Corrosion 6 (1950): pp. 235–244.
13. Anon, "Carpenter Stainless Steels," 4-99/12.5M (Reading, PA: Carpenter Technology Corp., 1999): 356 pp.
14. J. F. Grubb, "AL-6XN®," Edition no. 2 (Pittsburgh, PA: Allegheny Ludlum Steel Corp., 1995): 38 pp.
15. Anon, "Solutions to Materials Problems," CD (Huntington, WV: Inco Alloys International, 1997).
16. Anon, "Carpenter Alloys for Controlling Severe Corrosives" (Reading, PA: Carpenter Technology Corp., 1989).
17. C. W. Kovack, "High-Performance Stainless Steels," NiDI Reference Book Series No. 11 021 (Toronto, ON, Canada: NiDI, 2000): 96 pp.
18. Anon, "7-Mo PLUS® Duplex Stainless Steel Booklet" (Reading, PA: Carpenter Technology Corp.).
19. Anon, "Corrosion Handbook—Stainless Steels" (Sandviken, Sweden: AB Sandvik Steel, 1999): pp. I:I–II:88.
20. Anon, "General Corrosion" (Manchester, U.K.: Weir Material Services, 2003), http://www.weirmaterials.com/general_corrosion.htm.
21. Anon, "Hastelloy® G-30® Alloy," publication no. H-2028D (Kokomo, IN: Haynes International, 1997): 16 pp.
22. Anon, "High-Performance Alloys for Resistance to Corrosion," SMC-026 (Huntington, WV: Special Metals Corporation, 2000): 64 pp.

23. M. B. Rockel, "Nicrofer 5923 hMo—Alloy 59," VDM Case History No. 5 (Werdohl, Germany: Krupp VDM GmbH, 2000): 50 pp.
24. J. F. Smith, "Corrosion of Lead and Lead Alloys," J. R. Davis, ed., vol. 13, Corrosion, Metals Handbook, 9th Ed. (Metals Park, OH: ASM International, 1987): pp. 784–792.
25. Anon, "Dilute Sulfuric Acid," ChemCor 10 (St. Louis, MO: MTI, undated).
26. J. A. Mountford, "Titanium Meeting the Challenge of the New Millenium," paper no. 01329, NACE 2001 (Houston, TX: NACE International, 2001).
27. R. Porter, "Reaping the Rewards of Ti-Ru Alloys," Corrosion Processing, Feb (2002): pp. 40–42.
28. R. Graham, R. Sutherlin, "Niobium and Niobium Alloys in Corrosive Applications," Wah Chang TMS, Niobium 2001 Conference, Orlando, FL (2001).
29. Anon, Sulfuric Acid section, CD "Corrosion Handbook" (Frankfurt, Germany: Dechema eV, 2001).
30. B. J. Saldanha, M. A. Streicher, NACE International in Sulfuric Acid section, CD, "Dechema Corrosion Handbook" (Frankfurt, Germany: Dechema eV, 2001).
31. Anon, "Duriron and Durichlor 51M," Bulletin A/2j (Dayton, OH: Flowserve Corporation, 1998): 8 pp.
32. L. L. Shreir, ed., "Corrosion, Volume 1, Metal/Environment Reactions" (London: Newnham-Butterworth, 1976): pp. 3.52–3.53.
33. C. M. Schillmoller, "Selection and Performance of Stainless Steels and Other Nickel-Bearing Alloys in Sulphuric Acid," NiDI Technical Series no. 10 057 (Toronto, ON, Canada: NiDI, 1990): 10 pp.
34. K. N. Krishnan, K. Prasad Rao, "Corrosion Rates of Austenitic Stainless Steel Clad Metals in 5 N H_2SO_4 + 0.5 N NaCl," Corrosion 46, 10 (1990): p. 866.
35. ASTM A 393 Discontinued 1974, "Recommended Practice for Conducting Acidified Copper Sulfate Test for Intergranular Attack in Austenitic Stainless Steel," replaced by A262 (West Conshohocken, PA: ASTM).
36. ASTM A 262-98, "Standard Practices for Detecting Susceptibility to Intergranular Attack in Austenitic Stainless Steels" (West Conshohocken, PA: ASTM).
37. N. Sridhar and I. J. Storey, "Prediction of Corrosion Behavior in Acid Mixtures," Corrosion Prevention in the Process Industries, Proceedings of the First NACE International Symposium, R. N. Parkins, ed. (Houston, TX: NACE International, 1990).
38. Anon, "Alloy Data," CD-ROM version 2.0 (Reading, PA: Carpenter Specialty Alloys, 1998).
39. Scarberry et al. (1967) in A. J. Sedriks, "Corrosion of Stainless Steels" (New York: John Wiley & Sons, 1979): p. 220.
40. D. C. Agarwal, R. Behrens, "Cost Effective Alternatives to Tank Linings for Handling Corrosive Solutions," Corrosion 2003, paper no. 03054 (Houston, TX: NACE International, 2003): 12 pp.
41. Anon, "Trent SEA-CURE stainless steel for power generation and chemical processing," catalogue no. A18-7/00-5000 (East Troy, WI: Trent Tube, 2000): 20 pp.
42. Anon, "AL 29-4-2," brochure no. B153-Ed 1-10M-582P (Pittsburgh, PA: Allegheny Ludlum Steel Corp., 1982): 18 pp.
43. Anon, "Ferrallium® Alloy 255," AR/4K/LA19 (Slough, Berks, U.K.: Langley Alloys, 1991): 14 pp.
44. N. Sridhar, "The Effect of Velocity on Corrosion in H_2SO_4 + HF mixtures," Corrosion/90 paper no. 19 (Houston, TX: NACE, 1990): 13 pp.

45. Anon, "Hastelloy® B-2 Alloy," brochure no. H-2006E (Kokomo, IN: Haynes International, 1997): 122 pp.
46. M. Schussler, "Corrosion Data Survey on Tantalum" (Chicago, IL: Fansteel Inc., 1972).
47. P. Aimone, E. Hinshaw, "Tantalum Materials in the CPI for the next Millennium," paper no. 01330, NACE 2001 (Houston, TX: NACE International, 2001).
48. Anon, "Niobium," Technical Data Sheet NioNio-056 (Albany, OR: Wah Chang, 2003): 42 pp.
49. A. Robin, "Corrosion Behavior of Niobium, Tantalum and Their Alloys in Boiling Sulfuric Acid Solutions," Int J of Refractory Metals and Hard Materials 14, 6 (1997): pp. 317–323.
50. Anon, "Corrosion Handbook," Vol. 8, Sulfuric Acid (Frankfurt, Germany: Dechema eV, 1991).
51. Anon, "ULTIMET® Alloy," publication no. H-2087B (Kokomo, IN: Haynes International, 1992): 8 pp.
52. H. H. Uhlig, "Corrosion Handbook" (New York: John Wiley, 1948): p. 44.
53. G. A. Nelson, "Prevention of Localized Corrosion in Sulfuric Acid Handling Equipment," Corrosion 14, 3 (1958): p. 145t.
54. R. Covert, J. Morrison, K. Rohring, W. Spear, "Ni-Resist and Ductile Ni-Resist Alloys," NiDI publication 11 018 (Toronto, ON, Canada: NiDI, 1998): 42 pp.
55. Anon, "Corrosion Handbook," Sulfuric Acid section, CD (Frankfurt, Germany: Dechema eV, 2001).
56. Anon, "Incoloy® alloy 020," publication no. SMC-018 (Hereford, U.K.: Special Metals, undated): 4 pp.
57. Anon, "AL20® Nickel-Base Alloy," brochure no. B173/ED2/298/SW (Pittsburgh, PA: Allegheny Ludlum Corp., 1998): 6 pp.
58. D. C. Agarwal, et al. (1993) in I. A. Franson, J. F. Grubb, "Superaustenitic Stainless Steels," in CASTI Handbook of Stainless Steels, pp. 263–264.
59. Anon, "Nicrofer® 3127 hMo—alloy 31," material data sheet no. 4031 (Werdohl, Germany: Krupp VDM GmbH, 1997): 14 pp.
60. J. H. Potgieter, H. C. Brookes, "Corrosion Behavior of a High-Chromium Duplex Stainless Steel with Minor Additions of Ruthenium in Sulfuric Acid," Corrosion 51, 4 (1995): pp. 312–320.
61. Anon, "7Mo-PLUS Duplex Stainless Steel," brochure no. 4-89/7/5M (Reading, PA: Carpenter Technology, 1989).
62. P. Crook, personal communication (Kokomo, IN: Haynes International, 2003).
63. T. C. Spence, D. R. Stickle, "Corrosion-Resistant Casting Alloys," Advanced Materials & Processes 160, 1 (2002): pp. 51–54.
64. J. R. Crum, "Comparison of 5 mpy Isocorrosion Lines for Several Nickel Base Alloys in Sulfuric Acid" (Huntington, WV: Special Metals Corporation, 2004): 1 p.
65. Anon, Special Metals publication in DKL Engineering (2003), http://members.rogers.com/acidmanual/materials_metals.htm.
66. Anon, "Isocorrosion Diagram for the alloy Nicrofer 5923hMo (Alloy 59) in Pure Sulphuric Acid," Thyssen Krupp VDM web site (2004), http://www.wdisweb.de/wdisweb/wdis.
67. S. K. Brubaker, "Materials of Construction for Sulfuric Acid," Process Industries Corrosion—The Theory and Practice (Houston, TX: NACE, 1987).

68. D. R. Holmes, "Effect of Tin Content on the Corrosion of Zirconium 702 in Sulfuric Acid," NACE 2003, paper no. 03451 (Houston, TX: NACE International, 2001): 11 pp.
69. T-L. Yau, "Zirconium Meeting the Challenges of the New Millennium," NACE 2001, paper no. 01331 (Houston, TX: NACE International, 2001).
70. T-L. Yau, Presentation to MTI-TAC, St. Louis 910618 report by C. P. Dillon (1994).
71. B. Fitzgerald, "Issues around the Use of Zirconium in Alcohol Production," Wah Chang Zirconium/Organics Conference Proceedings, Sept. (1997), Salishan Lodge, Gleneden Beach, OR.
72. B. Fitzgerald, T-L. Yau, "The Mechanism and Control of Stress Corrosion Cracking of Zirconium in Sulfuric Acid," 12th International Corrosion Congress, Sept. (1993), Houston, TX.
73. B. Fitzgerald, T-L. Yau, R. T. Webster, "Stress Corrosion Cracking of Zirconium and Its Control in Sulfuric Acid," Corrosion '92, paper no. 154 (Houston, TX: NACE International, 1992).
74. T-L. Yau, "Methods to Treat Pyrophoric Film on Zirconium," ASTM STP 830 (West Conshohocken, PA: ASTM): pp. 124–129.
75. C. P. Dillon, "Pyrophoric Surfaces on Zirconium Equipment: A Potential Ignition Hazard," MTI Publication no. 19 (St. Louis, MO: MTI, 1986).
76. T. J. Glover, "Titanium and Zirconium Castings: An Economic Solution to Corrosion in the Process Industries," Corrosion Prevention in the Process Industries, Proceedings of the First NACE International Symposium, R.N. Parkins, ed. (Houston, TX: NACE International, 1990).
77. J. R. Davis, ed., "Corrosion," ASM Metals Handbook, Vol. 13, 9th ed. (Metals Park, OH: ASM International, 1987): p. 795.
78. W. Barker, T. E. Evans, K. J. Williams, "Effect of Alloying Additions on the Microstructure, Corrosion Resistance and Mechanical Properties of Nickel-Silicon Alloys," Br. Corr. J. 5, 3 (1970): pp. 76–86.
79. Various authors, in Sulfuric Acid section, CD, "Dechema Corrosion Handbook" (Frankfurt, Germany: Dechema eV, 2001).
80. DuPont (1986) in C. P. Dillon, ed., "Concentrated Sulfuric Acid and Oleum," vol. MS-1, Materials Selector for Hazardous Chemicals (St. Louis, MO: MTI Inc., 1997): 212 pp.
81. F. Todt (1961) in Sulfuric Acid section, CD, "Dechema Corrosion Handbook," (Frankfurt, Germany: Dechema eV, 2001).
82. Anon, "Zirconium in Sulfuric Acid Applications," brochure no. 5317 PDF (Albany, OR: Wah Chang, 2001): 9 pp.
83. Anon, "Zirconium in Sulfuric Acid Pickling Applications," brochure no. 5721 PDF (Albany, OR: Wah Chang, 2002): 5 pp.
84. F. J. Hunkeler (1986) in M. Renner, K. Andersson, D. Michalski-Vollmer, "Application Limits of Ta and Ta-2.5% W for Sulfur Acid Handling," Materials and Corrosion 49 (1998): pp. 877–887.

10

Corrosion in Contaminated Acid and Mixtures

Corrosion data obtained from laboratory tests in C.P. acid or from field storage of commercial grades usually reflect only the corrosivity of the sulfuric acid itself. However, corrosion in plant acids is often complicated by contaminants.

For example, in the manufacture of concentrated acid in which air combustion of feed stock is the source of sulfur trioxide, some nitrogen oxides may be formed (depending on burner design, temperature, etc.) that contaminate the product with nitric acid. Acid recovered from drying of chlorine contains chlorine (or hypochlorite ions, depending on acid concentration) and chloride ions. Drying of methyl chloride can introduce chloride ions. Unless steps are taken to preclude iron contamination, small amounts of ferric ion (Fe^{+++}) may be present due to air oxidation of ferrous sulfate ($FeSO_4$) removed from steel or iron surfaces.

Apart from these incidental cases of contamination, in some cases sulfuric acid may be deliberately mixed with other acids, such as nitric, for specific purposes. It may also become mixed with other chemicals as part of a particular process—for example, the introduction of fluorides during the production of phosphoric acid.

Organic contaminants may act as corrosion inhibitors (e.g., aryl sulfonates from petroleum operations) but others (e.g., diethyl sulfate) aggravate corrosion by solubilizing otherwise protective films (e.g., on lead). Inorganic contaminants (e.g., oxidants such as cupric or ferric ions or nitric acid, reducing cations such as stannous ions, halides such as chloride or fluoride ions) affect the redox potential of sulfuric acid. This may make some forms of localized corrosion such as pitting, crevice corrosion, and stress corrosion cracking more likely. Halides may also solubilize otherwise protective films that some metals (e.g., lead) depend on for corrosion resistance.

Nitric acid was an inherent constituent of 78% sulfuric acid in the now obsolete Lead Chamber Process because of the use of oxides of nitrogen to oxidize sulfur dioxide to SO_3. Today, when 78% acid is encountered, it may have been recovered from nitration processes involving mixed acids, and it may also be contaminated with nitric acid, although vacuum concentrators can be operated to eliminate NO_X. Nitric acid is also added as an antifreeze in oleum. Small amounts of oxides of nitrogen are introduced into strong acids by high-temperature oxidation of sulfurous materials (e.g., pyrites). High concentrations of nitric acid can form in a vapor-phase situation to cause stress corrosion cracking (SCC) of non-stress-relieved steel vessels handling mixed sulfuric/nitric acid liquors.

Effects of specific contaminants and mixtures on particular metals and alloys are discussed in this chapter. This will deal with contaminants of general interest and in any strength of acid. Other cases of incidental contamination in particular acid strengths or of special significance to specific alloys have been dealt with in the relevant preceding sections.

Chloride Ion Contamination

Halides in sulfuric acid make it difficult to achieve and maintain passivity on stainless steels and higher-nickel alloys. In 10M (approx. 64% acid) at room temperature, prepassivated type 316 (S31600) could remain passive indefinitely. With 5 ppm chlorides present, passivity was lost and active corrosion commenced.[1] Chlorides increase attack on stainless steels in the passive state but have an inhibiting effect on them when they are actively corroding. If an oxidizing agent is present together with chlorides, pitting or crevice corrosion may occur and corrosion rates may increase, Table 10.1.[2] This table also shows the beneficial effect of nitric acid on stainless steel and a slight but surprising detrimental effect on the higher-nickel alloys.

In a test in 10% acid at 80°C (176°F), alloy 904 showed variable resistance at a corrosion rate of <1 mpy (<0.025 mm/y), while alloy 20 corroded at 5 mpy (0.13 mm/y) or less. Under these conditions, the addition of 10,000 ppm chloride ion increased rates of attack about tenfold: to 30–50 mpy (0.76–1.27 mm/y) for alloy 20 and to >100 mpy (>2.54 mm/y) for alloy 904.[3] For alloy 20, rates >5 mpy (>0.13 mm/y) occur above a few hundred ppm chloride and are about 15 mpy (0.38 mm/y) with 1,000 ppm chlorides; alloy 904 was not tested in this range of chlorides.

Chloride ions in sulfuric acid can cause pitting, crevice corrosion, and SCC in stainless steels. Stress corrosion cracking tests (constant extension rate test; CERT) were used to test the relationship between SCC and pH in solutions of H_2SO_4 and NaCl. These tests showed that at 25°C (77°F) there was a transition between SCC and immunity for type 301 stainless steel that was directly related to solution pH in the range of 0–3, Figure 10.1.[4] The boundary between possible SCC and immunity on a pH vs. log [Cl^-] plot may be expressed as:

$$pH = k_1 \log [Cl^-] + k_2$$

Table 10.1 Effect of Oxidizing Agents and Chlorides on the Corrosion Rates in mm/y (mpy) of Various Alloys at 66°C (150°F)

Material	10% H_2SO_4	10% H_2SO_4 + 5% HNO_3	10% H_2SO_4 + 5% NaCl
Type 316L	0.74 (29)	0.13 (5)	17.04 (671)
Aloy 825	0.03 (1)	0.05 (2)	5.16 (203)
20Cb-3	0.03 (1)	0.05 (2)	4.60 (181)

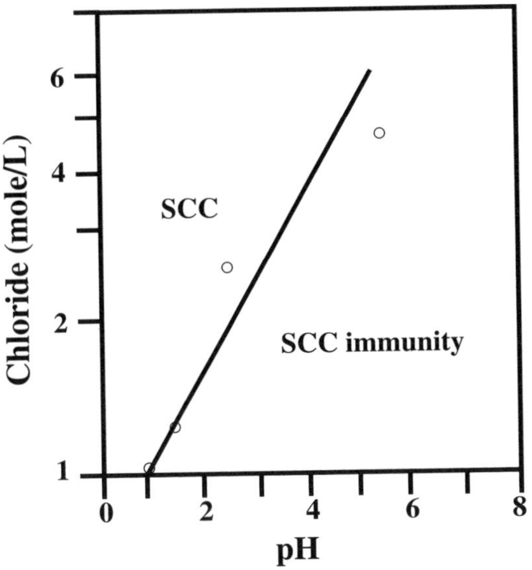

Figure 10.1 Effect of Chlorides and pH on SCC (CERT Tests) Resistance of Alloy 301 at 25°C (77°F)

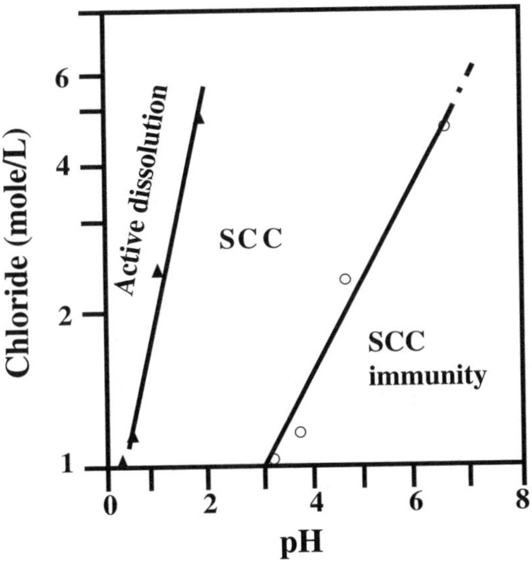

Figure 10.2 Effect of Chlorides and pH on SCC (SCG Tests) Resistance of Alloy 301 at 25°C (77°F)

where k_1 and k_2 are constants for a given alloy. The same alloy was also tested using an SCG (subcritical crack growth) technique under the same conditions, Figure 10.2. This technique also indicated the conditions under which active dissolution is taking place. Similar tests on type 310 stainless steel showed that SCC of this more alloyed steel was insensitive to pH.

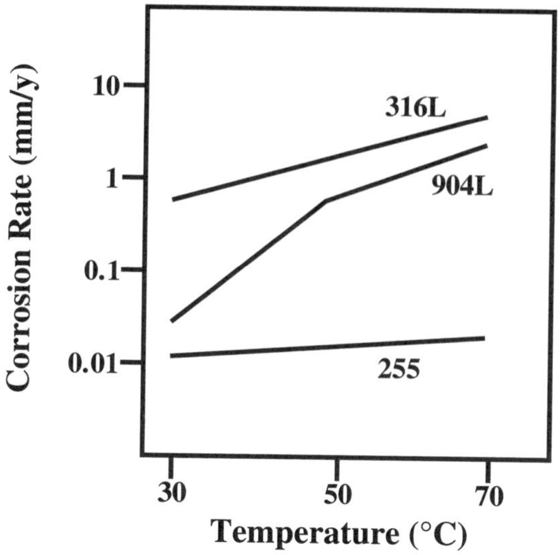

Figure 10.3 Effect of Temperature on Corrosion Rates of Stainless Steels in 2.5% H_2SO_4 + 3% NaCl

The corrosion behavior of various alloys has been tested in deaerated 2.5% H_2SO_4 + 3% NaCl, Figure 10.3.[5] These data showed that alloy 904L (N08904) was insufficiently alloyed to resist attack in this environment at 50 to 70°C (122 to 158°F) and that the duplex alloy 255 (S32550) was resistant under these conditions.

Data on the corrosion resistance of some of the superaustenitic alloys containing higher molybdenum levels than 904L (N08904) show the beneficial effects of increased molybdenum content in the presence of high chloride levels.[6] These high-molybdenum alloys all have good resistance in this aggressive environment (simulating some flue gas desulfurizing [FGD] conditions), with the 7% molybdenum alloy 27-7MO (S31277) performing slightly better than even alloy 276 (N10276) in these tests with 10,000 ppm chlorides, Table 10.2. These data were from 72-hour tests run at a temperature of 150°F (65°C).

Table 10.2 Corrosion Rates mpy (mm/y) in 10% H_2SO_4 + 10,000 ppm Chlorides

Material (UNS No.)	% Molybdenum	Trial 1	Trial 2	Average
254SMO (S31254)	6.3	32 (0.81)	38 (0.97)	35 (0.89)
25-6MO (N08926)	6.5	30 (0.76)	21 (0.53)	25.5 (0.65)
27-7MO (S31277)	7	0	0	0
Alloy C-276 (N10276)	16	0	1 (0.025)	0.5 (0.013)

In 10% H_2SO_4 with 2% HCl at 50°C (122°F), alloy 27-7MO (S31277) was uncorroded while alloy 25-6MO (N08926) had a corrosion rate of 29 mpy (0.74 mm/y).[6] Other trials in nominally the same environment reported a corrosion rate of 43 mpy (1.09 mm/y) for 27-7MO (S31277).[7] In 10% sulfuric at 65°C (149°F) without HCl neither of these alloys corroded.[6] In other trials simulating a more aggressive FGD environment, all of the alloys tested were severely attacked. This environment of 60% H_2SO_4 + 2.5% HCl + >2% HF + 0.5% flyash at 80°C (176°F) corroded alloy 27-7MO (S31277) at 153 mpy (3.91 mm/y) and alloy C-276 (N10276) at 28 mpy (0.71 mm/y).[7]

Other alloys of the "C" family, such as alloy 59 (N06059), have good resistance to weak sulfuric acid contaminated by chlorides. Alloy 59 (N06059) corrodes at 0.003–0.007 mm/y (0.12–0.28 mpy) in 20% acid with 1.5% Cl^- at 80°C (176°F), while in 50% acid with the same chloride level at 50°C (122°F) the corrosion rate increased to 0.42–0.75 mm/y (16.5–29.5 mpy).[8]

The effect of 2,000 ppm chlorides in 0–70% sulfuric acid on the corrosion behavior of duplex stainless steels and high-performance alloys has been summarized in Figures 10.4[9] and 10.5,[9,10] respectively. These figures (when compared with the nonchloride acid data, Figure 10.6[9]) show the benefit of molybdenum in chloride-containing acid solutions. The corrosion behavior of various nickel alloys in a mixture of 10% sulfuric acid and 5% hydrochloric acid has been tested at 80°C (176°F), Figure 10.7.[11] These data show that even the best of these alloys is substantially corroded in this aggressive environment.

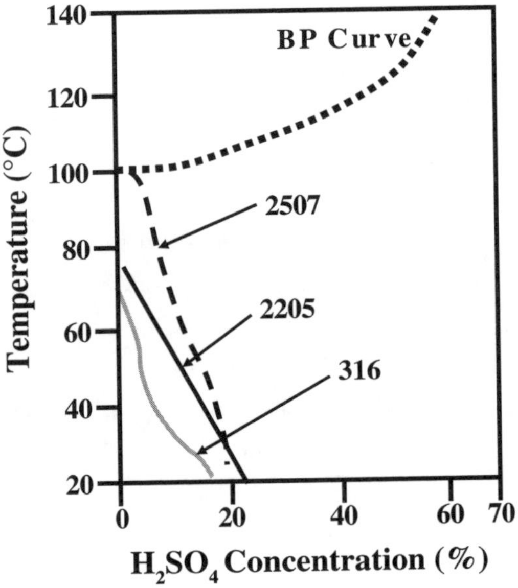

Figure 10.4 Isocorrosion Curves at 0.1 mm/y (3.9 mpy) of Duplex Stainless Steels Compared with 316 in Sulfuric Acid Containing 2,000 ppm Chlorides

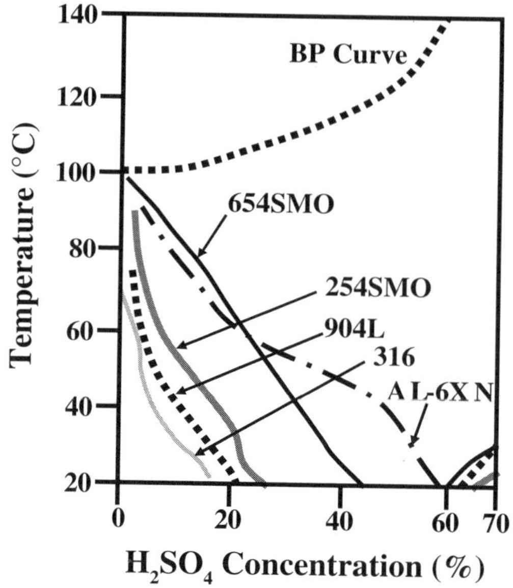

Figure 10.5 Isocorrosion Curves at 0.1 mm/y (3.9 mpy) of High-Molybdenum Stainless Steels Compared with 316 in Sulfuric Acid Containing 2,000 ppm Chlorides

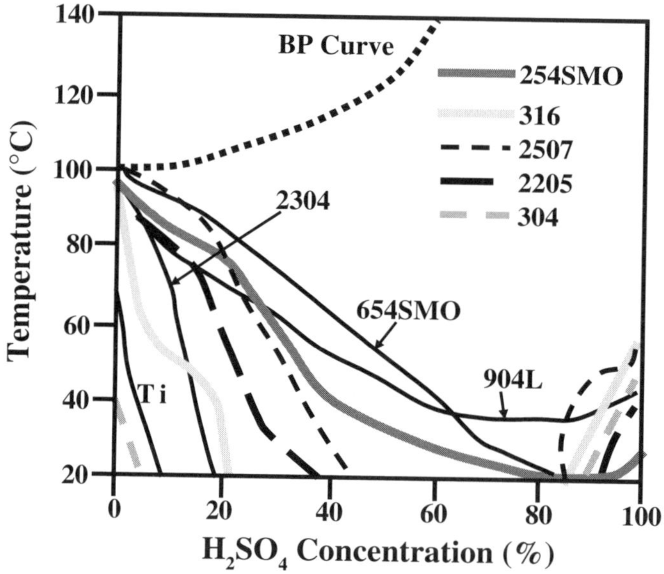

Figure 10.6 Isocorrosion Curves at 0.1 mm/y (3.9 mpy) of Stainless Steels and Titanium in Naturally Aerated 0–100% Sulfuric Acid without Chlorides

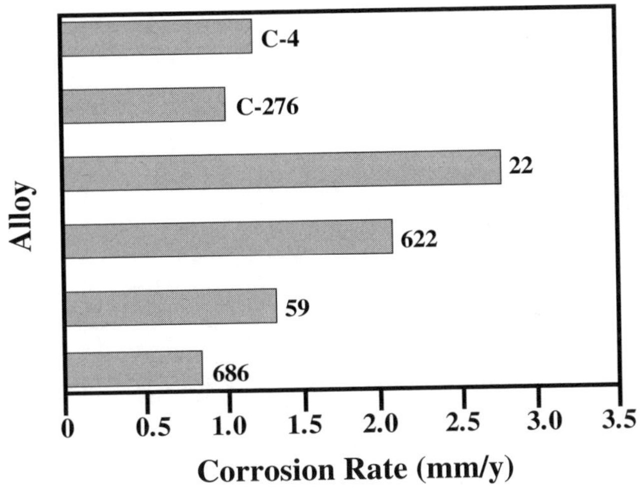

Figure 10.7 The Corrosion Resistance of Various Nickel-Based Alloys to 10% H_2SO_4 with 5% HCl at 80°C (176°F)

Fluoride Ion Contamination

Contamination of sulfuric acid by fluoride ions is common, and many materials are attacked by even small quantities of fluorides in acid solution. Fluoride contamination should be avoided if plant equipment contains glass, tantalum, zirconium, or silicon-containing alloys. The 5% silicon-containing austenitic stainless steels that are gaining favor in strong acid applications rely on a silica-rich surface to resist acid attack. This protective film is susceptible to attack by fluorides. Fluorides should be limited to the following levels to avoid attack of SX® alloy in 98–99.9% sulfuric acid:[12]

- <75°C (167°F), 5 ppm F^-
- 75–95°C (167–203°F), 2 ppm F^-
- >95°C (>203°F), 1 ppm F^-

The effect of fluorides in 93.5 and 98% acid on the corrosion of SARAMET® 23 was tested at different temperatures, Figure 10.8.[13] These data show a maximum level of fluoride ion, above which the corrosion rate decreases but is still at unacceptably high levels of attack. Corrosion of SARAMET® 23 was caused by fluoride contamination in a strong acid pipe distributor. Fluorocarbon refrigerant had leaked into the strong acid circuit for a considerable period of time. The acid brick of the tower was not attacked, implying relatively low levels of fluorides in the system.

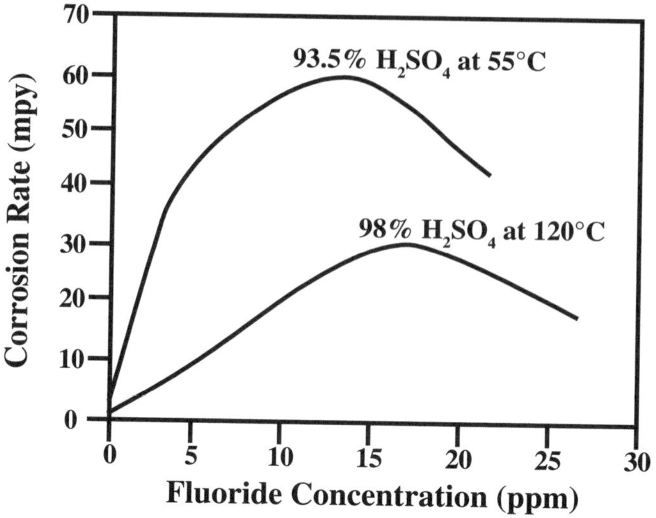

Figure 10.8 Effect of Fluorides on the Corrosion of SARAMET® 23 in 93.5% H_2SO_4 at 55°C (131°F) and 98% H_2SO_4 at 120°C (248°F)

Oxidizing Ion Contamination

The most common oxidant in strong sulfuric acid is nitric acid formed from nitrogen in air during the manufacturing process, from recovery of nitration acids, or added as an antifreeze in oleum. (See also section on mixed sulfuric/nitric acids, below).

Passivation of type 304 (S30400) begins in 70% acid at 60°C (140°F) when the nitric acid concentration is about 700 ppm (300 ppm for type 316 [S31600]). However, when type 304 (S30400) is activated in anhydrous mixed acids above about 50°C (122°F), nitric acid additions increase corrosion. Corrosion rates increase from about 14 mpy (0.35 mm/y) to 50–75 mpy (1.13–1.88 mm/y) at about 50°C (122°F) in mixed acids containing 0–20% nitric acid.[14] At higher temperatures, the rate increases from 40–60 mpy (1.0–1.5 mm/y) to several hundred mpy (>72 mm/y). In 76% sulfuric acid containing 2.3% nitric, type 304 (S30400) shows rates of <1 mpy (0.03 mm/y) at 45°C (113°F) in a dinitrotoluene process, while rates of 1.2 mpy (0.03 mm/y) are observed in a 90% mixed acid with slightly less than 2% nitric acid content.[15] Traces of chromic oxide (Cr VI, i.e., Cr with a valence state of 6^+) benefit type 304 grades in 70 to 100% sulfuric acid, raising the limiting temperature from 20°C (68°F) to 60 to 80°C (140 to 176°F).

Other oxidizing cations such as Fe^{+++} or Cu^{++} can drastically reduce corrosion in concentrated acid and even in less concentrated acid at elevated temperatures, as shown for 75 and 85% acid in Table 10.3.[16] It is, however, unwise to rely on passivation by oxidants to reduce corrosion, as crevices (in which an adequate concentration of inhibitor is difficult to maintain) are almost always present in equipment, and the oxidants may be inadvertently reduced, for example, by organic contaminants.

Similar data for weak and intermediate-strength sulfuric acid show that oxidizing contaminants are effective in reducing corrosion in this range of acid strength as well as in concentrated acid, Table 10.4.[16]

Table 10.3 Corrosion Rates in mm/y (mpy) of Various Stainless Steels in Concentrated Sulfuric Acid with Different Additions at 100°C (212°F)

	75%			85%		
% Oxidant	Type 316	Alloy 904L	Alloy 28	Type 316	Alloy 904L	Alloy 28
None	80 (3,150)	8 (315)	3 (118)	37 (1,460)	7.5 (295)	5 (197)
0.5% $CuSO_4$	8 (315)	6 (236)	0.4 (15.7)	—	—	—
1.0% $CuSO_4$	8 (315)	4 (157)	2 (79)	—	—	—
0.5% $Fe_2(SO_4)_3$	0.1 (3.9)	0.07 (2.8)	0.02 (0.8)	25 (984)	3 (118)	0.05 (1.97)
1.0% $Fe_2(SO_4)_3$	0.1 (3.9)	0.07 (2.8)	0.02 (0.8)	25 (984)	3.5 (138)	2 (79)

Table 10.4 Corrosion Rates in mm/y (mpy) of Various Stainless Steels in 20% and 65% Sulfuric Acid with Different Additions

	20% at 80°C (176°F)			65% at 100°C (212°F)		
% Oxidant	Type 316	Alloy 904L	Alloy 28	Type 316	Alloy 904L	Alloy 28
None	11 (433)	1.2 (47)	0.8 (31)	220 (8,660)	0.6 (24)	0.6 (24)
0.5% Fe2(SO4)3	0	0	0	—	0.1 (3.9)	0.04 (1.6)
1.0% Fe2(SO4)3	—	—	—	0.2 (7.9)	0.1 (3.9)	0.04 (1.6)
0.5% CuSO4	—	—	—	200 (7870)	—	0.02 (0.79)
1.0% CuSO4	—	—	—	—	0.07 (2.8)	0.02 (0.79)

The effect of oxidizing impurities on the molybdenum-rich nickel-based alloy B-2 (N10665) has been quantitatively defined. Table 10.5 shows the adverse effects of ferric ion and nitrate ion contamination on alloy B-2 (N10665) and the passivating influence on type 316L (S31603).[17]

Table 10.5 Effect of Oxidants on Corrosion Rate of Type 316L and Alloy B-2 in 96% Sulfuric Acid at 130°C (266°F)

	Oxidant			
Material	None	1,000 ppmFe^{+++}	10 ppm NO_3	1,000 ppm NO_3
Type 316L	72 (1.8)	60 (1.5)	82 (2.1)	5.2 (0.13)
Alloy B-2	6.6 (0.17)	9.8 (0.25)	23 (0.58)	148 (3.7)

Mixed Acids

Sulfuric/Nitric Acid Mixtures

In commercial processing, sulfuric and nitric acids are often combined in varying proportions with each other and with water. The effect of the combined acids on the corrosion rate of the containment materials varies, and whether the corrosion is retarded or accelerated depends on the mixture proportions and the composition of the alloy. Additions of nitric acid to sulfuric acid generally reduce the corrosion rate of stainless alloys. The corrosion rate of a low-chromium, high-molybdenum alloy such as alloy C-276 will not be appreciably affected. On the other hand, the corrosion rate of 316L (S31603), with only slightly more chromium but much less molybdenum, is greatly reduced by the addition of nitric to 30% sulfuric acid, Figure 10.9.[17]

Plastics such as PE and PVC are resistant to acid mixtures such as 54% H_2SO_4, 28% HNO_3, and 15% H_2O at room temperature.[18]

When sulfuric acid is added to nitric acid in amounts as small as 5–10%, a marked reduction in the corrosion rate of carbon steels is noted. At ambient temperatures, these mixtures are sometimes contained in carbon steel. Mixed acids with less than 15% H_2SO_4 and less than 20% H_2O do not severely attack carbon steel or cast iron.[18] Certain mixtures of nitric/sulfuric acids are permitted by the Interstate Commerce Commission (ICC) to be stored and transported in carbon steel, Table 10.6.[19]

Nitric acid additions to sulfuric acid are also beneficial for the 18% chromium austenitic grades, e.g., types 347 (S34700) and 304L (S30403), in amounts as small as 5–10%. For example, the corrosion rate for type 304 (S30400) in 65% sulfuric acid is about 120 mpy (3.05 mm/y) but this is reduced to <20 mpy (<0.51 mm/y) by the

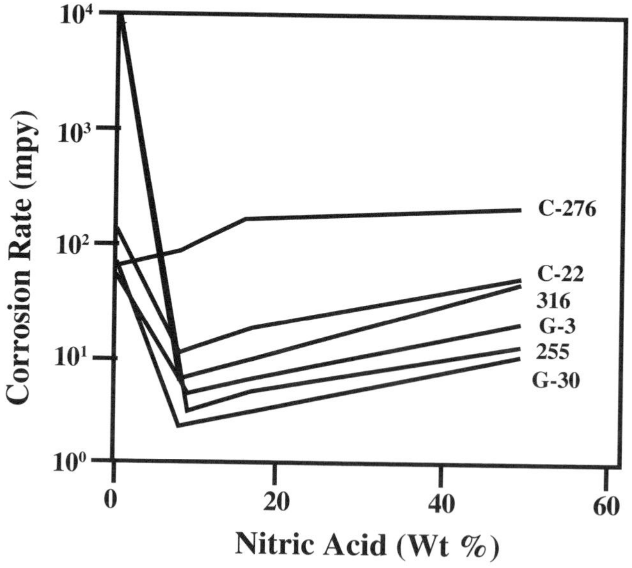

Figure 10.9 Effect of Nitric Acid Concentration on Corrosion of Various Alloys in Boiling 30% Sulfuric Acid

Table 10.6 Critical Water/Acid Values for Steel Containers

Water % Maximum	H_2SO_4 % Minimum	HNO_3
10	15	Balance
15	15	Balance
20	20	Balance
38	68	Balance

addition of about 5% nitric acid. Stainless steel 304L (S30403) is the preferred material of construction for process equipment other than tanks. The corrosion rate of types 304 (S30400) and 316 (S31600) is less than 0.11 mm/y (4.3 mpy) at 50°C (122°F) in typical sulfuric acid/nitric acid/water mixtures, Table 10.7.[19]

Results of other corrosion tests of 304 (S30400) in various mixtures of anhydrous sulfuric and nitric acids are shown in Table 10.8.[14] These data show somewhat higher corrosion rates for 304 (S30400) in the 50/50 mixture at 50°C (122°F) than did the previous table in the absence of water. They also show that the addition of nitric to sulfuric acid increases the corrosion rate up to a maximum at 20% nitric acid, after which the rate decreases. Type 304 (S30400) is not a suitable material for handling any of these mixtures of anhydrous acids even at 50°C (122°F).

Table 10.7 Concentrations of Sulfuric Acid/Nitric Acid/Water at 50°C (122°F) in Which the Corrosion Rate of 304 and 316 is < 0.11 mm/y (<4.3 mpy)

% H_2SO_4	% HNO_3	% H_2O
50	50	—
70	10	20
30	5	65
20	15	65
75*	25*	—

* Only type 316 (S31600)

Table 10.8 Corrosion Rates mpy (mm/y) of Type 304SS in Mixtures of Anhydrous Sulfuric Acid and Nitric Acid at Various Temperatures

Temperature °C (°F)	100% H_2SO_4 + 0% HNO_3	90% H_2SO_4 + 10% HNO_3	85% H_2SO_4 + 15% HNO_3	80% H_2SO_4 + 20% HNO_3	65% H_2SO_4 + 35% HNO_3	50% H_2SO_4 + 50% HNO_3
50 (122)	13.7 (0.35)	50.2 (1.3)	68.1 (1.7)	76.1 (1.9)	64.1 (1.6)	46.3 (1.2)
75 (167)	40.1 (1.0)	249 (6.3)	345 (8.8)	370 (9.4)	328 (8.3)	242 (6.2)
100 (212)	57.8 (1.5)	1,006 (25.5)	1,374 (34.9)	1,452 (36.9)	1,288 (32.7)	—

Various stainless steels were tested in nitrating acid at a range of temperatures. The results showed that at the lower temperature the 4% silicon alloy was the most resistant, while at higher temperatures the 304 (S30400) grades were better, Table 10.9.[20] Surprisingly, the 25% Cr, 20% Ni steel (alloy 25L) showed the highest corrosion rate at the elevated temperature.

Polarization studies on stainless steels 316 (S31600) and 316L (S31603) showed that the low carbon grade was highly resistant to nitrating acid (75% H_2SO_4, 25% HNO_3) at room temperature compared with 316 (S31600) with 0.08% carbon. The low-carbon grade retained passivation throughout the experiments, even when Cl^- and Cu^{++} ions were present. There was very little increase in passivation current, i.e., from 3,500 to 4,200 $\mu A/cm^2$, with an increase in sodium chloride concentration from 5 ppm to 200 ppm. At 5 ppm Cu^{++} ions, the passivation current decreased from 3,500 to 2,500 $\mu A/cm^2$, indicating inhibition.[21] In a TNT explosives plant, the reaction chamber is made from N08904 (2RK65®), handling 25% nitric and 75% sulfuric that heats up as the reaction proceeds.[22]

When the temperature is raised, mixed sulfuric/nitric acids become very aggressive and require alloys that are more resistant than standard stainless steels. Even below the boiling point, more highly alloyed steels or nickel-based alloys such as the special 310L (S31050 NAG), alloy 20Cb-3 (N08020), and CN7M (N08007) may be needed.

Chromium-bearing, nickel-rich alloys have been compared in one mixed acid comprised of 50% sulfuric plus 10% nitric acid at the boiling point, Table 10.10.[23] These data show that in this type of mixture, containing significant concentrations of nitric acid, the high-chromium, low-molybdenum alloys are more resistant.

Above the boiling point, the mixed nitric and sulfuric acids are very corrosive. Materials serviceable in this environment are zirconium alloy 702 (R60702), 14.5% silicon iron (F47003), and glass-lined steel, but even these materials can have significant rates of corrosion.

Corrosion rates of zirconium in a range of acid mixtures and temperatures are shown in Table 10.11.[24]

Table 10.9 Corrosion Rates $g\ m^{-2} h^{-1}$ (mm/y) of Various Stainless Steels in Nitration Acid, 60% H_2SO_4, 32% HNO_3

Temp. °C (°F)	25L	304L	304L-X (NAG)	4% Si Stainless
60 (140)	0.09 (0.10)	0.11 (0.12)	0.08 (0.09)	0.04 (0.04)
85 (185)	0.24 (0.26)	0.24 (0.26)	0.22 (0.24)	0.08 (0.09)
110 (230)	0.46 (0.50)	0.24 (0.26)	0.23 (0.25)	0.30 (0.32)

Table 10.10 Chromium-Nickel-Iron Alloys in 50% H_2SO_4, 10% HNO_3

Material	% Cr	% Mo	% W	Corrosion Rate mpy (mm/y)
Alloy C-22	22	13	3	70 (1.78)
Alloy G-3	22	7	1	30 (0.76)
Alloy G-30	30	5	3	16 (0.41)

Table 10.11 Corrosion Rates of Zirconium in Some Mixed Acids

Test Solution wt %	Temperature °C (°F)	mpy	mm/y
1% H_2SO_4, 99% HNO3	RT, 100 (212)	0.06	0.0015
10% H_2SO_4, 90% HNO_3	RT, 100 (212)	WG*	WG*
14% H_2SO_4, 14% HNO_3	Boiling	0.1	0.0025
25% H_2SO_4, 75% HNO_3	100 (212)	150	3.8
50% H_2SO_4, 50% HNO_3	RT	0.63	0.016
68% H_2SO_4, 5% HNO_3	Boiling	2,000	51
68% H_2SO_4, 1% HNO_3	Boiling	11	0.28
75% H_2SO_4, 25% HNO_3	RT	260	6.2

*WG = weight gain

Sulfuric/Hydrofluoric Acid Mixtures

Mixtures of sulfuric and hydrofluoric acids are common in many processes. Examples include the production of hydrofluoric acid and the manufacture of some organic fluorides.

A mixture of 20% H_2SO_4 with 3% HF at 70°C (158°F) is aggressive to most common alloys, particularly the low-alloy stainless steel 316L (S31603), Table 10.12.[25]

In other tests, 1–5% HF additions to various concentrations of sulfuric acid up to 60% at 79ºC (174ºF) increased the corrosion rate on alloy G-30 (N06030). The same additions of HF at higher sulfuric acid concentrations and the same temperature acted as an inhibitor and reduced the corrosion rate, Table 10.13.[26] The duplex alloy 255 (S32550) was similarly affected by HF in these tests. In 20% acid at 79°C (174°F) the corrosion rate for this alloy was <0.01 mm/y (<0.04 mpy), but this was increased by several orders of magnitude to >10 mm/y (>400 mpy) by the addition of 1% HF and to >100 mm/y (>4,000 mpy) with 3–5% HF. Corrosion rates in sulfuric acid strengths above 40% were reduced by HF additions.

For stainless steels such as alloy 255 (S32550) and copper-containing nickel-based alloys containing chromium, such as alloy 20 (N08020), alloy 825 (N08825), and alloy 630 (N06030), the addition of HF in the range of 1–3% acts as a corrosion accelerator up to about 60% sulfuric acid (the "switching concentration"). At concentrations above this it functions as a corrosion inhibitor. The inhibition effect is most pronounced at higher sulfuric acid concentrations where the corrosion rate is much higher in the pure acid.

Table 10.12 Corrosion Rates mpy (mm/y) in 20% H_2SO_4 with 3% HF at 70°C (158°F)

Material	Corrosion Rate
Type 316L (S31603)	13,600 (345)
Type 904 (N08904)	220 (5.6)
Alloy 28 (N08028)	120 (3.1)
Alloy G-30 (N06030)	26 (0.69)
Alloy 825 (N08825)	15 (0.38)
Alloy C-276 (N10276)	14 (0.36)
Alloy 20Cb-3 (N08020)	11 (0.28)

Table 10.13 Corrosion Rates mm/y (mpy) of Alloy G-30 (N06030) in Mixtures of H_2SO_4 and HF at 79°C (174°F)

% H_2SO_4	0% HF	1% HF	3% HF	5% HF
20	<0.003 (<0.12)	0.52 (20.5)	0.53 (21)	0.53 (21)
40	0.018 (0.71)	0.21 (8.3)	0.18 (7)	0.22 (9)
60	0.283 (11)	0.21 (8.3)	0.20 (8)	0.23 (9)
80	3.13 (123)	0.048 (1.9)	0.02 (0.79)	0.05 (2)
90	3.20 (126)	—	0.048 (1.9)	—

For alloys without copper additions such as alloy 22 (N06022), alloy C-276 (N10276), and alloy 625 (N06625), switching was at around 80% sulfuric acid. In all cases, the switching concentration was unaffected by the HF concentration.

In a question raised on the NACE Corrosion Network, it was reported that alloy 20 (N08020) lasted only 18 months in 67% sulfuric acid and 3% nitric at 100°C (212°F).[27] One respondent quoted rates of 5.7 mpy (0.14 mm/y) and 9.5 mpy (0.24 mm/y) for 304 (S30400) and 316 (S31600), respectively, for 60% sulfuric acid containing 2% nitric acid at 100°C (212°F). With 5% nitric acid in 60% sulfuric acid at the same temperature, corrosion rates were slightly higher, 6.3 mpy (0.16 mm/y) and 12 mpy (0.3 mm/y), respectively.

References

1. O. L. Riggs (1963) in Anon, "The Corrosion Resistance of Nickel-Containing Alloys in Sulfuric Acid and Related Compounds," CEB-1 (Suffern, NY: The International Nickel Co. Inc., 1983): 90 pp.
2. Anon, "The Corrosion Resistance of Nickel-Containing Alloys in Sulfuric Acid and Related Compounds," CEB-1 (Suffern, NY: The International Nickel Co. Inc., 1983): 90 pp.
3. N. Sridar, I. J. Storey, "Prediction of Corrosion Behavior in Acid Mixtures," Corrosion Prevention in the Process Industries, Proceedings of the First NACE International Symposium, R. N. Parkins, ed. (Houston, TX: NACE International, 1990).
4. H. K. Juang, C. Altstetter, "Effect of pH and Chloride Contents on Stress Corrosion Cracking of Austenitic Stainless Steels at Room Temperature," Corrosion 46, 11 (1990): pp. 881–887.
5. P. Amelot, J. C. Bavay, B. Baroux, "Behavior of a Duplex Stainless Steel in H2SO4 + NaCl Environments," Corrosion/91, paper no. 570 (Houston, TX: NACE International, 1991): 12 pp.
6. N. C. Eisinger, J. R. Crum, L. E. Shoemaker, "An Enhanced Superaustenitic Stainless Steel Offers Resistance to Aggressive Media," Corrosion 2003, Paper No. 03256 (Houston, TX: NACE International, 2003): 15 pp.
7. Anon, "Incoloy® alloy 27-7MO," publication no. SMC-092 (Huntington, WV: Special Metals Corporation, 2002): 4 pp.
8. M. B. Rockel, "Nicrofer 5923 hMo—alloy 59," VDM Case History No. 5 (Werdohl, Germany: Krupp VDM GmbH, 2000): 50 pp.

9. Anon, "Corrosion Handbook—Stainless Steels" (Sandviken, Sweden: AB Sandvik Steel, 1999): pp. I:I–II:88.
10. J. F. Grubb, "AL-6XN®," Edition no. 2 (Pittsburgh, PA: Allegheny Ludlum Steel Corp., 1995): 38 pp.
11. Anon, Special Metals publication on DKL Engineering (2003), http://members.rogers.com/acidmanual/materials_metals.htm.
12. Anon, "SX® Isocorrosion Curve" (Goteburg, Sweden: Edmeston Material System Engineering): 1 p.
13. G. Harding, "Developments in Sulphuric Acid Resistant Metal, 'SARAMET®' Technology" (Vancouver, BC, Canada: Aker Kvaerner Chemetics, 2003): 18 pp.
14. J. M. Viebrock, "Corrosion of Type 304 Stainless Steel in Mixed Anhydrous Nitric and Sulfuric Acids," Corrosion 25, 9 (1969): pp. 371–378.
15. L. L. Fairgold et al., "Influence of Added Nitric Acid on Corrosion of Stainless Steel in Sulfuric Acid," Protection of Metals 17, 3 (1981): p. 251.
16. G. Berglund, C. Martenson, "Applications of a High-Alloyed Stainless Steel in Sulfuric Acid Environments," Corrosion/87, paper no. 21 (Houston, TX: NACE International, 1987): 14 pp.
17. N. Sridhar, "Behavior of High Performance Alloys in Sulfuric Acid Streams," Corrosion '87, paper no. 19 (Houston, TX: NACE International, 1987): 16 pp.
18. E. Rabald, "Corrosion Guide" (Amsterdam, Netherlands: Elsevier Scientific Publishing Co., 1968): pp. 466–474.
19. C. P. Dillon, "Corrosion of Type 347 Stainless Steel and Aluminum 1100 in Strong Nitric and Mixed Nitric-Sulfuric Acids," Corrosion 12 (1956): pp. 47–50.
20. S. Nordin, "The Austenitic Stainless Special Steel UHB 25L for Nitric Acid Service" (Torshalla, Sweden: Uddeholm, undated), pp. 1–5.
21. I. Singh, A. K. Bhattamishra, D. K. Basu, "Electrochemical Behaviour of AISI-316 Stainless Steels in Sulphuric/Nitric Acid Mixtures," Anti-Corrosion Methods and Materials 44, 3 (1997): pp. 200–203.
22. Anon, Sulfuric Acid section, CD, "Dechema Corrosion Handbook" (Frankfurt, Germany: Dechema eV, 2001).
23. Anon, "Hastelloy G-30 Alloy," Brochure H-2028D (Kokomo, IN: Haynes International, 1997): 16 pp.
24. Anon, "Zirconium," Lecture notes, Zapp (undated).
25. P. A. Schweitzer, "Corrosion Resistance Tables—Metals, Plastics, Nonmetallics and Rubbers," Second Edition, Marcel Dekker, Inc., New York (1986).
26. N. Sridhar, "The effect of velocity on corrosion in H_2SO_4 + HF mixtures," Corrosion/90 paper no. 19 (Houston, TX: NACE, 1990): 13 pp.
27. T. Spence, private communication (NACE Corrosion Network, 1998).

11

Resistance of Nonmetallic Materials

Concentrated Acid and Oleum

Although nonmetallic materials do not corrode electrochemically, they do suffer specific types of degradation in concentrated sulfuric acid, including chemical stress corrosion cracking (CSCC), also known as environmental stress corrosion cracking (ESCC). The powerful oxidizing nature of hot strong acid limits many materials and can desiccate other materials containing hydroxyl groups (e.g., cellulose in wood). Plastics, elastomers, coatings, and carbon or graphite comprise the nonmetallics of an organic nature, while inorganic materials consist of ceramic-type materials.

Thermoplastics

In the form of plastic-lined pipe or valves, plastics may be used in concentrated sulfuric acid within certain limits. Solid plastics, plain or reinforced, can pose safety and reliability problems. Even when solid plastics are exposed to chemically compatible conditions, problems may arise from the high thermal expansion, notch sensitivity, creep characteristics, and other inherent weaknesses. Some process chemical companies simply forbid the use of solid plastics in strong sulfuric acid, regardless of chemical resistance, because of concerns over safety, reliability, and insurance aspects of their utilization. Also, bonded plastic linings, like elastomeric linings, are susceptible to blistering if the acid contains dissolved sulfur dioxide. Plastics other than fluorocarbons can be attacked by sulfur trioxide and oleum.

Polyolefins Polyethylene (PE) resists strong sulfuric acid up to 98% at ambient temperature. The high-density variety (HDPE) will tolerate up to about 49°C (120°F) and is used in the laboratory for beakers and bottles. Tanks up to 1,000-gallon (3,800-L) capacity have been used for up to 96% acid in ambient temperature service. Chlorosulfonated polyethylene has been successfully used in hoses handling up to 93% acid at ambient temperature. Braided stainless steel or alloy reinforcement is necessary.[1]

Polypropylene (PP) can be used, especially for lined steel pipe. It will withstand 98% acid at ambient temperatures and lower concentrations to 93°C (199°F), and it will resist intermediate strengths of about 70% to 85°C (185°F) for prolonged periods. Polypropylene immersion and external heat exchangers are used in the chemical-processing and metal-finishing industries to handle up to 80% acid to 85°C (185°F).

Both PE and PP can be subject to ESCC, their susceptibility increasing with increasing temperature and acid concentration.[1] To avoid this, the plastic is kept under compressive stress, if possible, or a copolymer or more resistant material is used.

Chlorocarbon Plastics Polyvinyl chloride (PVC, type 1, Grade I) is resistant to up to 93% acid at ambient temperatures. Type 2 is resistant to 90% acid at up to 54°C (129°F). Although useful for short piping runs, the chlorinated hydrocarbon plastics are impractical for longer runs because of the extensive support required and an irreversible expansion in service. However, FRP-reinforced PVC has been used in the United Kingdom for ambient 96% acid. Chlorinated PVC (CPVC), usually employed for its higher temperature capability (e.g., in hot water systems), will withstand slightly higher temperatures than PVC in sulfuric acid. A conservative figure for the limiting temperature is about 65°C (149°F).

Neither PVC nor CPVC should be exposed to 98% acid or greater concentrations under any circumstances.

Since about 1980, several major chemical companies have advised their customers against the use of solid PVC piping in sulfuric acid service, citing the possibility of failure from external mechanical damage or from fatigue (due to vibration) or thermal stress.[2] Small fatigue cracks can develop in piping that has aged or has low impact strength, or both, causing catastrophic failure. PVC is also subject to ultraviolet (UV) degradation by sunlight.

Polyvinylidene chloride (PVDC) is not recommended for concentrated sulfuric acid, but PVC and PVDC have been successfully used as piping for ambient-temperature acid up to 96% concentration.[1]

Solvent cement-jointing systems for PVC are normally as resistant as the base polymer. This is not the case, however, in sulfuric acid concentrations above 70%. This possibility of joint attack should be considered if this polymer is exposed to concentrated acid.[3]

Fluorocarbon Plastics The family of fluorocarbon plastics is satisfactory in concentrated acid within the concentrations and temperature limits given in Table 11.1. PTFE-lined steel pipe has incurred nitrate-induced stress corrosion cracking of the steel substrate from diffusion of nitric acid through the PTFE liner (e.g., in 72% sulfuric containing about 0.5% nitric acid).[4]

Fluorocarbons have a permeability problem because SO_3 can pass through PTFE, for example, to attack external metal. Stainless steel hose with PTFE lining is *not* suitable for oleum service for this reason. Similarly, PTFE spaghetti-type coolers are unsuitable for oleum.

FEP is inert in all concentrations and has been used for spaghetti-type heat exchangers in contact plants, although newer versions are made of PFA. FEP is *not* suitable for oleum service because of its permeability to sulfur trioxide. SO_3 then reacts with water exothermically, and temperatures can reach 427°C (800°F) and destroy the FEP. This mechanism has been responsible for failure of FEP-lined mist scrubbers on oleum storage tanks as well as FEP heat exchangers.[4]

Table 11.1 Thermoplastics Max. Temperature °C (°F) in Various Strengths of Concentrated Sulfuric Acid

	H_2SO_4 Concentration				
Plastic	<90%	93%	96%	98%	100%
Acrylic	—	—	—	20 NR	—
Acrylonitrile-butadiene-styrene (ABS)	—	20 (68) NR	20 (68) NR	20 (68) NR	20 (68) NR
Polyethylene (PE)	40 (104)*	40 (104)*	40 (104)*	20 (68) NR	—
Polypropylene (PP)	80 (176)	80 (176) 100 (212)*	20–60 (68–140)* 90 NR	20 (68)*	20 (68) NR
Polyvinyl chloride (PVC)	60 (140)	60 (140)	20 (68) 60 (140)*	20 (68)* 40 NR	20 (68) NR
Chlorinated polyvinyl chloride (CPVC)	60 (140) 80 (176)*	60 (140) 80 (176)*	20 (68) 40–60 (104–140)* 60–90 (104–194) NR	20 (68) 40 (104)* 60 (140) NR	20 (68) NR
Polyvinylidene chloride (PVDC)	20 (68)	20 (68)	20 (68)	20 (68) NR	20 (68) NR
Polyvinylidene fluoride (PVDF)	80–100 (176–212) 120 (248) NR	80–100 (176–212) 120 NR	60 (140) 90* 100 NR	52–60 (126–140) 100 NR	20 (68) NR
Ethylene chloro-trifluoroethylene and Ethylene tetrafluoroethylene (ECTFE and ETFE)	150 (302)	150 (302)	150 (302)	150 (302)	—
Fluorinated ethylene propylene (FEP)	205 (400)	205 (400)	205 (400)	205 (400)	20 (68) NR
Perfluoroalkoxy (PFA)	260 (500)	260 (500)	260 (500)	260 (500)	260 (500)
Polytetrafluoroethylene (PTFE)	260 (500)	260 (500)	260 (500)	260 (500)	260 (500)

* Conditionally resistant; medium can attack, cause swelling, and reduce life

NR: Not resistant at this temperature

The noncrystalline fluorocarbons, notably PFA, are useful for coatings and linings. Failures have occurred in stainless-steel-braided fluorocarbon plastic hose in oleum and SO_3 service. Permeation of the liner by SO_3 has occurred, forming dilute acid externally by reaction with atmospheric moisture and attacking the braid. The use of alloy 20Cb-3 (N08020) braid, instead of type 304 (S30400) or 316 (S31600), is effective in resolving this problem.[5]

At room temperature the following fluorocarbon polymers are resistant to concentrated acid, up to 100%: PTFE, PFA, ECTFE, FEP, and ETFE.[1] PVDF is resistant at room temperature in acid up to 98%.

Comparison of Thermoplastics Temperature and acid concentration limits are published by polymer producers, piping suppliers, and others. There are often discrepancies in published limits that may reflect differences in behavior of different formulations but may also indicate differences in testing methods and evaluation of performance. The temperature and acid-strength limits for various thermoplastics shown in Table 11.1 were compiled from a number of sources. The stated maximum temperatures presuppose that the mechanical properties are also adequate and probably mean they would only be used as fully supported linings at these elevated temperatures. There would also be more risk of permeation through the polymer as temperature increased.

In the form of plastic-lined pipe or valves, plastics may be used in concentrated sulfuric acid within certain limits, Table 11.2.[6] Most thermoplastics used in strong sulfuric acid are used in this way so that the steel outer pipe can provide the strength and protection for the plastic liner. PVDF is attacked by free sulfur trioxide so that 98% sulfuric acid should be considered the absolute maximum strength of acid to be used in PVDF-lined pipe. Other applications where free SO_3 is present should be avoided.

Table 11.2 Liner Selection Guide for Thermoplastic-Lined Steel Pipe

	Maximum Temperature °F (°C)				
H_2SO_4	450 (232)	200 (93)	175 (79)	150 (66)	Not resistant
85%	PTFE	PVDF	PP	—	SARAN®
93%	PTFE	PVDF	—	—	SARAN®
96%	PTFE	—	PVDF	—	SARAN®
98%	PTFE	—	—	PVDF	SARAN®
Fuming	PTFE	—	—	—	PVDF, PP, SARAN®

SARAN®: Polyvinylidene chloride resin
PP: Polypropylene
PVDF: Polyvinylidene fluoride resin
PTFE: Polytetrafluoroethylene

Table 11.3 Temperature Limits °C (°F) for Various Plastics in Dual-Laminate Construction in Sulfuric Acid

	H_2SO_4 Concentration				
Plastic Liner	Up to 80%	90%	96%	97%	98%
PVC	60 (140)	60 (140)	20 (68) 60 (140)*	—	20 (68) NR
CPVC	60 (140) 80 (176)*	60 (140) 80 (176)*	40 (104)* 60 (140) NR	40 (104)	20 (68) NR
PE	40 (104) 60 (140)*	20 (68)*	20 (68) NR	—	—
PP	40 (104) 60 (140)*	40 (104) 60 (140)*	20 (68) NR	—	20 (68) NR
PVDF	100 (212)* 120 NR	80 (176) 120 NR	60 (140) 100 NR	20 (68)* 40 (104) NR	20 (68) NR
ECTFE	100 (212) 120*	100 (212)	60 (140)	—	40 (104) 60 (140) NR
ETFE	—	—	—	—	150 (302)
FEP	60 (140)	60 (140)	150 (302)	—	150 (302)
PFA	150 (302)	150 (302)	150 (302)	—	150 (302)

* Conditionally resistant; medium can attack, cause swelling, and reduce life
NR: Not resistant at this temperature

Thermoplastic linings as part of dual-laminate construction with FRP reinforcement are also used in some sulfuric acid applications within certain limits, Table 11.3.[7]

Thermosetting resins

Fiberglass-reinforced plastics (FRP), also called RTP (reinforced thermoset plastics) and GRP (glass fiber-reinforced plastics) in the United Kingdom, are based on conventional polyesters and vinyl esters. They are limited to <90% sulfuric acid at ambient temperatures and to 75% acid at up to 65°C (149°F); actual limits depend on the resin type and formulation, Table 11.4. This table was compiled from various sources. Much of the variation of resistance between nominally similar resins is due to different formulations used to achieve different properties that also affect chemical resistance. For example, resins may be formulated to be flexible, be fire retardant, have good high-temperature properties, etc.

The reinforced thermosetting resins have a higher strength and lower coefficient of thermal expansion than the thermoplastics. Both formulation and compounding affect their chemical resistance.

Table 11.4 Temperature Limits °C (°F) for Thermosetting Resins in Various Strengths of Sulfuric Acid

Material	H_2SO_4 Concentration		
	75%	80%	93%
Vinyl ester	43–49 (110–120)	NR	NR
Epoxy novolac vinyl ester	120 (248)	NR-38 (NR-100)	NR
Chlorendic polyester	—	66 (150)	NR
Bis-A Fumarate polyester	43 (110)	NR-80 (NR-176)	NR
Furane	66 (151)	38 (100)	NR
Brominated isophthalic	—	NR	—
Brominated terephthalic	—	NR	—
Isophthalic	NR	NR	NR

NR: Not resistant

FRP is also used as the reinforcing, structural element with a thermoplastic liner in dual-laminate construction. The corrosion resistance depends on the resistance of the liner, although resistant resins are often used in the construction of the FRP reinforcement in case of permeation and leaks in the thermoplastic liner.

A proprietary brand of phenol-formaldehyde piping reinforced with silica is rated to 70°F (27°C) in 70–93% sulfuric acid.[8]

Plastic Coatings

The only organic coatings useful in concentrated sulfuric acid are those based on fluorocarbons (e.g., PFA and ECTFE) and baked phenol-formaldehyde types. It is axiomatic that coatings should only be used when corrosion of the metallic substrate at defects or "holidays" would be low.

The spray-and-bake PFA-type coatings are applied at about 10 mils (0.25 mm) thickness per coat, usually to a final thickness of about 40 mils (1.0 mm). They may be applied to complex shapes such as valve bodies, dust collectors, agitator blades, and pump components, as well as to tanks or vessels. Temperatures up to 232°C (450°F) can be tolerated, according to the manufacturer.

Coatings based on ECTFE are similarly resistant to all concentrations but have a lower temperature limit of about 140°C (300°F). PFA/PTFE coating applied at 60-mil (0.14-mm) thickness to a surface to which stainless-steel mesh has been spot/seam-welded is very resistant to disbonding.

Baked phenol-formaldehyde coatings, cured at about 225°C (437°F), are routinely used for steel tanks in strong acid service, within specific limits at ambient temperatures. These coatings are approved for food-grade chemicals such as 1.835 Sp. Gr. Battery-Grade acid.

Details concerning the use of these coatings in specific industrial grades of acid are given in other chapters. Briefly, they are widely used in 93% acid and, with a shorter life, in 98% acid. The reduced life in nominal 98.5% acid may be due to traces of nitric acid contamination. Fluoroplastic coatings are largely replacing the baked phenolic coatings in this latter service. Baked phenolic coatings are not used with oleum.

Elastomers

Elastomeric products can be successfully used as sealing components at various temperatures and concentrations of sulfuric acid. The performance of elastomeric products in a sulfuric acid medium is dependent on the following variables (and perhaps even more):

- acid concentration
- application temperature
- application pressure
- the presence of other chemicals in the acid stream
- specific service conditions: static, dynamic, compression on seal, etc.
- specific family and grade of elastomer
- the compound formulation
- crosslinking chemistry of compound and compatibility with medium

Table 11.5 presents information on elastomer performance in sulfuric acid.[9] However, material formulation and cure chemistry can lead to different performance ratings. It is always recommended that elastomeric materials be tested for suitability before use.

The performance of seals made of FKM (e.g., Viton® fluoroelastomer) in sulfuric acid is dependent on the variables listed previously. Broadly speaking, seals made of FKMs with higher fluorine content and using a peroxide/coagent vulcanization system will perform better than alternatives. Suggested grades include Viton® B, Viton® GF-S, and Viton® ETP-S, listed in order of increasing performance.

Table 11.5 Compatibility of Elastomers with Sulfuric Acid (0–98% Concentration)

Elastomer Family	Immersion Test Conditions	Performance Ratings[a]		
		Suitable	Satisfactory[b]	Not Suitable
Buna-N / HNBR	≤ 23°C (73°F)	X	23°C (73°F) to 50°C (122°F)	—
Butyl	≤ 23°C (73°F)	X	—	—
EPR, EPDM	< 50°C (122°F)	—	X	—
Fluorohydrocarbon (FKM, Viton®, etc.)	< 100°C (212°F)	X	—	—
Fluorosilicone (FVMQ)	< 23°C (73°F)	—	—	X
Chlorosulfonated polyethylene (CSM, Hypalon®, etc.)	< 50°C (122°F)	X	—	—
Natural rubber—soft	≤ 30°C (86°F)	—	≤ 70 %	—
Natural rubber—hard	≤ 30°C (86°F)	—	≤ 80%	—
Perfluoroelastomer (FFKM, Kalrez®, etc.)	≤ 150°C (302°F)	X	> 150°C (302°F)	—

[a] Ratings up to service temperature of elastomer unless otherwise noted
[b] Testing recommended

A fully fluorinated elastomer, such as Kalrez® perfluoroelastomer, has chemical resistance similar to PTFE. Kalrez compounds 7075 and 4079 are reinforced with carbon black and will resist up to 100% acid. A titanium-dioxide-reinforced grade, Kalrez 1045, is suggested for oleum up to 260°C (500°F). Kalrez 7075 and 4079 *may* be satisfactory for lower strengths of oleum if the reinforcing agent is suitable for the exposure. The upper temperature limits in air alone are about 325°C (627°F) and 315°C (600°F) for compounds 7075 and 4079, respectively. Kalrez products are available as O-rings or specialty shapes; they are not suitable for sheet lining.

Finally, it should be noted that elastomers can suffer severe attack if even parts per million of certain organic compounds are present in the acid (e.g., chlorinated solvents and aromatic solvents). These can be preferentially absorbed over prolonged periods to attack either the elastomer itself or any adhesive employed in its application.

Carbon and Graphite

Carbon and graphite are allotropic forms of the same element. The graphite form, with its ordered crystalline structure, has better electrical conductivity and heat-transfer properties. Commercially, graphite is used as "impervious graphite," the pores having been filled with a resin impregnant (e.g., phenolic, epoxy, furan, or tetrafluoroethylene). Carbon has somewhat better oxidation resistance than impervious graphite, while the latter material provides more effective heat transfer.

Carbon is used primarily for carbon brick linings in strong acid services to about 315°C (599°F), using acid-resistant cements. It is used where fluorides are present and must not be placed directly in contact with lead, e.g., lead lining of a steel shell. Carbon in contact with lead in sulfuric acid causes major galvanic corrosion of the lead. A barrier layer of acid-resistant brick must always be placed between the carbon and the lead-lined vessel.

Impervious graphite heat exchangers are used up to 93% concentration. Acid concentration and temperature limitations are given in Table 11.6.

The 170°C (338°F) limit is due to the phenolic impregnant. Actually, this is somewhat higher than the resin alone would tolerate, but the microscopic pockets are physically supported and protected by the appropriate surrounding graphite structure.[10]

The development of tetrafluoroethylene impregnants (e.g., Carbone® Type TH) has greatly extended the usefulness of impregnated graphite. It is recommended to 175°C (347°F) in up to 80% sulfuric acid and to 110°C (230°F) in 95% acid.[11] It is reported to have a unique applicability in hot acids contaminated with sulfur dioxide. In tests in 96 and 98% sulfuric acid, the presence of 400 ppm of nitric acid caused major attack of this grade of graphite at 150°C (302°F).[12] Fluoroplastic bonded graphite (e.g., Diabon® F 100) is used for plate exchangers in 80% sulfuric acid up to

Table 11.6 Temperature/Concentration Limits for Impervious Graphite Heat Exchangers with Phenolic Impregnant

H_2SO_4 Concentration %	Temperature °C (°F)
70–85	170 (338)
85–90	150 (302)
90–96	70 (158)

120°C (248°F) and in 90% acid up to 100°C (212°F). The synthetic resin grade, Diabon® NS 2, is resistant to 95% sulfuric acid at up to 160°C (320°F).[13]

Block-type exchangers are used up to 300 psig (2.1 MPa). Shell and tube exchangers are limited to about 75 psig (520 kPa). As a practical matter, the latter should be limited to perhaps 50 psig (340 kPa) saturated steam, i.e., about 150°C (302°F), because of cemented tube-to-tubesheet joints.

Graphite fiber cloth is available and is used in gasketing and packing applications.

Ceramic Materials

Glass Borosilicate glass pipe with PTFE-gasketed flexible joints has been used in 93% acid up to the temperature limit of the plastic. Borosilicate glass is also used to concentrate sulfuric acid in single- or multi-stage vacuum evaporator units and for acid dilution equipment.[14,15] Glass-lined steel equipment is used in all concentrations of acid up to about 200°C (392°F) in commercial practice but can be attacked, particularly at elevated temperatures, Figure 11.1.[16] This figure also shows the beneficial effect of silica inhibition in this environment.

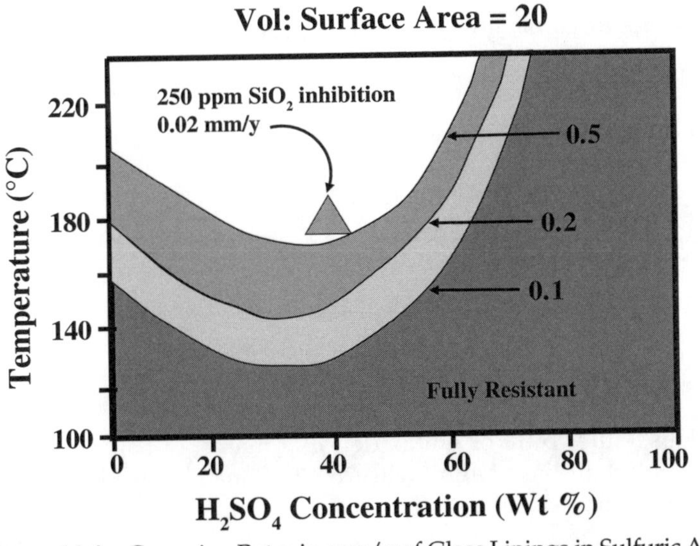

Figure 11.1 Corrosion Rates in mm/y of Glass Linings in Sulfuric Acid

Glass-lined steel pipe and vessels must be protected against external corrosion by acids because hydrogen generated on the metal surface can cause internal failure of the glass lining.

Stoneware Alumina (Al_2O_3) and Silica (SiO_2) Earthenware is the term used to describe equipment made from fired clays. Chemical stoneware describes equipment, made from selected clays, that is stronger and more chemically resistant than earthenware. Chemical stoneware has substantially the same resistance as glass but is more susceptible to mechanical shock. Porcelain is a dense product with comparatively poor mechanical properties and thermal shock resistance. It is made from silica and alumina and has good resistance to all acids except hydrofluoric, but generally little resistance to alkalies. These ceramics are typically glazed for chemical application to assist in cleaning and to seal any surface porosity.[17]

Alumina (aluminum oxide) is attacked by concentrated acid at temperatures above about 140°C (285°F). Silica (silicon dioxide) is used in preference to alumina for more severe acid services.

Fused silica (typically 99.6% silica) is available as dense shapes or insulating foam bricks and has excellent corrosion resistance to sulfuric acid. This material is used in specific lining applications, such as in drum concentration plants (see Chapter 12).

Silicon Carbide (SiC)

Alpha silicon carbide (e.g., Metaulic Systems Hexoloy®) can be used for heat exchangers up to about 175°C (347°F).

The rate of attack of silicon carbide (Hexoloy®) and other ceramics is usually stated in a weight-loss rate, mg/cm^2 yr, Table 11.7. A material with a rate of 0.3–9.9 mg/cm^2 yr is considered suitable for long-term service.[18] In tests in 30% oleum at 100°C (212°F), the weight loss of Hexoloy SA was acceptable (0.04 mg/cm^2 yr), but some loss of bend strength was noted. The silica phase of the Hexoloy KT grade was attacked by hot oleum, causing weight loss >90 mg/cm^2 yr and a loss in hardness and strength.[19]

Table 11.7 Corrosive Weight Loss (mg/cm^2 yr) of Various Ceramics in 98% H_2SO_4 at 100°C (212°F)

Hexoloy SA (no free Si)	Hexoloy KT (12% Si)	Tungsten Carbide (6% Co)	Aluminum Oxide (99.0%)
1.8	55.0	>1,000	65.0

Wear-Resistant Coatings

Ceramic coatings, either pure or embedded in a matrix of corrosion-resistant alloys (e.g., Stellite® 6, a cobalt-based alloy with 26% Cr), may be used for mechanical seals in pumps for concentrated acid service.

Acid-Brick Lining

Acid-brick linings are widely used to handle hot, concentrated acid. This type of construction can be subject to long-term "growth" of the acid brick (unless fired at high temperature) and to problems with cemented joints. It is desirable to have a membrane between the metallic shell and the brick lining because of incursions of acid at defects existing (or developing) in the lining. Such membranes may be metallic (e.g., lead) or nonmetallic (e.g., asphaltic coatings, rubber, fluoroplastic linings or coatings). The temperature limit for the brick itself is about 315°C (600°F), and the role of the traditional double course of brick is to reduce the temperature of the leaking acid to a tol-

Figure 11.2 Acid Tower Brick Lining in Progress

erable level by the time it reaches the membrane. A typical tower lining consists of a modified asphalt, Pecora® mastic layer next to the steel shell, a PTFE membrane next to the mastic, and multiple layers of fireclay brick exposed to the acid, Figure 11.2.

Typical nonmetallic materials used for membranes behind brick in sulfuric acid applications include the following:[20]

- Modified asphalt mastic (Pecora® A-103). This is trowelled onto the prepared steel shell and is followed by a layer of mortar into which the first brick layer is bedded. It must cover the entire surface to be effective and can break down or become damaged with time.
- Modified asphalt mastic followed by overlapping sheets of 10-mil-thick PTFE. This provides a more impermeable membrane, but it can also be mechanically damaged.
- 10-mil-thick PTFE sheets attached to a prepared steel shell using epoxy resin. The epoxy resin can be damaged thermally and break down over time.
- Rhepanol® ORG sheeting is a graphite-filled polyisobutylene rubber sheet product applied directly to the steel shell. This material is commonly used as a membrane behind brick in Europe.
- Ceramic paper soaked in potassium silicate is generally used in dry service rather than immersion.

Some naturally occurring clays are especially corrosion resistant and are used to make acid-resistant brick. Once these silica-alumina shales have been fired, they are particularly resistant to most chemicals, to abrasion, and to high temperatures. Acid-resistant bricks are normally made from red shale or fireclay, both alumino-silicate mixtures.

Common red shale or fireclay brick can cause iron contamination of acid until the surface contamination has been leached out by contact with the acid. Red-shale brick typically has an iron content of 6.5% compared to fireclay, which has only about 2.5% iron, so iron contamination is more likely if red shale is use. Fireclay bricks are preferred in areas exposed to high temperatures, such as gas inlets on absorber towers. They are less brittle and have better thermal shock resistance than red-shale bricks.[21] Carbon bricks are used as facing bricks in applications, such as scrubbers, in which the sulfuric acid contains fluorides. If carbon facing bricks are used and the steel shell is lead-lined, then an interlayer of acid-proof brick must be installed between the carbon and the lead to avoid severe galvanic corrosion.

Acid-brick linings are used extensively to resist hot, concentrated sulfuric acid in manufacture, utilization, and storage at elevated temperature. In the United Kingdom and elsewhere in Europe, design standards for acid-proof linings have long been available, resulting in wide use of this economical and long-lasting approach to corrosion control. Brick linings are also frequently employed in North America, even though there are no such standards for quality control in their application.[22]

It is now possible to engineer linings, improving performance and designing to more sophisticated requirements. New materials such as special membranes and highly insulating block linings (permitting reduced lining thickness), and new mortars with improved physical properties, improve lining performance and conserve energy.

Standards ASTM C 279[23] is the U.S. standard for acid brick, and Type III (previously designated Type L) brick in this standard is preferred for acid-proof linings. Both fireclay and red-shale types have been satisfactory, the former offering an advantage in availability and shapes and the latter in reduced cost. The brick selected should have a history of prior successful service in sulfuric acid, or it should be tested for a minimum of eight weeks in the acid concentration at temperature. When both weight change and changes in compressive strength have been recorded every two weeks during the exposure, the results can then be quantitatively evaluated.

The mortars are at least as important as the bricks in lining. Furan mortars (carbon-filled, *not* silica-filled) will resist 50% acid to 100°C (212°F) but are not resistant above 60% concentration. Standard phenolic mortars are resistant to 50% acid at the boiling point. Mortars currently used in concentrated acid are usually based on chemical or self-hardening silicates. These mortars have porosity that is 3 to 10 times less and strength that is 1 to 2 times greater than the original air-hardening sodium silicates. Most acid-brick work now uses potassium silicate mortars such as HB®[24] or Corlock B®[25] or modified silicate mortars such as Pennwalt HES®,[26] which are resistant to all concentrations of sulfuric acid and can be used up to 900°C (1,650°F). The cement or mortar may be critical, and the fabricator of brick-lined vessels should be approached regarding the possible use of silicate or aluminate mortars.

Use of a compressible ceramic fiber-matte material has replaced the original compressible white asbestos or ceramic fiber rope, with greatly reduced acid solubility of joint packing.

The objectives of brickwork are to reduce temperature and to ensure stagnant conditions beneath it. Modern construction has made a tremendous improvement over older designs and materials.

Design Brick linings can now be designed to known stresses with known safety factors. One of the methods is to ensure that, at ambient air temperatures, more than one-half of the lining is in compression and the substrate steel is in tensile stress. The same design technology can be used for concrete vessels. Although there is no North American standard for brick-lined steel vessels, the German DIN 28061 and 28062 standards are available in English and cover many concerns, such as out-of-roundness and bottom deflection.

Specific details for typical process equipment, design of internal partitions, temperature gradients, hot-gas inlets, relative merits of dished versus flat bottoms, and heat-flow calculations are available.[27,28]

Existing storage tanks may be brick-lined to simultaneously reduce corrosion rates (by limiting access to dissolved air and lowering the temperature at the steel surface) and prevent the acid from freezing. The steel is first sandblasted, then pretroweled with 1/16 to 1/8 in (1.6 to 3.2 mm) of potassium silicate mortar of the polyaluminum phosphate-catalyzed variety, followed by about 4 in (10 cm) of acid brick.

Nonmetallic Materials in Weak and Intermediate-Strength Acid (0–70%)

The use of nonmetallic materials in weak and intermediate-strength sulfuric acid is more often limited by temperature rather than by concentration. The more strongly oxidizing nature of the acid can influence some polymers at the higher end of this concentration range, approaching 70%. The various types of nonmetallic materials that can be used in this range of acid strengths will be discussed together in this chapter with specific concentration limits indicated where appropriate.

Thermoplastics

In general, most nonfluorinated thermoplastics can be used up to about 80°C (175°F) in the 5–25% sulfuric acid range.

Polyolefins Low-density polyethylene (LDPE) and ABS are resistant to 25% acid to about 60°C (140°F). High-density material (HDPE) is good to about 70°C (160°F). Polyethylene (both cross-linked and linear) is resistant at room temperature over the entire 0–70% concentration range.

Polypropylene (PP) is resistant to about 80°C (175°F) unsupported and to about 95°C (200°F) properly supported (e.g., in polypropylene-lined steel pipe). PP is resistant to about 85°C (185°F) in the weaker range of acid strengths and up to around 60°C (140°F) in 70% acid.

Chlorocarbon Plastics Polyvinylidene chloride (PVDC) and PVC Grades I and II are resistant to about 54°C (130°F) in 10% acid when supported. Temperatures to 80°C (176°F) are acceptable for 10% acid when the plastic piping system is fully supported. PVC is resistant at room temperatures for 0–70% concentration. Chlorinated polyvinyl chloride (CPVC) is resistant to about 80°C (175°F) in this range of acid strengths when properly supported.

It should be noted that, despite the temperature limits indicated, PVC is quite easily hydrolyzed at temperatures above about 40°C (104°F) and is a potential source of chloride-ion contamination. The more temperature-resistant CPVC is probably a better choice if the solution has any contact with stainless steels, which are subject to pitting and SCC by chlorides.

Fluorocarbon Plastics Polyvinylidene fluoride (PVDF) is resistant to about 120°C (250°F) unsupported, while the higher-fluorinated plastics are resistant to their normal limiting temperatures of 177°C (350°F). As *linings*, FEP is limited to about 120°C (250°F) and PTFE to 160°C (320°F).

Comparison of Thermoplastics An acetal copolymer based on trioxane resists 3% acid at room temperature. In 3% acid, cellulose acetate (CA) is nonresistant, but cellulose acetate butyrate (CAB) and cellulose acetate propionate (CAP) are resistant at room temperature.

Silicones are resistant at room temperature over the entire 0–70% concentration range.

Temperature and acid concentration limits are published by polymer producers, piping suppliers, and others. There are often discrepancies in published limits that may reflect differences in behavior of different formulations but may also indicate differences in testing methods and evaluation of performance. The temperature and acid-strength limits for various thermoplastics shown in Table 11.8 were compiled from a number of sources.

Many thermoplastics materials can be used unsupported (e.g., in pumps) but require continuous support in piping systems. Some thermoplastic materials tend to "grow" in sulfuric acid services. For this reason, and in view of their poor mechanical properties as compared with metals, many process industry companies avoid using thermoplastic materials for piping except in underground drainage systems in which the pipe is both supported and protected.

In the form of plastic-lined pipe or valves, plastics may be used in a range of sulfuric acid concentrations within certain limits, Table 11.9.[6] PVDC-lined steel pipe suffered slow carbonization over a 2–3 year period in 65% acid at 85°C (185°F). It was replaced with PP-lined steel, which greatly extended the service life.

Many common thermoplastics are used in dual-laminate construction where the structural strength is provided by FRP (fiber-reinforced plastic) and the corrosion barrier is thermoplastic. This type of construction is commonly used in weak sulfuric acid applications within certain limits, Table 11.10.[7]

Table 11.8 Thermoplastics Max. Temperature °C (°F) in Various Strengths of Sulfuric Acid

Plastic	H_2SO_4 Concentration				
	10–25%	25%	50%	60%	70%
Acrylic	10%: 20 (68) 50 (122) NR	—	—	—	—
Acrylonitrile-butadiene-styrene (ABS)	15%: 49 (120)	—	49 (120)	—	23 (73)*
Polyethylene (PE)	60 (140)	60 (140)	25 (77)	60 (140)	25 (77)
Polypropylene (PP)	80 (176)	93 (200)	80 (176)	60 (140) 80 (176)*	66 (151)
Chlorinated polyether	—	115 (239)	115 (239)	—	—
Polystyrene	—	82 (180)	71 (160)	—	25 (77)
Polyvinyl chloride (PVC)	60 (140)	60 (140)	55 (130)	60 (140)	60 (140) 55 (130)
Chlorinated polyvinyl chloride (CPVC)	82 (180)	82 (180)	82 (180)	82 (180)	82 (182)
Polyvinylidene chloride (PVDC)	—	80 (175) 82 (180)	70 (158)	—	60 (140)
Polyvinylidene fluoride (PVDF)	120 (248)	115 (239)	104 (220)	100 (212) 120 (248)*	93–100 (200–212) 100–120 (212–248)*
Perfluoroalkoxy (PFA)	260 (500)	—	—	260 (500)	260 (500)
Polytetrafluoroethylene (PTFE)	260 (500)	—	—	260 (500)	260 (500)

* Conditionally resistant; medium can attack, cause swelling, and reduce life
NR: Not resistant at this temperature

Table 11.9 Liner Selection Guide for Thermoplastic-Lined Steel Pipe

	Maximum Temperature °F (°C)						
% H_2SO_4	450 (232)	250 (121)	225 (107)	200 (93)	125 (52)	75 (24)	Not Resistant
10	PTFE	PVDF	PP		SARAN®		
16	PTFE	PVDF		PP	SARAN®		
30	PTFE	PVDF		PP		SARAN®	
60	PTFE	PVDF		PP			SARAN®

SARAN®: Polyvinylidene chloride resin
PP: Polypropylene
PVDF: Polyvinylidene fluoride resin
PTFE: Polytetrafluoroethylene

Table 11.10 Temperature Limits °C (°F) for Various Plastics in Dual-Laminate Construction in Sulfuric Acid

Plastic	H_2SO_4 Concentration	
	Up to 40%	Up to 60%
PVC	40 (104) 60 (140)*	60 (140)
CPVC	80 (176)	80 (176)
PE	60 (140)	60 (140)
PP	60 (140)	60 (140)
PVDF	120 (248)	120 (248)
ECTFE (ethylene chlorotri-fluoroethylene)	100 (212) 120 (248)*	100 (212) 120 (248)*
ETFE (ethylene trifluoroethylene)	150 (302)	150 (302)
FEP (fluorinated ethylene propylene)	150 (302)	100 (212)
PFA	150 (302)	150 (302)

* Conditionally resistant; medium can attack, cause swelling, and reduce life.

Thermosetting Resins

Fiberglass-reinforced plastics (FRP), also called RTP (reinforced thermoset plastics) and GRP (glass fiber-reinforced plastics) in the United Kingdom, are based on conventional polyesters and vinyl esters. FRP with polyester resins is unattacked in dilute acid up to 90°C (195°F) and can be used in up to 75% concentration at <65°C (<150°F).[29] Other modified resins can be used at higher temperatures in this range of acid strengths but become increasingly attacked as strength approaches or exceeds 70% (see Table 11.11).

The reinforced thermosetting resins have a higher strength and lower coefficient of thermal expansion than the thermoplastics. Both formulation and compounding affect their chemical resistance. For example, crystalline forms of polyethylene terephalate (PETP) resist up to 30% acid to 60°C (140°F) when unfilled and to 80°C (175°F) when glass-filled.

Some of the structural-grade polyester resins are not resistant to sulfuric acid at concentrations above a few percent at room temperature. However, all chemically resistant polymers resist 0–25% acid; some formulations are resistant to weak acid (~5%) to the atmospheric boiling point. All of the chemically resistant thermosetting resins may be considered applicable in this range of acid strength, at least at ambient temperature, in the absence of other aggressive species.

It should be noted that FRP constructions are subject to a gradual degradation of properties in most chemical environments. Allowance must be made for this aspect of their design and application, e.g., by use of a more chemically resistant thermoplastic liner such as dual-laminate construction (see above). The corrosion-resistance of dual-laminate vessels depends on the resistance of the liner, although resistant

Table 11.11 Temperature Limits °C (°F) for Thermosetting Resins in Various Strengths of Sulfuric Acid

Material	H_2SO_4 Concentration		
	25%	50%	70%
Vinyl ester and brominated vinyl ester	82–99 (180–210)	82–93 (180–200)	77–88 (170–190)
Epoxy novolac vinyl ester	99–104 (210–220)	99–104 (210–220)	49–82 (120–180)
Epoxy	59 (122)	38 (100)	25 (77)
Bis-A Fumarate polyester	93–99 (200–210)	93–99 (200–210)	Ambient to 82 (77–180)
Chlorendic polyester	93–121(200–250)	93 (200)	66–79 (150–17)
Chlorinated polyester	88 (190)	88 (190)	60 (140)
Brominated THPA	88 (190)	60 (140)	71– 82 (160–180)
Isophthalic	49–77 (120–170)	NR-60 (NR-140)	NR-27 (NR-80)
Furane	99–121 (210–250)	93–121 (198–250)	NR-88 (NR-190)
Phenol-Formaldehyde	150 (300)	150 (300)	30 (86)

NR: Not resistant

resins are often used in the construction of the FRP reinforcement in case of permeation and leaks in the thermoplastic liner.

The maximum recommended service temperature for thermosetting resins in a range of sulfuric acid concentrations depends on resin type and formulation, Table 11.11. This table is a compilation of data from various sources. Much of the variation in resistance between nominally similar resins is due to different formulations used to achieve different properties that also affect chemical resistance. For example, resins may be formulated to be flexible, be fire retardant, have good high-temperature properties, etc.

Silicate-filled phenolic filament-wound tanks (Haveg® 7 have been used for alum boil tanks at 50% sulfuric acid strength. Strongly oxidizing contaminants, such as nitric acid, would attack phenolic resins. Haveg® 61NA is a furfuryl alcohol-formaldehyde grade and Haveg® 41NA is based on phenol formaldehyde. This latter grade is rated for use in up to 50% sulfuric acid and has only fair resistance to 70% acid at 100°C (212°F). Phenolic piping reinforced with silica is resistant in 0–50% sulfuric acid at 250°F and in 50–70% acid at 200°F.[30]

In one case, two FRP storage tanks made from Bis-A Fumurate resin failed within two years when handling 65–70% sulfuric acid with 0.1–0.2% nitric acid, nominally at ambient temperature. The failures were found to be due to a combination of stress and the environment, although the strains incurred were below those that would damage this material. The tanks were replaced with PVC/FRP dual-laminate ones.

It is essential that proper manufacturing procedures be specified and employed. FRP tanks and vessels for this service must be built to ANSI/ASTM RTP-1, "National Standard for Fabrication of FRP Tanks and Vessels." Off-the-shelf, filament-wound tanks and vessels are not satisfactory.[31]

Polymer Coatings

Polymer coatings on steel or stainless steel are used in this range of acid strengths. Phenolic and baked phenolic coatings are used at ambient temperatures for sulfuric acid up to 98% concentration.

Vinyl coatings can be used for up to 50% sulfuric at up to 150°F (66°C); some neoprene-based and epoxy coatings can be used for 50% acid at room temperature and conditionally up to 150°F (66°C).[32]

Elastomers

Both natural and synthetic elastomeric products can be successfully used as sealing components at various temperatures and concentrations of sulfuric acid. Many can be used in 0–98% sulfuric acid within their thermal limits, Table 11.5. Product formulations can have a major impact on product performance over this range of sulfuric acid concentration. Careful material selection is critical to maximize product performance.

Some elastomers have been used at higher temperatures than those stated for the full range of acid strengths. Examples of weak-acid use include the following:

- Polysulfide (to 10%) and polybutadiene elastomers are resistant only at room temperature.
- Butadiene-styrene (GRS) and butadiene-nitrile (Buna N) elastomers are resistant to 60°C (140°F) in 5% acid. Buna N is resistant to 10% acid at room temperature but is attacked in 25% acid.
- Butyl rubber is resistant to 82°C (180°F). It was resistant to 10% acid at 100°C (212°F) in a 24-hour test but became tacky during a 6-month exposure at room temperature.
- Hard natural rubber and Neoprene are resistant to 93°C (200°F).
- Polyurethane elastomers are resistant to 10% acid at 20°C (68°F) but are attacked at 60°C (140°F).
- EPDM (e.g., Nordel®) should not be used above 10% acid.
- CSPE is useful to at least 120°C (250°F).

It should be noted that elastomers can suffer severe attack if even parts per million of certain organic compounds are present in the acid (e.g. chlorinated solvents and aromatic solvents). These can be preferentially absorbed over prolonged periods to attack either the elastomer itself or any adhesive employed in its application.

Carbon and Graphite

Carbon and graphite are resistant to the atmospheric boiling point in the entire range of dilute sulfuric acid, unless significant amounts of strongly oxidizing contaminants (e.g., nitric acid) are present. The higher acid concentrations pose more danger than the very dilute acid. Materials resistant to the higher concentrations are without exception resistant to the more dilute acid.

Carbon and graphite are inherently resistant to the entire 0–70% range to the atmospheric boiling point. Unfortunately, in some equipment, the cement (e.g., in impervious graphite tubular heat exchangers) or mortar (e.g., for carbon brick) may be a limiting factor.

Standard phenolic mortars reportedly resist 50% (but not 70%) acid to the atmospheric boiling point. However, some proprietary variants will resist the boiling 70% acid. A rigid polyester mortar, combined with a reactive cross-linking polymer, will resist 70% acid to about 120°C (250°F).

Impervious graphite bursting discs are suitable for all strengths of sulfuric acid up to 85%; some grades, depending on the impregnant used, are also satisfactory in sulfuric acid with nitric acid present.[33]

Wood

Highly resinous woods (e.g., cyprus, larch, pitch pine, and redwood) can be used in up to at least 8% acid at room temperature.[34] Wooden tanks resisted up to 20% acid at 16–49°C (61–120°F) for more than 10 years when internally protected with an asphalt/tar lining. Pitch pine tanks have been used to handle pickling solutions containing 15% sulfuric acid.[35]

Ceramic Materials

Ceramic materials (e.g., glass, masonry, and acid-resistant silicate and aluminate materials) are resistant to dilute sulfuric acid to the atmospheric boiling point. Dilute sulfuric acid is usually less corrosive to ceramic materials than is boiling water.

Glass, stoneware, and acid-proof brick are resistant to acid of these concentrations to the atmospheric boiling point. Acid-proof brick is more resistant than the other ceramic materials at the high end of this concentration range.

The cement or mortar may be critical, and the fabricator of brick-lined vessels should be approached regarding the possible use of silicate or aluminate mortars. Furan mortars (carbon-filled, *not* silica-filled) will resist 50% acid to 100°C (212°F) but are nonresistant above 60% concentration. Standard phenolic mortars are resistant to 50% acid at the boiling point. Potassium silicate mortars such as HB®[24] or Corlock B®[25] and modified silicate mortars such as HES® are resistant to all concentrations of sulfuric acid and can be used up to 900°C (1,650°F).[27]

References

1. Anon, "Materials for the Handling and Storage of Concentrated (90 to 100%) Sulfuric Acid at Ambient Temperatures," RP 0391 (Houston, TX: NACE International, latest edition).
2. Anon, Materials Engineering Tech Alert, Vol. 3, No. 2, DuPont Engineering Department, DuPont Company, Wilmington, DE.
3. Anon, "Corrosion Resistance Guide" (Tustin, CA: George Fischer Signet Inc., 2003), www.georgefischer.com, 80 pp.
4. Private communication, in C. P. Dillon, ed., "Concentrated Sulfuric Acid and Oleum," vol. MS-1, Materials Selector for Hazardous Chemicals (St. Louis, MO: MTI Inc., 1997): 212 pp.
5. Anon, "Dilute Sulfuric Acid," ChemCor 10 (St. Louis, MO: MTI, undated).
6. Anon, "Sulfuric Acid Piping Systems," form no. 178-105-389 AMS (Midland, MI: Dow Chemical USA, 1989): p. 6.

7. Anon, "Chemical Resistance of Thermoplastics Used in Dual Laminate Constructions," DLFA (2002), http://www.dual-laminate.org/html/corrosion_guide.html, 143 pp.
8. Anon, "Chemtite® Type 'SP' Piping," brochure HA-22-81-5M/P (Wilmington, DE: Ametek Haveg Div., 1982): 14 pp.
9. R. W. Schnell, private communication (Newark, DE: DuPont Dow Elastomers LLC, 2004).
10. J. R. Schley, "Impervious Graphite for Process Equipment," Materials Engineering, 1—Selecting Materials for Process Equipment, Chemical Engineering, 1980.
11. Anon, "Corrosion Chart for Grafilor," brochure GC 5 FED 7821 (Moselle, France: Le Carbone-Lorraine, undated): 22 pp.
12. J. Rodda, unpublished report (1991).
13. Anon, "The Chemical Industry Builds on Graphite," PE-200/01 (Meitingen, Germany: SGL Carbon Group, 2003): 24 pp.
14. Anon, "Sulfuric Acid Concentration," Process Profile 12-2 (West Union, NJ: De Dietrich Process Systems, Inc., 1995): 2 pp.
15. Anon, "Sulfuric Acid Dilution," Process Profile 13-2 (West Union, NJ: De Dietrich Process Systems, Inc., 1995): 2 pp.
16. Anon, "Worldwide GLASTEEL 9100," Brochure no. SB95-910-5 (Rochester, NY: Pfaudler Reactor Systems, 2000): p. 2.
17. H. H. Uhlig, "The Corrosion Handbook" (New York: John Wiley & Sons, 1948).
18. Anon, "Hexoloy SA Corrosion Test in Liquids" (Niagara Falls, NY: Saint-Gobain Advanced Ceramics, 2003), http://www.carbo.com/datasheets/corrosiontest.html.
19. Anon, "Corrosion Test Results by Carborundum of Hexoloy® Silicon Carbide" (Niagara Falls, NY: Carborundum, 1981): 30 pp.
20. Anon, "Acid Resistant Linings," DKL Engineering (2003), http://members.rogers.com/acidmanual/materials_linings_acidlinings.htm.
21. Anon, "Acid Brick," DKL Engineering (2002), http://members.rogers.com/acidmanual/materials_linings_acidbrick.htm.
22. R. R. Pierce, W. G. Carpenter, "Sulfuric and Phosphoric Acid Plant Linings Systems," CEP, March (1982): pp. 156–160.
23. ASTM C 279, "Specification for Chemical-Resistant Masonry Units" (West Conshohocken, PA: ASTM).
24. Anon, "HB® MORTAR," brochure CE-207, Henkel Surface Technologies (1999): www.tufchem.com, 2 pp.
25. Anon, "CORLOCK B® MORTAR," brochure CE-281, Henkel Surface Technologies (1999): www.tufchem.com, 2 pp.
26. Anon, "HES® CEMENT," brochure CE-204, Henkel Surface Technologies (1999): www.tufchem.com, 2 pp.
27. W. G. Carpenter, R. R. Pierce, "Linings for Sulfuric and Phosphoric Acid Plant Process Vessels," Corrosion/83, paper no. 95 (Houston, TX: NACE International, 1983).
28. W. L. Sheppard, "Chemically Resistant Masonry" (New York: Marcel Dekker Inc., 1982): 286 pp.
29. J. R. Davis, ed., "Corrosion—Understanding the Basics" (Metals Park, OH: ASM International, 2000): p. 225.
30. Anon, "Chemical Equipment," brochure no. CED/65-5 (R-6/73) (Wilmington, DE: Ametek Haveg Div., 1981): 8 pp.

31. ASME RTP-1, "Reinforced Thermoset Plastic Corrosion Resistant Equipment (Fiberglass construction)" (New York: ASME International, latest edition.).
32. Anon, "Atlas Protective Coating Systems," bulletin 7-1 (Mertztown, PA: Atlas Minerals & Chemicals, Inc., 1982): 8 pp.
33. Anon, "Impervious Graphite Rupture Discs," ref. no. 2-2204-2 (Liberty, MO: Continental Disc Corp., 1998): 12 pp.
34. Anon, "Corrosion Handbook," Vol. 8, Sulfuric Acid (Frankfurt, Germany: Dechema eV, 1991).
35. E. Rabald, "Corrosion Guide" (Amsterdam, Netherlands: Elsevier Scientific Publishing Co., 1968): pp. 466–474.

12

Specific Production Equipment

The corrosion resistance of various materials under a variety of conditions has been discussed in previous chapters. Materials of construction that are or have been used for specific production equipment are detailed in this chapter. This chapter also includes aspects of design and operation that are relevant to particular types of equipment. One critical aspect of design is the specification of suitable materials of construction that will adequately resist the acid environment. Mechanical design is the other essential design step; its basic requirements are described in several standards, e.g., API 650 for Storage Tanks and ASME Section VIII, Division 1 for Pressure Vessels.[1,2] All pressure vessels should be built and tested to the requirements of the construction code to which they are built.

Process equipment for producing sulfuric acid by the contact process tends to be relatively simple but usually quite large for economic reasons. The original sulfuric acid plants were constructed of brick-lined equipment and thick-walled cast iron or carbon steel. As stainless steels became available, the early 15–18% chromium (similar to type 430 [S43000]) and 18% Cr, 8% Ni (type 304 [S30400]) steels began to be used. The standard stainless steels are still extensively used in acid plants, in some cases with the help of anodic protection. As operating temperatures and production rates have increased, more highly alloyed materials have become more commonly employed.

Contact Acid Plants

The equipment and materials used in the production of concentrated sulfuric acid vary depending on the source of sulfur. The sulfur can come from elemental sulfur, from waste acids such as alkylation processes, or from hydrogen sulfide from metallurgical processes or other sources. The equipment and process used to convert sulfur into sulfur dioxide varies depending upon the sulfur source. Once sulfur dioxide has been produced and cleaned, its conversion into sulfur trioxide and then sulfuric acid is largely independent of the original source of sulfur. Waste acids can also be cleaned and concentrated directly without first being made into sulfur dioxide; see the section on acid concentration plants later in this chapter.

Production of Sulfur Dioxide from Elemental Sulfur

Sulfur Melter The sulfur melter is made from carbon steel, cast iron or stainless, or sometimes from alloy 600 (N06600). Note, however, that alloy 600 is subject to liquid metal cracking (LMC) if it is welded without cleaning after having been exposed to sulfur. A thorough review of corrosion mechanisms involving sulfur has been published.[3]

Steel steam coils for providing the sensible heat may have only two to eight months' life due to moisture in the sulfur, and iron corrosion products from this source can plug the converters. Of the stainless steels, type 316L (S31603) is preferred, although higher alloys have been used, e.g., 4.5% molybdenum grades such as alloy 904L (N08904).

Sulfur Pipelines Pipelines are heated to keep the sulfur molten, >~130°C (266°F). Materials used include carbon steel, galvanized carbon steel, and alloy steels. Pipelines should be kept full; if pipelines are drained and allowed to cool, sulfuric acid can form and cause corrosion.

Valves are usually steam-jacketed plug, gate, or globe made from carbon steel, cast iron, stainless steel, alloy 400 (N04400), nickel, or nickel-based alloy.

Sulfur pumps are centrifugal, rotary, or reciprocating piston made from carbon steel or cast iron, with harder materials used for parts subject to wear.

Molten sulfur storage tanks are usually brick-lined carbon steel with internal or external steam coils. Sometimes molten sulfur is stored in pits made from similar materials.

Sulfur Furnace The furnace in which sulfur is burned is made from carbon steel and lined with refractory brick with or without insulating brick depending on whether the furnace is to be operated with a hot or cold shell. Castable refractory linings are an option.

Sulfur Burner The sulfur burner is usually carbon steel, with type 309 (S30900) or 310 (S31000) burner nozzles.

Waste-Heat Boilers Waste-heat boilers are typically made of 2¼% Cr, 1% Mo (K21590), with types 309 (S30900) or 310 (S31000) tube inserts at the hot end.

Production of Sulfur Dioxide from Wet Gas

So called "wet-gas" processes produce sulfur dioxide by the combustion of one of several sulfur-containing materials and include the following:

- Metallurgical (e.g., pyrite combustion): Gas is cooled, cleaned, and dried.
- Spent acid regeneration (e.g., alkylation sludge): Acid is burned with fuel to form gas that is cooled, cleaned, and dried.
- H_2S burning: Does not usually require fuel. Gas is then cooled and cleaned as above.

The sulfur dioxide gas stream from these operations contains significant amounts of moisture. The use of these sulfur sources causes more severe problems in terms of SO_2 generation than the use of elemental native sulfur. Waste acids are more dilute and corrosive and contain entrained carbon and other particles that contribute to erosion-corrosion. Various configurations and processes are used for these wet gases, usually involving some combination of waste-heat boiler, quench, cyclones, hot ESP, wet ESP, heat exchangers, and packed towers. The circulating weak acid is an integral part of the gas-cleaning process, and the materials and equipment used largely depend on the impurities present. While there is a huge variation in the processes used, typical equipment and materials that might be encountered are described in this section.

Furnace The furnace used to burn the sulfur-containing feed is usually steel-lined with either refractory brick or a castable refractory. The off-gases from metallurgical smelting operations go straight to the waste-heat boiler.

Waste-Heat Boilers Regardless of the feedstock, wet-gas processes share a common need for heat recovery by waste-heat boilers, which must be kept at elevated temperatures to avoid dew-point corrosion. Waste-heat boilers operate with the combustion gas products above 315°C (599°F) and are often fabricated from 2¼% Cr, 1% Mo steel (K21590), while the outlet piping is steel, PTFE-lined steel, or dual-laminate FRP/PTFE.

Alternatively, high-temperature stainless steels such as type 310 (S31000) or nickel-rich alloys such as alloy 800 (N08800) may be used depending on the composition of the feed. Bimetallic tubes in a fluidized-bed pyrite roaster, type 329 (S32900)/carbon steel (ASTM 210,[4] Grade A1 [K02707]) have been used in this service.[5]

Gas Cleaning Wet gases often contain entrained solids and fine particulate matter that must be removed before the SO_2 drying operation. Also, the higher combustion temperatures, as compared with those for burning sulfur, produce some sulfur trioxide ahead of the converter and small amounts of oxides of nitrogen, causing nitric acid contamination. Fluoride and/or chloride contamination may be present in the gases, resulting in excessive corrosion of stainless-steel components. Arsenic and mercury constituents can degrade catalyst and contaminate the product acid. Mercury can be removed by conversion to chloride via treatment with hydrochloric acid, or to sulfide with hydrogen sulfide, or by reduction with thiosulfate in 70–85% sulfuric acid.[6]

Wet gases must have entrained solids and mists removed prior to entering the acid converter. Different systems are used for wet gas with and without volatile metals present, Figure 12.1 and Figure 12.2, respectively.[7]

The wet gas from the furnace operations must be kept above 315°C (599°F) to minimize dew-point corrosion before being cooled in a gas-saturation or scrubbing tower. In some cases the gas is first cleaned in a hot, dry electrostatic precipitator (ESP). The quench or spray tower is usually a brick-lined carbon-steel vessel with a fluorocarbon, rubber, or lead membrane behind the brick. The outer face of the lining must be carbon brick if substantial amounts of fluorides are present, as is often the case with smelter gases. The spray nozzles can be PTFE, silicon iron, or metal ranging from 316L (S31603) to alloy C-276 (N10276) depending on the weak acid being sprayed.

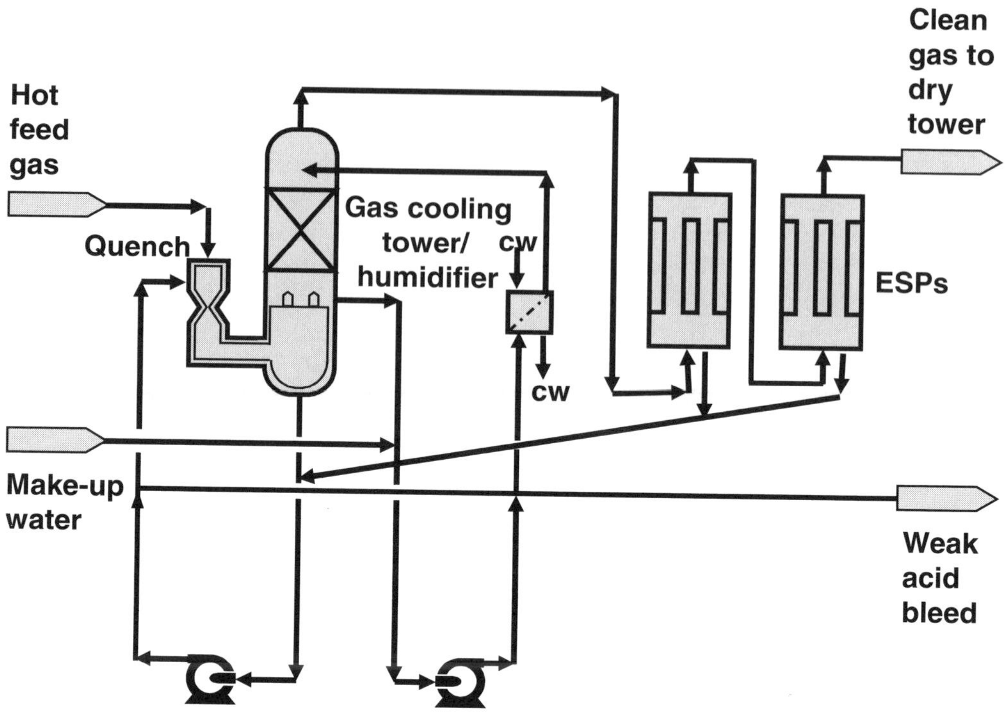

Figure 12.1 Schematic Flow Diagram for a Simple Gas-Cleaning System

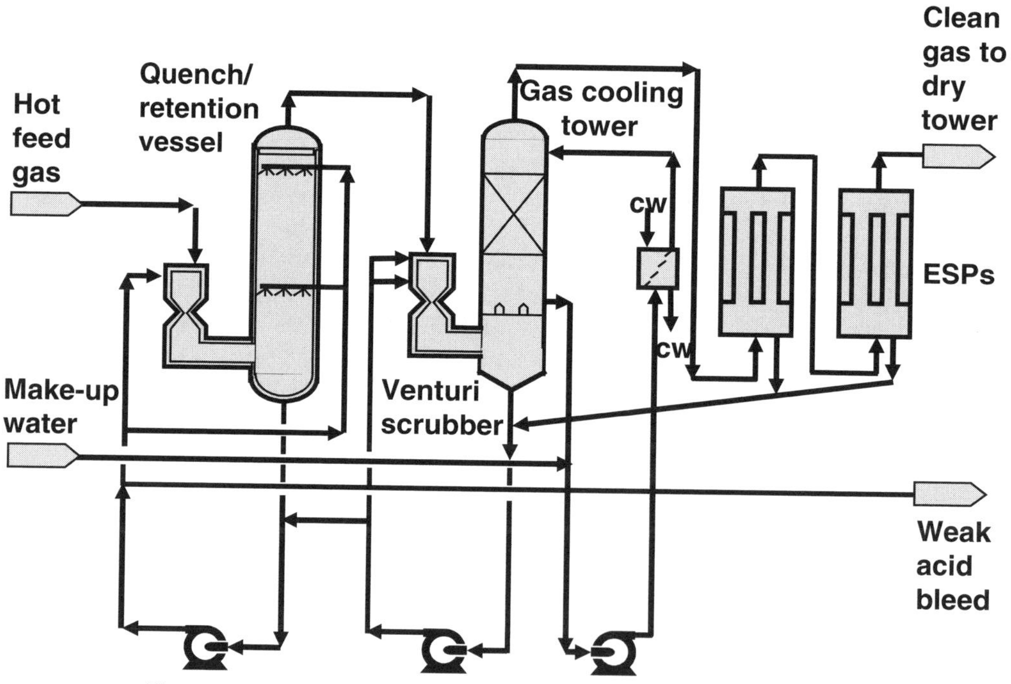

Figure 12.2 Schematic Flow Diagram for a Gas-Cleaning System with Volatile Metals Present

The quench system is usually operated with a weak-acid concentration of less than 10% H_2SO_4. Occasionally, the system is operated at concentrations up to 30% H_2SO_4 in order to reduce the weak-acid effluent flow to the treatment system and reduce the size of the equipment. Weak-acid circulation piping is PTFE- or PVDF-lined steel, FRP, or rubber-lined steel below about 50°C (122°F). The weak-acid pumps may be impervious graphite; PTFE-lined steel; plastic such as high-molecular-weight, high-density polyethylene (HMWHDPE) to resist abrasion; or 14% silicon cast iron. The weak-acid cooler may be impervious graphite shell and tube exchangers or plate coolers made from various alloys ranging from type 316L (S31603) to alloy C-276 (N10276) or zirconium, depending on operating conditions.

From here, the gas goes through cooling and cleaning sections which may be surface condensers, plate towers, or packed towers. Various materials are used in this service depending on the design of equipment and operating conditions. Graphite shell and tube condensers are used, as are lead star coolers. Alloy tubing can be used in very dilute acid conditions. In the case of plate towers, typically 316L (S31603) is used for the shell and alloy 20Cb-3 (N08020) or a similar alloy is used for the plates. Plastic packing, typically polypropylene saddles, is often used in packed towers with FRP-lined steel tower shells. Alternatively, the towers can be made from FRP or dual laminate, e.g., FRP/PVC.

Sulfuric acid mist—formed by contact between sulfur trioxide and water—and remaining particulates must be removed before the gas is dried and fed to the acid plant. Mist removal is normally carried out in wet electrostatic precipitators (WESPs). These are typically made from lead and/or lead-coated steel or FRP.[6] Dual-laminate thermoplastic/FRP is becoming more common with an electrically conductive veil incorporated into the collecting tube surface. There are also WESP units, made from PVC tubes in a lead- or rubber-lined steel shell, or FRP or PVC/FRP shell that are being used in cleaning wet gas for acid production. Alloy units are also used with, for example, alloy 904L (N08904), 254SMO (S31254), or alloy C-276 (N10276) in severe conditions. An alternative design is based on an FRP electro filtering venturi.

The wet SO_2 is then heated to about 440°C (824°F) before entering the contact process converter. Such heaters usually contain type 310 (S31000) or bimetallic tubes, e.g., type 310 (2½% Cr, 1% Mo). Subsequently, drying of air, conversion, and absorption follows normal contact process procedures (see Figure 5.1), as described below.

Conversion of Sulfur Dioxide into Sulfuric Acid

The catalytic conversion of sulfur dioxide into sulfur trioxide and absorption to form sulfuric acid uses similar equipment and materials regardless of the original source of sulfur. Similarly, the cooling and transfer of the acid during and after production uses identical equipment and materials.

A modern double-absorption sulfuric acid plant in Sweden was designed to take 250,000nm^3/h of smelter off-gas, Figure 12.3. Typical flows and items of equipment in a sulfur-burning double-absorption acid plant are shown in Figure 12.4 for gas streams and in Figure 12.5 for acid streams. Similar diagrams illustrate gas and acid flows in a metallurgical acid plant, Figure 12.6 and Figure 12.7, respectively.[8] The major items of equipment and typical materials of construction have been summarized, Table 12.1. These equipment items are discussed and described in detail below.

Figure 12.3 A Double-Absorption Metallurgical Acid Plant at Ronnskarr, Sweden (Courtesy of Chemetics, a Division of Aker Kvaerner)

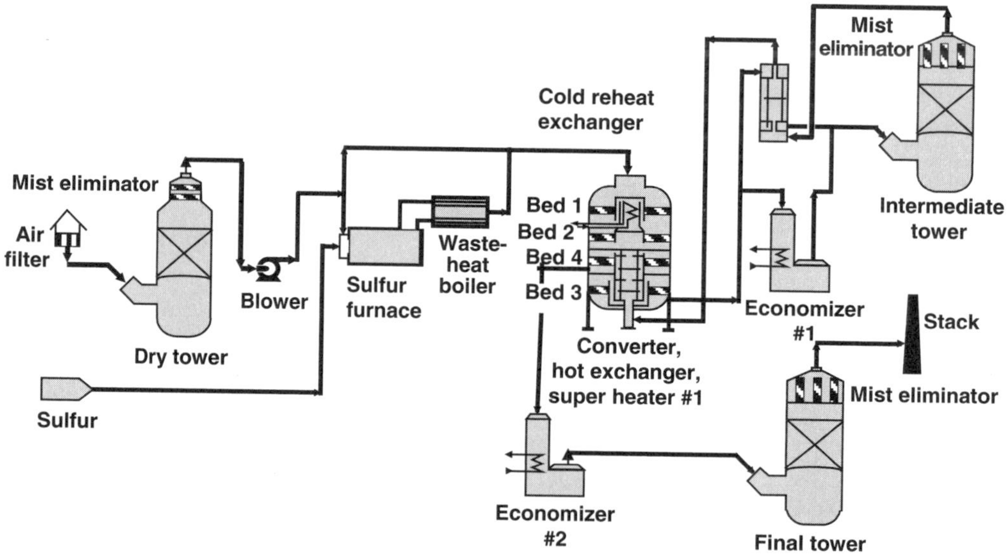

Figure 12.4 Typical Gas Flows in a Sulfur-Burning Double-Absorption Acid Plant

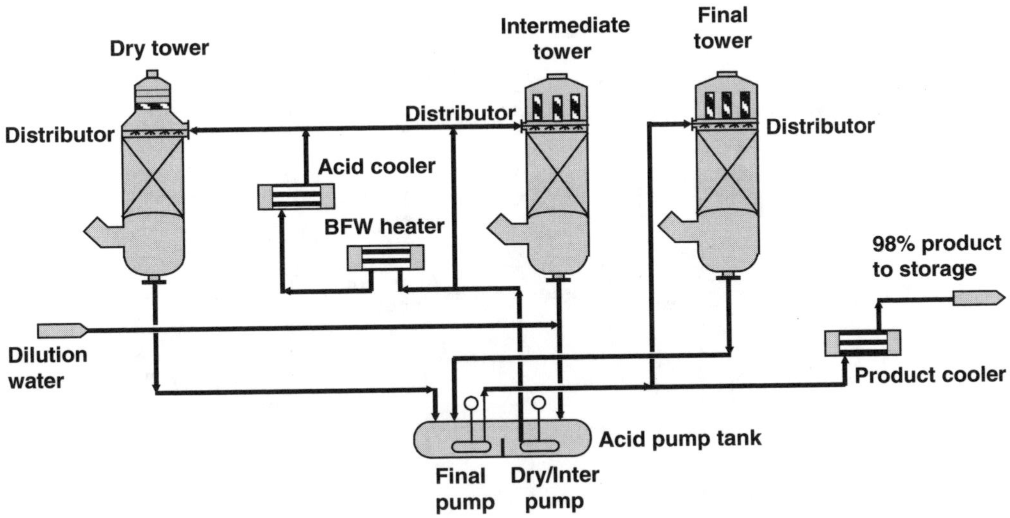

Figure 12.5 Typical Acid Flows in a Sulfur-Burning Double-Absorption Acid Plant

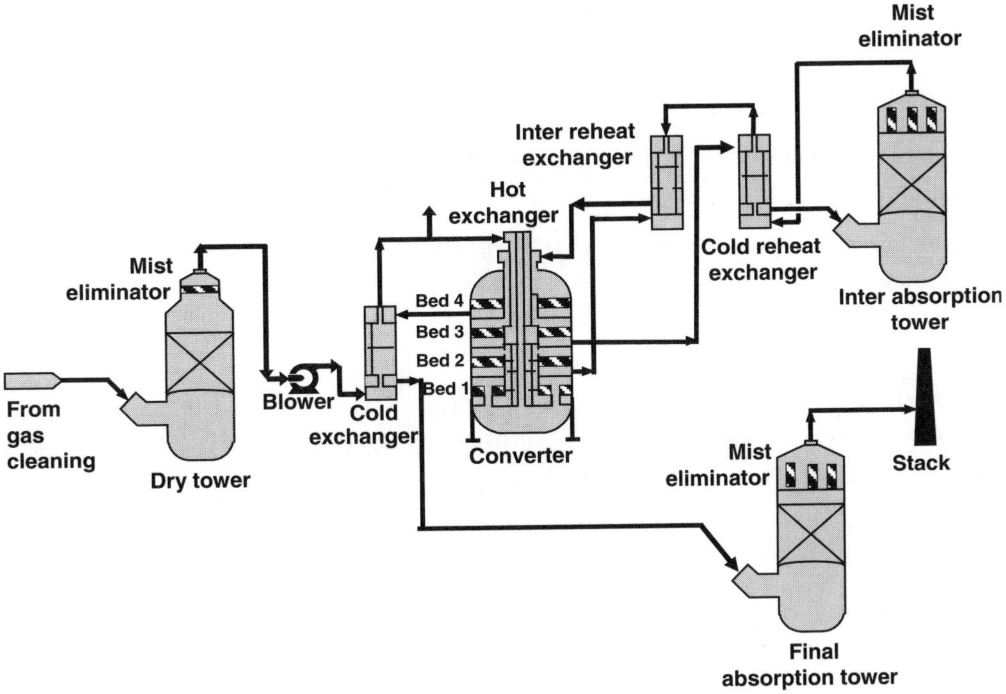

Figure 12.6 Typical Gas Flows in a Metallurgical Double-Absorption Acid Plant

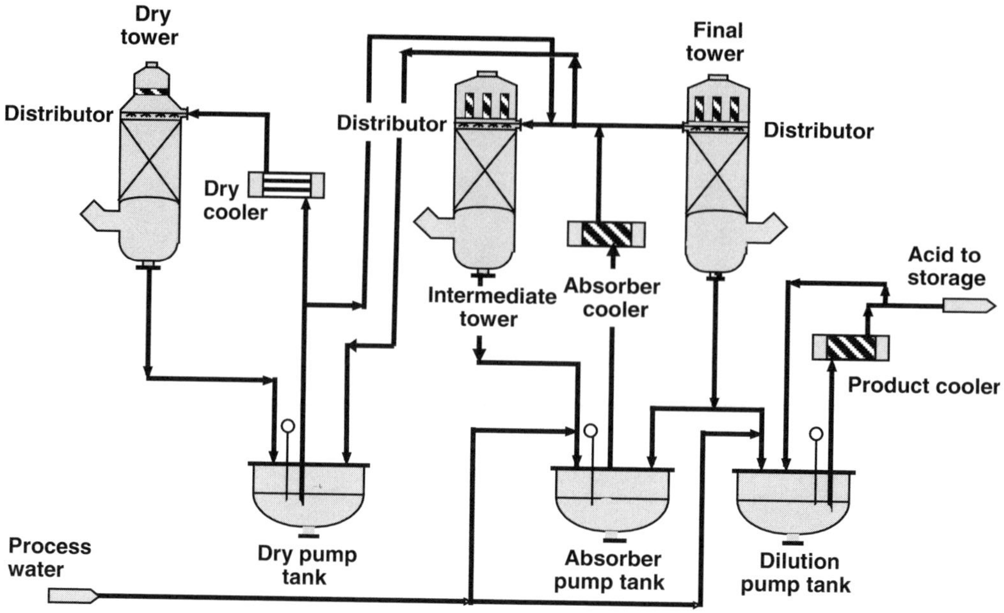

Figure 12.7 Typical Acid Flows in a Metallurgical Double-Absorption Acid Plant

Table 12.1 Equipment and Materials of Construction for Sulfuric Acid Production

Specific Equipment	Materials of Construction
Converters	304H SS
Drying towers	Brick-lined CS, high-Si SS
Absorbing towers	Brick-lined CS, high-Si SS
Oleum towers	CS
Tower packing	Ceramic saddles
Acid distributors	High-Si SS, CI, DI, alloy CI
Gas-to-gas heat exchangers	CS up to 482°C (900°F), 304H, metallized CS at higher temps
Strong-acid coolers	CI, AP SS 316 or Mo austenitic, high-Si SS
Weak-acid coolers	Impervious graphite, high-performance stainless alloys, PTFE
Pump tanks	Brick-lined CS, high-Si SS
Acid pumps	Alloy castings, e.g., Lewmet® with alloy SS shafts
Gas ducting	CS, lined CS, 304H, alloy SS
Concentrated-acid piping	DI, CI, alloy CI, high-Si or other alloy SS, AP 304, PTFE-lined pipe
Weak-acid piping	FRP, plastic-lined pipe
Storage tanks	CS, AP CS, SS

Converters

The converter operates at high temperature (up to 625°C [1,157°F]) in an oxidizing atmosphere. Sulfur dioxide from the burners enters the top of the converter through Al-coated steel or type 304 ducting. Traditionally made of brick-lined carbon steel or cast iron, converters today are usually fabricated from type 304H (S30409), often with a minimum 0.04% carbon to provide increased high-temperature strength. Type 304L (S30403) does not have adequate strength at the temperatures used in modern converters. Sensitization of high-carbon stainless steel will occur throughout the structure during service, not just at the welds. This is not a problem, however, since the stainless steel is never in contact with wet chlorides, except possibly from the outside during the very short startup and shutdown periods.

Recent work has shown that type 304H (S30409) stainless steel exhibits accelerated corrosion at temperatures above 630°C (1,166°F) in converter atmospheres. The high-silicon stainless steels were found to be more resistant under these conditions, but the effect on mechanical properties of silicon-rich phases was thought to need further study. If converters are operated at temperatures consistently above 630°C (1,166°F), it is likely that materials that have better high-temperature corrosion resistance than 304H (S30409) will be needed.[9]

In the past, the catalyst beds and supports were normally made from various grades of alloy cast irons, such as Meehanite®. Today, catalyst beds are normally placed on perforated 304H (S30409) supported from the converter shell. Other designs still use posts and grids, but these are also usually made from stainless steel. In some designs the hot gas-to-gas exchanger is installed within the central core of the converter, thus avoiding the need for high-temperature gas ducting.

Gas-to-Gas Heat Exchangers

There are a series of large gas-to-gas shell-and-tube heat exchangers associated with the converter. These are either carbon steel or stainless steel depending on the duty, particularly the operating temperature. In some cases, brick lining is used to protect parts of the exchanger from condensation corrosion. The break point between using carbon steel and stainless steel is about 482°C (900°F). Metallized and Alonized® (carbon steel surface-treated to produce an aluminum-rich surface) tubes are still used in some cases above this temperature. In some designs carbon steel, with or without surface treatment, is used at the cold end of the exchanger with stainless steel at the hot end. Although this saves money, the differences in thermal expansion properties can lead to mechanical design difficulties.

Various stainless steel grades are being used for waste-heat boilers, superheaters, economizers, and interchangers including types 304 (S30400), 316 (S31600), 321 (S32100), 309 (S30900), and 310 (S31000) for higher temperature resistance. The dry air leaving the drying tower can be preheated in type 321 (S32100) or type 310 (S31000) heat exchangers or in alloy 253 MA® (S30815) for higher temperatures.

Drying Towers

The gas feed to the converter is dried with 98% acid, producing 93% sulfuric in the process. Traditionally, the drying tower was brick-lined steel with a fluorocarbon or other membrane behind the brick. The packing was ceramic saddles supported on ceramic piers and beams or ceramic domes. The distributors were generally Lewmet® 66 or similar alloy cast-iron troughs. Mist eliminators were, and still are, made from stainless steel or more resistant alloy, PTFE, glass, or ceramic fiber depending on the design and operating parameters of the tower.

This type of tower design is still common, but all-metal towers are also now being constructed. These are typically made from one of the 5% silicon austenitic stainless steels, which are also being used to make acid distributors, either trough or pipe design. The packing is usually still ceramic saddles, although the packing supports may be metallic.

All metal towers are sensitive to excursions from the normal operating acid concentrations and temperatures. Localized hot spots, non-wetted areas, and weak-acid formation have all resulted in tower failures. The gas inlet area is subject to weak-acid formation due to the lack of acid flow and the wet gas entering the tower. Special attention must also be paid to the area under the packing support to ensure that acid flows evenly across the surface of the metal to avoid dry areas.

The high-silicon stainless steels (SX®, SARAMET®, and ZeCor®) are sensitive to weak-acid formation and the resulting higher corrosion rates. A grade of SARAMET® has now been produced that is more tolerant of somewhat lower acid strengths (see Figure 8.16). A nickel-based alloy such as alloy C-22 (N08022) is more forgiving if weak acid forms inside the tower, but it is also more expensive. All-metal towers are generally not recommended for metallurgical or acid-regeneration plants.[10]

Absorption Towers

Sulfur trioxide formed by catalytic oxidation in the converters is absorbed in 93% acid to produce 98.5% product.

The absorbers have traditionally been brick-lined carbon steel, using a membrane behind the brick and ceramic packing. Asbestos paper saturated with sodium silicate solutions was used to replace lead membranes in acid-brick linings for contact plants. This aggravated the formation *in situ* of ferrous sulfate, causing bulging of steel walls and arching of steel bases of towers.

To alleviate these problems, trowellable asphalt-based membranes such as Pecora® were developed. Although not acid-resistant, this membrane works in practice when mortar joints in the brickwork are completely filled, the brickwork uncracked, the mortar always solid against the membrane, and the temperature under 57°C (135°F) versus air at 21°C (70°F). Membranes behind brick lining now usually consist of a mastic, usually Pecora®, covered with overlapping 5-mil (0.13-mm) thick PTFE film. In Europe it is common to line the steel tower shell with Rhepanol® ORG, which is a graphite-filled polyisobutylene rubber sheet, beneath the brick lining.

There is also increasing use of high-silicon stainless steels to produce all-metal absorption towers. Mist eliminators use alloy 20Cb-3 (N08020) or alloy 825 (N08825), or alloy C-276 (N10276) when chlorides are present. Orifice plates, troughs, pipe distributors, nozzle-liners, in-line filters, and so on, are Lewmet® 66 or high-silicon stainless steel.

If oleum is to be made, additional towers are added to the double-absorption contact plant. Sulfur trioxide formed by catalytic oxidation in the converter is absorbed in 98% acid to produce 35% oleum product. The absorption tower is carbon steel with ceramic packing and type 304 (S30400) internals.

Pump Tanks

Pump tanks for 98.5% product acid are typically brick-lined with a membrane of high-silicon austenitic stainless steel.

Pump tanks for oleum may be steel or type 304 (S30400) stainless steel, depending on velocity and operating temperatures.

Pumps

Submerged pumps in pump tanks may be CF-3M (J92800) or CN-7M (N08007) with alloy 20Cb-3 (N08020) shafts and Lewmet® 55 impellers. Oleum pumps may be CF-3M (J92800), or Process Iron (specially formulated to contain small graphite flakes in isolated clusters) castings with Lewmet® 55 impellers and alloy 20Cb-3 (N08020) shafts.

In some cases, the shafts of submersible pumps are protected at the liquid-vapor interface and in the vapor phase by a heat-shrinkable fluorocarbon plastic sleeve.

Valves

Throttling control valves may be cast CN-7M (N08007) with Lewmet® 55 seats, or Lewmet® 55 throughout. Shut-off valves of CN-7M (N08007) or CF-3M (J92800) are suitable. Oleum valves are usually CF-8M (J92900).

Heat Exchangers

Concentrated sulfuric acid may need to be cooled to remove or recover the heat of reaction and to control corrosion, or heated to prevent freezing and to lower viscosity. Various methods and types of equipment are used in acid heat exchangers.

Coolers The original strong-acid coolers were made from cast-iron straight sections and return bends bolted together to make a serpentine structure. Hot acid flows through the cast iron and water flows over the outside to remove heat. These cast-iron serpentine coolers are still in use but have largely been replaced with more corrosion-resistant and energy-efficient designs.

Problems with this type of cast-iron exchanger are common. For example, cast-iron coolers using seawater were replaced due to high annual maintenance costs and loss of cooling efficiency caused by heavy marine fouling and rusting. Individual cast-iron pipe sections had been periodically replaced. Replacement shell and tube coolers were made of alloy 28 (N08028), the 98.5% sulfuric acid on the shell side being cooled from 95°C (203°F) to 57°C (135°F) by Mediterranean seawater. The seawater at 26°C (79°F), flowing at 6 ft/s (2 m/s) and intermittently chlorinated, is warmed, in turn, to 37°C (99°F). This change in materials provided 20% higher heat efficiency than design rate, while showing no evidence of either localized corrosion or erosion-corrosion effects.[11]

Seawater corrosion of another cast-iron cooler necessitated replacement at three-year intervals. The external attack was aggravated by flange leaks that introduced acid into the cooling water medium.

The service conditions involved exposure to 98.5% acid of 105°C (221°F) inlet temperature, cooled to 75°C (167°F) by seawater. The water at 29°C (84°F) was heated to

40°C (104°F) in the exchange of sensible heat. This cooler was replaced with an alloy 28 (N08028) serpentine cooler fabricated from 3-in (76-mm) Schedule 10 pipe. Designed to fit the existing rack, the new cooler was a low-cost, corrosion-resistant replacement.[11]

There are many similar examples of the improvement in thermal efficiency and extended maintenance-free life provided by the replacement of cast-iron coolers by anodically protected shell-and-tube exchangers, which are now standard for most strong-acid coolers. A typical design has the hot acid on the shell side and cooling water flowing inside the tubes. Inert metal cathodes, such as a nickel-chromium-molybdenum alloy (e.g., alloy C-276 [N10276]), are inserted through the water boxes into the tube bundle (see Figure 7.5). These cathodes provide the current to raise the potential of the outsides of the tubes and the inside of the shell into the passive region, i.e., anodic protection (AP). AP has also been applied to stainless-steel exchangers in which the acid is on the tube side, and also to plate-type exchangers, but performance and durability have been poorer than with the acid on the shell side of shell and tube units.

Coolers for concentrated acid are made from anodically protected type 316L/304L (S31603/S30403) or 6% molybdenum superaustenitic alloys—e.g., alloys 254 SMO (S31254), AL6-XN (N08367), and 926 (N08926)—when using seawater or other high-chloride cooling waters. High-silicon austenitic stainless-steel heat exchangers, without AP, are also now being used for cooling hot acid. In some instances these are bimetallic tubes incorporating a more chloride-resistant grade of stainless steel on the inside of the tube or providing a high-silicon shell and chloride-resistant tubes, e.g., alloy 28 (N08028).[12] Plate-type coolers of alloy C-276 (N10276) have also been used for seawater cooling of concentrated acid.

Impervious graphite coolers are used for weak acid (for example, in wet gas acid plants) and may also be used for some acid concentrations and temperatures within the strong acid range. Glass exchangers and fluoroplastic spaghetti-type exchangers (in which the acid flows through narrow-bore fluorocarbon tubing in a cooling tank) are also employed in specific applications.

Oleum coolers may be carbon-steel spiral heat exchangers or type 304L (S30403) with or without AP depending on operating conditions.

Heaters Hot-wall effects due to heating by steam or other heat-transfer media or by process streams demand more resistant materials than do coolers. Above about 80°C (176°F), Lewmet® or high-silicon stainless steels offer a distinct advantage over alloy C-276 (N10276), which is also sometimes used.

Acid Piping

Gray cast iron was the standard piping material for concentrated acid but is now rarely used because of safety concerns, although its corrosion resistance in strong acid is normally better than the mechanically superior ductile iron. Ductile iron pipe is widely used in 98% acid at up to 65°C (149°F) at <2 m/sec velocity and up to 85°C (185°F) at <1.3 m/sec. A modified ductile iron pipe, Mondi®, is also used for acid piping systems, operating at velocities from 1.2 to 3.0 m/s (4 to 10 ft/s) for 3-in to 30-in line sizes.[13]

Some of the Meehanite® range of cast alloys are also used in acid piping, particularly as inserts or thimbles. Ductile irons with small quantities of chromium are commonly used in Europe for hot, turbulent areas such as at the bottom of absorber towers. Steel is rarely used because of its susceptibility to erosion-corrosion but is occasionally used at low temperatures and velocities (0.6–0.9 m/sec for 98% acid depending on temperature).[14]

For piping of less than 1-in (25-mm) size, screwed type 304 (S30400) or 316 (S31600) is appropriate at ambient temperatures. Pipe of NPS 4 (4 in [100 mm]) or less may be polypropylene-lined within its acid-concentration and temperature limits or, preferably, fluoroplastic-lined steel or type 304L (S30403). The 300 stainless steels should be limited to <45°C (<113°F) and <2 m/sec in 98% acid and to <45°C (<113°F) and 1 to 2 m/sec velocity for 93% acid. The molybdenum-free grade is generally preferred above about 93% concentration, but 316L (S31603) is somewhat more resistant to velocity.

Anodically protected stainless-steel piping has been used in some cases at production temperatures. At higher temperatures and velocities, more highly alloyed stainless steels or nickel alloys are sometimes used, such as type 310M (310S; S31008 modified) stainless steel and 20Cb-3 (N08020), and the 5%-silicon-containing, austenitic stainless steels are now routinely being used for piping in sulfuric acid production plants. High-silicon stainless steels (e.g., alloy SX® [S32615]) are available to ASME B31.3[15] in sizes from DN 25 (1 in) to DN 600 (24 in) with appropriate sizes of standard flanges.[12]

The molybdenum-containing austenitic stainless steel, alloy 926 (Cronifer® 1925 hMo[N08926]), has also been used in a number of concentrated-acid systems operating at 98.5% H_2SO_4 at 100°C (212°F).[16] In one case this alloy was uncorroded over a period of 3 years, while in another case rapid failure occurred. This rapid corrosion was thought to be due to traces of strong reducing agents and local turbulence.

Type 304 (S30400) piping is usually specified for oleum service. The L grade is not needed, as intergranular attack (IGA) is not a problem unless dilution occurs. The velocity should be limited to 4 ft/s (1.3 m/s), as higher velocity can increase attack, even in the 20 to 60°C (68 to 140°F) temperature range. Liquid oleum lines need to be heat-traced to avoid freezing, and they should be maintained at least 15°C (27°F) above the freezing point at a minimum of 35°C (95°F). Vapor lines should be held at 90°C (194°F) and preferably monitored and alarmed to avoid freezing of sulfur trioxide. In vapor lines that occasionally see liquid oleum—for example, lines that also act as vessel overflow—the temperature should be maintained as for liquid lines. If these lines are held at 90°C (194°F), excessive corrosion can occur when liquid oleum is flowing. The heating system used on this type of dual-function line should, however, be capable of heating to 90°C (194°F) to clear any blockages that occur.[17]

Weak and intermediate acid piping is typically plastic-lined pipe—for example, PVDF-lined—or FRP, thermoplastic/FRP dual-laminate within their temperature limitations.

Metallic Piping Design

A disadvantage of ductile iron pipe is that it is heavy, requiring large supports and robust design of surrounding equipment, which increases secondary costs. It is also quite susceptible to erosion-corrosion, requiring long-radius ells and extra thickness of reducers. Regardless of the specific metallic material selected, piping velocities are usually limited to about 6 ft/s (2 m/s) by flow resistance within the system. Piping needs to be sized accordingly.

Pipe bends or ells should be long-radius to minimize erosion-corrosion. Provision must be made for thermal expansion: Austenitic stainless steels have about a 35% higher coefficient of thermal expansion than do iron or steel. As previously noted, oleum and some concentrations of strong acid may require heating to reduce viscosity and/or prevent freezing. It is essential that heat tracing on piping be physically separated from contact with the external pipe wall to prevent hot-wall effects. Pumping schedules are important in the design of such systems, as a stagnant line more readily overheats.

Electrical tracing is generally preferred to steam tracing because it is more amenable to sophisticated control. Thermal cements must be carefully selected for types 304L (S30403) and 316L (S31603) piping, as chlorides in the contact cement may cause external stress corrosion cracking (ESCC).

Piping of welded construction should have full-penetration welds, with an inert-gas root pass or a consumable insert to minimize turbulence over the interior. Weld-neck flanges are preferred.

Acid Concentration Plant

Various processes and types of equipment are used to clean and/or increase the concentration of weak and intermediate-strength acids. The sources of these acids are usually waste products from other processes that use sulfuric acid, contaminating and diluting them while doing so. Typical sources for these waste acid streams include nitrations, alkylations, sulfonations, and titanium dioxide production. Waste acids used to be dumped, neutralized and dumped, thermally decomposed to produce sulfur dioxide that was fed to a contact acid plant, or fed to a concentrator. The dumping option is now largely unacceptable or unavailable.

Pot Concentrators

The original process for concentrating and/or cleaning weak or waste acid was by boiling it in cast-iron pots set in refractory frames and heated from beneath, usually by open gas or oil flame. Acid is fed to this type of pot concentrator (or Plinke or Pauling concentrator) through a packed column countercurrent to the vapors leaving the pot. It operates at the atmospheric boiling point of around 300°C (572°F) to produce 95–96% acid in which much of the organic material has been thermally decomposed. Pots are heavy-wall cast iron, partly ceramic-tile-lined. The column is cast iron with ceramic or silicon iron packing. Pot life can be very short, a year or less, and occasionally pot failure can be sudden, producing vast quantities of acid vapor.

Drum Concentrators

Drum concentrator plants originally consisted of horizontal, cylindrical steel vessels with a lead membrane behind two or three thick layers of acid-brick, bonded with air-setting or chemically hardening sodium silicate mortars. The brick thickness was designed to maintain the temperature at <74°C (<165°F) at the lead surface when the outside air temperature was 21°C (70°F). The mortar joints had to be crack-free to limit acid penetration and prevent the formation of hot spots. Often, the vessel was divided into compartments and had to be designed to withstand high-tensile stresses, particularly in the winter, as the compartment-wall brickwork would be hotter than the vessel-lining brickwork.

This type of design is no longer practical. Drum concentrators now incorporate silica or borosilicate glass block within the lining. This type of material has much better thermal insulating properties, thus reducing the need for very thick, multilayer brick linings. Polyisobutylene rubber lining now replaces the lead membrane, although the same temperature limits are required as for the lead membrane. Mortars used are potassium or sodium silicate types, hardened with a heat-condensed polyaluminum phosphate catalyst.

In a drum concentrator, weak acid is fed through a brick-lined cooler drum to a brick-lined concentrator drum with lead or rubber membranes, where it is blown with hot air to remove water. The hot air is introduced using a type 410 (S41000) stainless-steel blower through type 409 (S40900) or aluminized steel piping and introduced below the acid surface through 14% silicon cast-iron dip-tubes. The venturi scrubber on the cooler drum is alloy 20Cb-3 (N08020). An acid mix tank and strong-acid pump tank are brick-lined, with PTFE-lined or 14% silicon cast-iron pipe and Lewmet® pumps. The product cooler may be type 316L (S31603) with anodic protection or a 6% molybdenum, superaustenitic alloy, depending on cooling water chemistry. Downstream of the cooler, piping may be a high-silicon stainless steel or fluoroplastic-lined steel, with appropriate valves.

Vacuum Concentrators

In this type of concentration plant, the feed acid is preheated in the heat-recovery exchanger (a tantalum U-bundle in a glass-lined steel shell) before entering a series of concentrator/separators, usually three. A typical plant can treat, for example, a number of spent acids including acid from DNT (dinitrotoluene) and MNT (mononitroluene) production, Figure 12.8. This plant, built in 1987 in the United States, operates on a batch basis and processes approximately 700 tonnes per day (as 100% H_2SO_4) of concentrated acid from about 70% up to 93%.

The separators are now usually glass-lined steel, replacing lead-lined steel and the now-obsolete 14% silicon-nickel cast alloy. The reboilers are tantalum bayonet heaters, and the surface condensers are type 316 (S31600), alloy 20Cb-3 (N08020), or 6% molybdenum superaustenitic alloy tubes (e.g., alloys 254 SMO [S31254], AL-6XN [N08367], 926 [N08926], and 31 [N08031]). The makeup cooler is tantalum in a glass-lined shell, while the weak-acid coolers are impervious graphite up to 150°C (302°F) and the concentrated-acid coolers are type 316 (S31600) with anodic protection or 6% molybdenum superaustenitic alloys. The tanks are glass-lined steel, while pumps are PTFE-lined steel or acid-resistant alloys (e.g., alloy 20Cb-3 [N08020], CN-7M [N08007] casting, 14% silicon cast iron or a Lewmet® alloy). Valves are PTFE- or PFA-lined or specialty alloys, while piping may be glass-lined steel, PTFE-lined steel (of special design intended for vacuum service), or 14% silicon cast iron.

Vacuum concentrators are also available in glass, with tantalum steam heating tubes. These glass units can be single- or multi-stage and can be designed for

Figure 12.8 Multiple-Effect Vacuum Concentration Plant Treating Waste Toluene-Based Chemicals on a Batch Basis (Courtesy of Chemetics, a Division of Aker Kvaerner)

throughputs of up to 6,600 lb/hr (expressed as 100% H_2SO_4) in a single stream. For higher capacities, the use of multiple streams is recommended.[18]

Titanium Dioxide Acid Concentrator

Waste acid (typically 15–22% H_2SO_4) from the production of titanium dioxide can be burned with natural gas in a steel combustor lined with a castable refractory. The acid is then concentrated by passing through a series of towers. The towers are brick-lined steel with carbon-block heat exchangers in steel shells. Piping is 14% silicon cast iron; pumps may be the same or of a Lewmet® composition. Mist eliminators are PVDF mesh and grid, and the candle demister is an alloy 20Cb-3 (N08020) shell with glass filament packing. Where a multiple-effect evaporator is used, it may be of either brick-lined steel with a lead membrane or glass-lined steel.

This type of concentration plant normally produces acid at around 70% concentration, but it has been shown to be capable of producing 96% acid from this titanium dioxide waste acid.[19]

References

1. API 650 Welded Steel Tanks for Oil Storage (Washington, DC: API, latest edition.)
2. ASME Section V111, Pressure Vessels (New York: ASME International, latest edition.)
3. G. Schmitt, "Effect of Elemental Sulfur on Corrosion in Sour Gas," Corrosion 47, 4 (April 1991).
4. A210/A210M-96, "Standard Specification for Seamless Medium-Carbon Steel Boiler and Superheater Tubes" (West Conshohocken, PA: ASTM, 1996).
5. T. Odelstam, G. Bergland, "Sandvik 10RE51/4L7 Composite Cooling Tubes in Fluidized Beds for Pyrites Roasting—Technical and Economical Advantages," February 1982, AB Sankvik, Sankvikan, Sweden.
6. L. J. Friedman, "The Metallurgical Sulfuric Acid Plant—Design, Operating and Metallurgical Considerations," Corrosion/87, paper no. 18 (Houston, TX: NACE International).
7. J. Thomson, "Flow Diagrams for Gas Cleaning Systems" (Vancouver, BC, Canada: Chemetics, A division of Aker Kvaerner Canada Inc., 2004).
8. J. Thomson, "Typical Gas and Acid Flows in Double Absorption Acid Plants" (Vancouver, BC, Canada: Chemetics, A division of Aker Kvaerner Canada Inc., 2004).
9. M. B. Ives, K. S. Coley, J. Rodda, "Corrosion in Hot Gas Converters of Sulphuric Acid Plant," JCSE, 6 Paper C042 (2003): 15 pp.
10. Anon, "Drying Towers," DKL Engineering (2002), http://members.rogers.com/acidmanual/materials_drytowers.htm.
11. G. Berglund, C. Martenson, "Applications of a Highly-Alloyed Stainless Steel in Sulphuric Acid Environments," Corrosion/87, paper no. 21 (Houston, TX: NACE International, 1987): 14 pp.
12. Anon, "The Edmeston SX System—for the Sulphuric Acid Industry" (Goteborg, Sweden: Edmeston AB, 2003): 8 pp.
13. Anon, "Materials of Construction—Mondi®" DKL Engineering (2003), http://members.rogers.com/acidmanual/materials_mondi.htm.
14. M. Davies, "Materials for Piping in Sulfuric Acid Production Plants," MP 30, 9 (1990): pp. 57–59.
15. ASME/ANSI B31-3, "Chemical Plant and Petroleum Refinery Piping," ASME Code for Pressure Piping (1990).
16. Private communication in E. Altpeter, R. Kirchheiner, F. E. White, "Sulphuric Acid Corrosion of Some Special Stainless Steels and Nickel Alloys: Laboratory Tests and Plant Experience," NACE Conference, Grado, Italy (1995): 9 pp.

17. Anon, "Recommended Safe Practices and Emergency Procedures for Sulfur Trioxide, Oleum and Chlorosulphonic Acid," The Soap and Detergent Industry Association, April (1979): 39 pp.
18. Anon, "Sulfuric Acid Concentration," Process Profile 12-2 (West Union, NJ: De Dietrich Process Systems, Inc., 1995): 2 pp.
19. M-C. Simonetti, "Waste Acid Recovery Ends Industrial Emissions," Process Industry Canada, Aug/Sept (1987): 2 pp.

13
Shipping, Handling, and Storage

The materials and equipment described in Chapter 12 are used during the production of sulfuric acid and in chemical processes using sulfuric acid. This chapter deals with the shipping, handling, and storage of sulfuric acid, usually concentrated, and oleum at temperatures up to about 50°C (120°F). There is, however, some overlap in these areas—for example, in transfer piping from production to storage, venting loading and storage facilities, etc. Guidance is also available in a NACE standard that relates to the materials to be used in handling and storage of concentrated sulfuric acid.[1]

Sulfuric acid is a hazardous substance under the Federal Water Pollution Control Act. It is also regulated under the Federal Insecticide, Fungicide, and Rodenticide Act, under which it is exempt from a tolerance for pesticide chemicals in or on raw agricultural commodities, and from the requirement of a tolerance when used in accordance with good agricultural practice as a herbicide in the production of garlic, onions, and potatoes.

The EPA Offices of Solid Waste and Toxic Substances also regulate sulfuric acid. The Comprehensive Environmental Response, Compensation, and Liability Act, and the Emergency Planning and Community Right-to-Know Act of 1986 also regulate sulfuric acid. Under Section 302 of this act, sulfuric acid is listed as an extremely hazardous substance and has a threshold planning quantity of 1,000 lbs. Also under this act, releases of more than one pound of sulfuric acid into the air, water, or land must be reported annually and entered into the Toxic Release Inventory (TRI).[2]

Shipping of Sulfuric Acid

Shipping of sulfuric acid is regulated in the United States by the Department of Transportation (DOT).

The various commercial grades of acid have DOT identification numbers and hazard classification, Table 13.1.

Table 13.1 Commercial Grades of Various Strengths of Sulfuric Acid

Shipping Name	Hazard Class	Grade	ID No.
DOT: Sulfuric acid	Corrosive; class 8	0.1 to 5N H_2SO_4	UN1830
International IMO: Sulfuric acid		10N H_2SO_4 2% H_2SO_4	UN1760
Corrosive liquid		Sulfuric acid <51% Sulfuric acid 51–100% Fuming sulfuric acid Spent sulfuric acid	UN2796 UN1830 UN1831 UN1832

Currently, the relevant government regulations for both land and sea transport of sulfuric acid are detailed in Title 49 of Code of Federal Requirements (CFR), Parts 100–177.

Sulfuric acid of greater than 65.25% may be shipped in the following DOT-approved tank cars and tank trucks.[3]

Tank Cars	Tank Trucks
103A-103AW-103CW	MC-307
105A300W	MC-312
111A60W2-111A100W2	MC-407
111A100W6-111A100F2	MC-412

Shipping of 93% Sulfuric Acid

Bulk shipment of 93% acid may be handled in tank trucks or railroad tank cars or by marine transport (e.g., barges, tanker ships).

The safe unloading of 66°Bé sulfuric acid by the application of air pressure or pump from tank cars and tank trucks requires careful attention. Detailed sulfuric acid unloading procedures and safety regulations can be found in the following industry and government publications.

49 CFR 171-181 Code of Federal Regulations, Department of Transportation (DOT)
29 CFR 1910 Code of Federal Regulations, Department of Labor (OSHA)

Recommended procedures for specific installations can also be obtained from acid suppliers and shippers. These procedures include recommendations relating to safety equipment and operation of equipment as well as advice in cases of frozen acid, blocked vents, leaks, spills, and so on.

Tank Trucks Steel containers are still used where iron contamination is permissible. They should be painted white on the outside to minimize the effect of solar heating on internal corrosion. Stainless-steel containers are becoming more commonly used for shipping 93% sulfuric acid. The molybdenum-bearing grades such as types 316

(S31600) and 316L (S31603) are preferred over types 304 (S30400) and 304L (S30403), which can lose passivity at about 45–50°C (113–122°F), temperatures that can be reached by direct-sunlight heating. Type 316L (S31603) is the preferred shipping-container material, because of diminished iron contamination as compared with steel and because of versatility (i.e., it does not need to be restricted to specific services). This grade also provides added protection against weaker-acid "heels" left in the tank. The meticulous grades must be shipped in inert materials.[4]

Baked Phenolic-Coated Steel Tank Trucks Phenolic-coated cars must be used with caution because high rates of attack may be experienced at pinholes in the coatings when the tanks are empty and exposed to moisture. Such cars should be flushed quickly with at least three volumes of fresh water after the 93% acid has been discharged. Under no circumstances should neutralizing agents (e.g., soda ash) be added to the wash water because alkaline solutions attack phenolic coatings.

Discharge nozzles should have a stainless-steel internal surface (solid, clad, or weld overlay; type 316L [S31603] preferred). All valves should be at least CN-7M (N08007) castings because of possible dilution effects at the external surfaces.

All vessels of carbon-steel construction are potentially subject to brittle fracture at low ambient temperatures (see section on brittle fracture, Chapter 8). Proper design must take this potential problem into consideration, especially with regard to the danger of impact when vessels are carrying acid (e.g., in the event of a collision).

Steel Railroad Tank Cars The original double-riveted steel cars have long since been replaced with fusion-welded cars of modern materials, design, and fabrication.[5]

US DOT Specification 111A100W2[6] covers the domeless 13,000-gal (5,000-L), 100-ton (90,000-kg) car, vent-equipped, with a rail load of 263,000 lbs (120,000 kg). The cars are fabricated from ASTM A 516-70[7] steel, fusion-welded, stress-relieved, radiographed to specification, designed for 240 psi (1,600 kPa) burst pressure, and tested at 100 psi (6,900 kPa). ASTM A 515[8] is not permitted. The shell is usually 9/16 in (14.2 mm) thick. The potential use of ASTM A 808[9] steels is being investigated for improved low-temperature properties and weldability.[10]

Since 1978, the Association of American Railroads (AAR) has mandated the "Canadian design car," which has long operated in the United States under DOT exemption E8931. This design entails the use of a bottom-outlet ball valve (or specially designed butterfly valve) protected by a skid device or other special design to minimize damage (as in derailment).

Prohibitions have been invoked relative to bleed-holes in siphon pipes, friction securement of rupture disks, and lead gaskets in acid-loading covers. Current requirements include bottom protection for sulfuric acid wash-outs, three-bolt fill-opening covers, shelf couplers, and AAR performance and qualification standards for rupture disks. The safety-vent settings have been changed to tank test pressure, and the outage for cars has been revised to two percent.

Increased corrosion may be anticipated because of lower ferrous-ion concentration in acid as a result of using more corrosion-resistant alloys in the manufacturing process. Localized corrosion (e.g., hydrogen grooving and erosion-corrosion) is a problem, especially in the newer and larger cars. When traces of nitrosulfuric acid ($NOHSO_4$) are present, ferrous ions become oxidized to ferric sulfate, precipitation of which causes cloudiness in the final product. External monitoring by nondestructive examination, such as ultrasonics and acoustic emission, is favored to minimize the dilution and washing required for internal inspection.[11]

The possibility of brittle fracture must be considered in cold weather. Railroad cars are subject to possible impact and hydraulic jolting during routine maneuvering. The specific changes in railroad tank-car design over several decades are intended to minimize leaks, spill, and splashes from top fittings under non-accident conditions.[5] Faulty rupture disks have been the major leaking problem. Currently, stainless-steel disks with a 1/8-in (3.2-mm) breather hole are used to dissipate any hydrogen buildup within the cars.

Although uncoated steel cars are still employed to convey 93% acid, they have largely fallen into disuse because of iron contamination problems. Most new cars are either stainless steel or phenolic-coated steel.

Phenolic-Coated Tank Cars High-baked pigmented phenolic coatings are widely used for railroad tank cars handling 93% sulfuric acid. A NACE standard provides detailed recommendations for this application.[12] The inherent problems involved with inadvertent dilution and the prohibition against alkaline neutralization apply as previously described concerning over-the-road tankers.

Up to 10 years of service can be expected in 93% acid at ambient temperatures before coating repair is required. The maximum recommended service temperature is 50°C (122°F).[13] In shipments of up to five days' duration, no iron pickup has been observed.

These coatings will react with traces of nitric acid to form nitrophenols, which cause slight yellow color on first exposure. However, "seasoned" coatings produce no further discoloration.

Marine Transport Marine transport of 93% sulfuric acid can be effected in suitable barge or tanker compartments (e.g., baked phenolic-coated, austenitic stainless steel) or in modularized containers of similar construction. Marine transport is governed by U.S. Coast Guard regulations.[4]

Problems have been encountered in barges under tow due to capsizing and ingress of seawater through the weighted pressure-vacuum valves of the inverted tank. Designing for a low center of gravity is an important consideration for such vessels. Regulations for marine transport have changed over the years, requiring increased thickness to cope with corrosion, banning electrical equipment in adjacent spaces, and requiring devices to detect acid leakage from the tanks.

The current regulations for shipping sulfuric acid state that the containment system may be:[14]

1. Made of unlined steel if the cargo composition is between 70 and 80% or between 90 and 100% acid by weight;
2. Lined with lead if the cargo composition does not exceed 96% acid by weight; or
3. Lined with natural rubber or neoprene if the cargo composition does not exceed 51% acid by weight.

In multicargo sea tankers, type 317L (S31703) compartments have been specially strengthened to carry sulfuric acid cargos up to full deadweight.[15]

Commercial-grade sulfuric acid is normally shipped as 93%, 96%, or 98%. Below about 95%, the standard 300-grade stainless steels have better corrosion resistance by about 5°C than the duplex 2205 (S31803). If a marine chemical tanker is dedicated to sulfuric acid and is normally used for the less concentrated acid, then 316 (S31600) would normally be used. The duplex alloy has a good advantage over 316 (S31600) when frequent tank cleaning takes place, especially if seawater is used. Alloy 2205 (S31803) is being used for marine chemical tankers to give a margin of safety from corrosion after cleaning operations.[16]

Shipping of 96–99% Sulfuric Acid

Bulk shipment of 96–99% acid may be effected in tank trucks or railroad tank cars or by marine transport (e.g., barges, tanker ships). Steel containers are still used where iron contamination is permissible.

Tank Trucks Carbon steel over-the-road tanks may be employed where iron contamination of the acid is permissible.

Where iron contamination is not acceptable, types 304 (S30400) and 316 (S31600) and L-grade stainless-steel tank trucks are suitable. In fact, most new tank trucks for this service are made from 316L (S31603).[1]

Phenolic-coated steel cars must be used with caution because high rates of attack may be experienced at pinholes in the coatings when the tanks are empty and exposed to moisture. Such cars should be flushed quickly with at least three volumes of fresh water after 98.5% acid has been discharged. Under no circumstances should neutralizing agents (e.g., soda ash) be added to the wash water because alkaline solutions attack phenolic coatings.

Discharge nozzles should have a stainless-steel internal surface (solid, clad, or weld overlay, type 316L preferred). All valves should be CN-7M (N08007) minimum, because of possible dilution effects at external surfaces. All vessels of carbon-steel construction are potentially subject to brittle fracture at low ambient temperatures, as discussed in Chapter 8, although the hazard is considerably less with the warmer temperatures required for this acid strength than for 93% acid. Proper design must take this potential problem into consideration, especially with regard to the danger of impact when the vessel is carrying acid (e.g., in the event of a collision).

Railroad Tank Cars Steel tank cars are satisfactory if iron contamination of the acid is acceptable. The Association of American Railroads (AAR) permits the use of any approved grade of carbon steel. In the past, ASTM A 515, Grade 70,[8] and the previously equivalent ASTM A 212,[17] Grade B, have been the most popular grades.

Because of the greater danger of impact in railroad transport, steel should be specified to ASTM A 516[7] (see section on shipping of 93%). This steel grade, in the normalized condition, is required for all railroad tank cars ordered after 1988. Tank-car plates are normally 9/16 in (14 mm) thick, including a ⅛-in (3.2-mm) corrosion allowance.

Tank cars for 98% acid are insulated to retain heat and do not require the white external coating recommended for the 93% grade.

Most new tank cars for this service have a baked phenolic lining or are stainless steel. If iron contamination is to be avoided, type 304 (S30400) cars are preferred, but type 316 (S31600) and type 316L (S31603) are also acceptable.

Marine Transport Marine transport of 98.5% sulfuric acid is suitable in barge or tanker compartments (e.g., baked phenolic-coated, austenitic stainless steel) or in modularized containers of similar construction, with appropriate provisions for heating the acid.

The current regulations for shipping sulfuric acid state that the containment system may be:[14]

1. Made of unlined steel if the cargo composition is between 70 and 80% or between 90 and 100% acid by weight;
2. Lined with lead if the cargo composition does not exceed 96% acid by weight; or
3. Lined with natural rubber or neoprene if the cargo composition does not exceed 51% acid by weight.

Shipping of Oleum

Bulk shipment of oleum may be handled in carbon-steel tank trucks or railroad tank cars or by marine transport in steel barges. A shipping containment system for oleum may be of unlined steel if the concentration of free sulfur trioxide in the oleum exceeds 20% by weight.[14]

Shipping of Other Concentrations

Because of the effects of motion as well as often undefined contaminants, the noncommercial concentrations in the 20–100% range are usually transported in corrosion-resistant containers. DOT regulations forbid the transportation of spent sulfuric acid on passenger-carrying aircraft or railcars.

Steel Tank Trucks Although carbon steel trailers have been used (depending on the sulfuric acid and nitric acid content), a corrosion rate of about 30–40 mpy (0.75–1.0 mm/y) must be expected. Substantial iron contamination would also occur.

Baked phenolic coatings are unsuitable if nitric acid contamination is present but may otherwise be suitable.

Steel trailers are often lined with five coats of baked PVDF, which provides a surface amenable to spark-testing to ensure there is no steel substrate exposed at holidays. This lining is highly resistant to mechanical damage.

Glass-lined steel trailers may be available as specialty items. Mechanical damage can be repaired with tantalum or high-alloy patches.

Stainless-Steel Tank Trucks Stainless-steel trailers, preferably type 316L (S31603), may be employed if the nitric acid content is at least 1%. It should be noted, however, that the nitric acid content may diminish with time if organic contaminants are also present, leading to increased corrosion attack. Chloride contamination may cause pitting, crevice corrosion, or stress corrosion cracking.

Railroad Tank Cars Carbon-steel tank cars may be used, subject to limitations mentioned previously. They should be thermally stress-relieved and weld hardnesses should not exceed 285 HBN (Brinell), namely HRC (Rockwell C) 30.

Stainless-steel tank cars should be type 316L (S31603). They are limited to acids containing at least 1% free nitric acid.

Marine Transport Steel barges may be used within the limitations previously described. Stainless-steel compartments or containers may be used with mixed acids containing at least 1% nitric acid. Fluorocarbon-lined containers are suitable. Glass-lined containers require padding protection against damage in rough weather.

Storage and Handling of Sulfuric Acid

The next sections provide summaries for the storage and handling of various concentrations of sulfuric acid and oleum. These summaries are followed by details of the tanks and other equipment used in storage and handling. Further information about storage and handling of sulfuric acid can be found in the NACE Standards on the subject[1,18] and in an informative summary.[19] Manufacturers of sulfuric acid also provide information about the safe storage of their products.

Storage and Handling of 93%

Materials used for the storage and handling of 93% sulfuric acid have been summarized for various temperature ranges, Table 13.2. These ranges are for storage at ambient temperatures (0 to 40°C [32 to 104°F]), moderately elevated temperatures that might occur in hot climates (40 to 60°C [104 to 140°F]), and at still higher temperatures (60 to 80°C [140 to 176°F]), which would only be encountered in day tanks or receivers of warm acid from the production process. The footnotes to this table relate to specific considerations, and a primary distinction is made between acid storage tolerant of iron contamination versus very low iron requirements.

Because 93% acid has a low freezing point of –28°C (–18°F), 93% acid may be stored conveniently in steel at ambient temperature. When there are restrictions on permissible iron content, control of contamination may be effected by nitric acid inhibition or anodic protection (see Chapter 7), baked phenolic coatings, or use of austenitic stainless-steel containers.

Minor dilution (for example, to about 90%) by water ingress will bring the freezing point dangerously near to that of water, i.e., -6°C (21°F). Inadvertent use of higher-concentration acid can also permit freezing. Also, there are special considerations regarding nil-ductility transition temperatures of steel vessels that apply widely but are a major concern in this service because of the personnel hazards with brittle fracture of acid tanks (see Chapter 8).

Certain distinctions must be made relative to temperature, pressure, velocity/turbulence, and heat transfer effects regarding different types of equipment. Also, certain distinctions must be made for ordinary commercial-quality or process acid. The meticulous grades require almost totally resistant materials of construction to prevent any product contamination.

Table 13.2 Materials for Handling 93% Sulfuric Acid

	Equipment	0–40°C (32–104°F)	40–60°C (104–140°F)	60–80°C (140–176°F)
Tanks[a,b,c,d]	Fe permitted	Carbon steel with discharge nozzles; stainless-or-alloy lined, clad, or overlaid (316L, 20Cb-3)	High-Si SS; brick-lined; glass- or FEP- or PFA-lined	High-Si SS; brick-lined; glass- or PFA-lined
	Low-Fe	Steel with AP (>1,000 tons); steel with baked phenolic coating;[e] 316L	High-Si SS; brick-lined; glass- or FEP- or PFA-lined	High-Si SS; brick-lined; glass- or PFA-lined
Piping	(Continuous flow) <1 in (2.5 mm)	Stainless (304 or 316), screwed[o]	Alloy B-2; C-276;[i] high-Si SS	Alloy B-2; C-276
	1 to 3 in (2.5 to 7.5 mm)	PVDF-lined; 304;[f] Sch 10 316L[f]	14% Si-iron; B-2; C-276; glass; PVDF- or PTFE-lined; high-Si SS; Lewmet®	14% Si-iron; glass; PTFE- or PVDF-lined; Lewmet®
	>3 in (7.5 mm)	DI; Sch 10 304 or 316; PVDF-lined steel	304L with AP; 14% Si-iron; B-2; C-276; glass; PVDF- or PTFE-lined; high-Si SS; Lewmet®	14% Si-iron; glass; PTFE- or PVDF-lined; Lewmet®
	(Intermittent flow)[g] >3 in (7.5 mm)	Carbon steel;[h] 304 or 316; PVDF-lined steel	Does not apply	Does not apply
Valves	Shutoff	CF-8M; CD-4MCu; CN-7M; PTFE	Glass- or PTFE-lined	Glass- or PTFE-lined
	Throttling	CN-7M; PTFE[j]	Si-iron[l] or Lewmet®	Lewmet 55®
	Pumps	CN-7M; Lewmet;[k] Si-iron	Lewmet®; Si-iron; glass- or PTFE-lined; graphite; solid alumina	Lewmet®; Si-iron; glass- or PTFE-lined; graphite
	Heat exchangers	Impervious graphite; 316L (AP); tantalum[n]	Impervious graphite; 316L (AP); tantalum;[n] B-2; C-276; glass	Graphite; PFA; 5% Si SS; tantalum;[m,n] SiC; alloyed SS[p]

(continued)

Table 13.2 Materials for Handling 93% Sulfuric Acid *(continued)*

Equipment	0–40°C (32–104°F)	40–60°C (104–140°F)	60–80°C (140–176°F)
Gaskets	Spiral SS/PTFE; PVDF/hexafluoro-propylene; silica or aluminum silicate-filled PTFE; envelope PTFE/SBR; CTFE; graphite	Spiral SS/PTFE; PVDF/hexafluoro-propylene; silica or aluminum silicate-filled PTFE; envelope PTFE/SBR; CTFE; graphite	Graphite or fluorinated complex
Instruments	304L/316L; ceramics; fluoroplastics	High-performance SS;[p] ceramics; fluoroplastics	High-performance SS;[p] ceramics; fluoroplastics

[a]Add impact requirements for cold climates (Brittle fracture, Chapter 8).

[b]Horizontal tanks should be designed and built in accordance with Section VIII, Division 1 of ASME Boiler & Pressure Vessel Code for Unfired Pressure Vessels. Vertical atmospheric tanks should be designed and built in accordance with API Standard 650, Welded Steel Tanks for Oil Storage," 14, with modifications to design and fabrication as required, e.g., nozzle placement. Vessel walls will be thicker than those for oil storage because of the greater density of acid. Welding slag and mill scale should be removed prior to service exposure. Small tanks and columns should be constructed in accordance with ASME B&PV Code.

[c]Corrosion allowance should be based on calculated anticipated corrosion, min. 6 mm (0.25 in).

[d]For high-humidity climates, use desiccating vents to minimize ingress of atmospheric moisture (or dry gas atmosphere blanketing) to minimize corrosion in vapor phase and at liquid level.

[e]Baked phenolic coatings must not be exposed to alkaline neutralizing washes unless flushed quickly with at least three volumes of fresh water.

[f]Acceptable to 10 ft/s (3.3 m/s) velocity.

[g]Drain back all metal piping under nitrogen purge.

[h]Acceptable maximum flow rate is 2 ft/s (0.6 m/s); up to twice this rate can be tolerated if pumping schedule is <8 hrs/day.

[i]Use alloy C-276 (N10276) when oxidizing contaminants are present.

[j]Where no abrasive particles are present.

[k]Seal hard-facing chromium oxide.

[l]Si-Cr cast iron (ASTM A 518, Grade 2) may be used in lieu of conventional 14% Si iron.

[m]Maximum tube-wall temperature of 200°C (392°F).

[n]Tantalum must be electrically isolated from other metals to prevent hydrogen pickup and embrittlement.

[o]Only dry air should be used to blow out stainless piping; otherwise, dilute acid will cause severe localized corrosion.

[p]6% Mo superaustenitic alloys with AP for aggressive cooling waters.

Table 13.3 Materials for Handling 96–99% Sulfuric Acid

Equipment	0–40°C (32–104°F)	40–60°C (104–140°F)
	Tanks[a,b,c]	
Fe permitted	Steel tanks with discharge nozzles; stainless-lined, clad, or overlaid (304L, 310L) [d]	Glass-lined steel; brick-lined[l] steel or high-Si SS
Low-Fe	Steel with AP; 304L; glass-lined steel for special requirements[e]	Glass-lined steel; brick-lined[l] steel or high-Si SS
	Piping	
(Continuous flow) <1 in (2.5 mm)	304[f]	High-Si SS; alloy 625
1 to 3 in (2.5 to 7.5 mm)	304, Sch 40	High-Si SS; alloy 625; Pyrex[m]
>3 in (7.5 mm)	DI; 304, Sch 10	High-Si SS; alloy 625; ASTM F47003; FEP-lined steel (Std) or heavy PTFE-lined steel[g]
(Intermittent flow)[h]	Carbon steel, Sch 80;[i] DI; 304L, Sch 10; 20Cb-3; FEP- or PVDF-lined steel or heavy PTFE-lined steel	High-Si SS; alloy 625; ASTM F47003; FEP-lined steel (Std) or heavy PTFE-lined steel[g]
	Valves	
Shutoff	CF-8M; CN-7M; fluoroplastic-lined steel	CW-12MW; CN-7M; fluoroplastic-lined steel
Throttling	CN-7M	CN-7M; High-Si SS; Lewmet®
Pumps[j]	CN-7M	Lewmet® 55

(continued)

Table 13.3 Materials for Handling 96–99% Sulfuric Acid *(continued)*

Equipment	0–40°C (32–104°F)	40–60°C (104–140°F)
Heaters	304L or external plate coils	304L or external plate coils
Gaskets	Spiral-wound 304/PTFE; PTFE (silica or aluminum silicate-filled); PTFE (multiple braid, 25% glass or CAF filler); envelope PTFE/SBR-bonded compressed asbestos;[k] PVDF/hexafluoropropylene; fluoroplastics; CTFE	Spiral-wound 304/PTFE; PTFE (silica or aluminum silicate-filled); PTFE (multiple braid, 25% glass or CAF filler); envelope PTFE/SBR-bonded compressed asbestos;[k] PVDF/hexafluoropropylene; fluoroplastics; CTFE
Instruments	304; 316; ceramics; fluoroplastics	304; 316; ceramics; fluoroplastics

[a] Horizontal tanks should be designed and built in accordance with Section VIII, Division 1 of ASME Boiler and Pressure Vessel Code for Unfired Pressure Vessels. Vertical atmospheric tanks should be designed and built in accordance with API Standard 650, "Welded Steel Tanks for Oil Storage," 14, with modifications to design and fabrication as required, e.g., nozzle placement. Vessel walls will be thicker than those for oil storage because of the greater density of acid. Welding slag and mill scale should be removed prior to service exposure. Small tanks and columns should be constructed in accordance with ASME B&PV Code.
[b] Corrosion allowance should be based on calculated anticipated corrosion, min. 6 mm. (0.25 in).
[c] For high-humidity climates, use desiccating vents to minimize ingress of atmospheric moisture (or dry gas atmosphere blanketing) to minimize corrosion in vapor phase and at liquid level.
[d] Type 316L and alloy 20Cb-3 would give added protection against inadvertent dilution or loadings with weaker acid, but they are less resistant to 98.5% sulfuric acid.
[e] Any spillage of acid on the exterior of bare glass-lined steel tanks will cause internal failure of the glass lining due to ingress of atomic hydrogen generated by corrosion. A protective physical barrier (e.g., types 304L or 316L sheathing) is recommended
[f] Type 304L is an acceptable alternative. Type 316L is also used to resist corrosion during water washing.
[g] PVDF-lined acceptable to 52°C (126°F).
[h] Use nitrogen purge, not air, with all metallic materials.
[i] Acceptable maximum flow rate is 2 ft/s (0.6 m/s); up to 4 ft/s (1.2 m/s) can be tolerated if pumping schedule is <8 hrs/day.
[j] Seal hard-facing of chromium oxide.
[k] Envelope PTFE/flexible graphite in United States.
[l] Low iron content may not be achieved using brick-lined steel if common shale or fire-clay brick is used, or at least not until surface iron contamination on the brick has been leached out.
[m] Pyrex pipe is potentially hazardous if support is constrained, due to possibility of delayed fracture ("static fatigue").

Storage and Handling of 96–99%

Materials used for storing and handling this range of acid strengths have been summarized, Table 13.3.

There are minor differences in the handling of 98.5% acid as compared with 93% grade because the higher concentration is somewhat less corrosive to common materials of construction. Also, the solubility of ferrous sulfate is substantially lower in 98.5% acid.[20] Further, the freezing point of the 98.5% acid is higher, approximately 0°C (32°F).

For conventional shipment and storage, the temperature for 98.5% acid should be kept between 5 and 40°C (41 and 104°F), although bulk temperature should not exceed 32°C (90°F) for significant periods. Steel tanks may be insulated, the acid temperature being controlled either by plate exchangers on the tank or by circulation through an external heater.

It is necessary to make certain distinctions relative to temperature, pressure, velocity/turbulence, and heat-transfer effects regarding different types of equipment, for example, tanks (receivers, day tanks, feed tanks, field storage), piping, valves, pumps, and heat exchangers. Iron contamination is not normally a problem because of the diminished solubility of iron sulfate at this acid concentration.

Storage and Handling of Oleum

Commercial grades of oleum may be stored in steel at ambient temperatures, with corrosion rates on the order of 3–5 mpy (0.8–0.13 mm/y). Above 60°C (140°F), type 304 is the preferred material. Suggested materials of construction for 0 to 40°C storage and also for temperatures above 40°C are given in Table 13.4. Suggested materials for piping at elevated temperatures are shown in Table 13.5.

Liquid lines should be heat-traced to avoid freezing of oleum. They should be maintained at least 15°C (27°F) above the freezing point with a minimum of 35°C (95°F). Vapor lines should be held at 90°C (194°F) and preferably monitored and alarmed to avoid freezing of sulfur trioxide. In vapor lines that occasionally see liquid oleum (for example, lines that are vapor lines but also act as vessel overflow), the temperature should be maintained at least 15°C (27°F) above the freezing point, with a minimum of 35°C (95°F). If these lines are held at 90°C (194°F), excessive corrosion can occur when liquid oleum is flowing. The heating system used on this type of dual-function lines should be capable of heating to 90°C (194°F) to clear any blockages that occur.[21]

Engineered oleum vent packages are available, designed to control SO_3/H_2SO_4 emissions from oleum storage tanks, tank trucks, railcars, and loading stations. Units are available in flows from 100 to 1,500 cfm; they hydrolyze SO_3 vapor into sulfuric acid mist and then remove the mist in a high-efficiency mist eliminator.[22,23]

Table 13.4 Materials for Handling Oleum

Equipment	0–40°C (32–104°F)	40–60°C (104–140°F)
Tanks[a,b,c]	Carbon steel	Steel to 60°C (140°F); 304 >60–80°C (>140–176°F)
Piping	Carbon steel;[d] 304L[e]	See Table 13.5
Valves	CF-8, CF-8M, or CN-7M	CF-8M or CN-7M to 80°C (176°F); Glass-filled PTFE-lined[f]
Pumps[g,h]	Process iron; CH-20; CK-20; CN-7M	Process iron; CF-8 to 90°C (194°F); CF-8M to 80°C (176°F); CH-20; CK-20; CN-7M
Coolers[i]	Does not apply	Steel[j] to 70°C[k] (158°F); 304 to 90°C (194°F); 309 >90°C (194°F); alloy 800 to 90°C (194°F)
Gaskets[l]	Spiral-wound 304/PTFE; PTFE (silica or aluminum silicate-filled); PTFE (multiple braid, 25% glass or CAF filler); CTFE; blue african asbestos[m]	Spiral-wound 304/PTFE; PTFE (silica or aluminum silicate-filled); PTFE (multiple braid, 25% glass or CAF filler); CTFE; blue african asbestos[m]
O-rings	FFKM, Kalrez®, etc.	FFKM, Kalrez®, etc.
Instrumentation[l]	304; ceramics; fluoroplastics	304; ceramics; fluoroplastics

[a] Vents, mist eliminators, etc., should be of type 304L construction due to possible moisture ingress and attendant problems with dilution effects. The L grade is protection against the possibility of IGA due to uncontrolled dilution.
[b] Some users have changed to type 304L tanks to minimize iron pickup and reduce corrosion during clean-out. (External ultrasonic thickness measurements are preferred to reduce the incidence of clean-out corrosion when preparing for internal inspection.)
[c] Tanks must not be left empty because of potential problems from drip or dilution effects.
[d] Maximum velocity is 6 ft/s (2 m/s); anticipate 5 mpy (0.13 mm/y).
[e] Where Fe pickup is objectionable, type 304L is acceptable.
[f] Carbon-filled PTFE is unsuitable.
[g] Seal hard-facing of chromium oxide.
[h] Submersible pumps are preferred to minimize drip and dilution.
[i] May be modified for cooling water chemistry and corrosivity. Molybdenum-rich alloys are less resistant than low-molybdenum alloys and should be avoided.
[j] The water side of spiral heat exchangers may be coated with baked phenolic to resist water side corrosion.
[k] Maximum tube wall temperature allowed.
[l] Retorquing requirements are critical to prevent moisture ingress and localized dilution. Do not use unfilled PTFE.
[m] Where permissible; other types of asbestos are unsuitable.

Table 13.5 Piping for Oleum > 40°C (104°F)

Material	Concentration	Temp. Limit °C (°F)[b]
Carbon steel[a]	5–30%	60 (140)
	35–65%	70 (158)
Type 304	5–30%	90 (194)
	35–65%	110 (230)
Type 316	All	80 (176)
Alloy 20Cb-3	All	80 (176)
Types 309 and 310[c]	All	130 (266)
PTFE- or PFA-lined steel	All	175 (347)

[a] To 5 ft/s (1.7 m/s).
[b] There is a potential problem with solidification of SO^3 in a vapor space, which may cause blockage, followed by rapid vaporization when the blockage is cleared. There should be provisions to heat vapor lines and vents to 90°C (194°F) to prevent blockage.
[c] Usually available only in mill lots.

Storage of Other Concentrations

In the static storage of noncommercial grades at ambient temperatures (as for receivers, day-tanks, feed tanks to concentrators, etc.), temperature, pressure, velocity/turbulence (such as for piping, valves, and pumps), and heat-transfer effects must be distinguished. At moderately elevated temperatures, e.g. >40°C (>104°F), more corrosion-resistant materials must be used and contamination effects may be more pronounced. Materials for storage of these miscellaneous concentrations are suggested in Table 13.6.

Table 13.6 Materials for Handling Noncommercial Concentrations of Sulfuric Acid

Equipment	0–40°C (32–104°F)	40–60°C (104–140°F)
	Tanks[a]	
Fe permitted	Carbon steel[b]	Brick-lined steel; lead-lined (<80%);[c] glass-lined steel; fluoroplastic lining
Low-Fe	Lead-lined (to 90%) 316L;[d] 5% Si SS; alloy 904L	Brick-lined steel; lead-lined (<80%);[c] glass-lined steel; fluoroplastic lining

(continued)

Table 13.6 Materials for Handling Noncommercial Concentrations of Sulfuric Acid *(continued)*

	Equipment	0–40°C (32–104°F)	40–60°C (104–140°F)
Piping	(Continuous flow) <1 in (2.5 mm)	Type 904L or 20Cb-3; 304L; 316L;[e] PVDF;[g] PP[g] (to 93%); lead (to 90%)	Alloy 625 or C-276;[f] PVDF; PFA;[g] PTFE
	(Continuous flow) 1 to 3 in (2.5 to 7.5 mm)	PP-lined (93% max.); lead or lead-lined 316L, Sch. 10[f]; Type 904L; 20CB-3; 304L; 316L; PVDF	Glass;[g] 20Cb-3;[f] Lewmet® 66; fluoroplastic-lined[h]
	(Continuous flow) >3 in (7.5 mm)	DI (>90%); Glass;[g] 20Cb-3;[f] Lewmet® 66; fluoroplastic-lined[h]	Glass;[g] 20Cb-3;[f] Lewmet® 66; fluoroplastic-lined[h]
	Intermittent flow	Glass;[g] 20Cb-3;[f] Lewmet® 66; fluoroplastic-lined[h] ;Type 904L; 20CB-3; 304L; 316L; PVDF	Glass;[g] 20Cb-3;[f] Lewmet® 66; fluoroplastic-lined[h]
Valves	Shutoff	CN-7M;[e] fluoroplastic- and glass-lined steel	CN-7M;[e] fluoroplastic- and glass-lined steel
	Throttling	Lewmet® 55 or 66; N-12MV or CW-12MW[f]	Lewmet® 55 or 66; N-12MV or CW-12MW[f]
	Pumps[i]	CN-7M; Lewmet® 55; impervious graphite[h] (93% max.)	ASTM F47003; Lewmet® 55 or 66
	Heat exchangers	Impervious graphite;[g] PFA or glass;[g] alloy 625 or C-276[f]	Impervious graphite;[g] PFA or glass;[g] alloy 625 or C-276[f]
	Gaskets	Flexible graphite; PTFE (silica or aluminum silicate-filled); PTFE (multiple-braid, 25% glass or CAF filler); envelope PTFE/SBR-bonded compressed white asbestos;[j] PVDF/ hexafluoropropylene	Flexible graphite; PTFE (silica or aluminum silicate-filled); PTFE (multiple-braid, 25% glass or CAF filler); envelope PTFE/SBR-bonded compressed white asbestos;[j] PVDF/ hexafluoropropylene
	Instrumentation	High-performance SS; ceramics;[g] fluoroplastics	High-performance SS; ceramics;[g] fluoroplastics

[a] See below for comments on ASME B&PV Code and API constructions, desiccating vents, inert-gas blankets, and nondestructive examination.
[b] Subject to nitric acid content (see Chapter 9). Based on anticipated corrosion rate of 10 mpy (0.25 mm/y), which will vary with nitric acid content.
[c] Sensitive to nitric and chloride content.
[d] With 1% nitric acid minimum.
[e] Where nitric acid or other oxidants can maintain passivity.
[f] Use nickel-chromium-molybdenum (e.g., alloy C-276 [N10276]) when oxidants >1%; beware of even ppm with nickel-molybdenum (e.g., alloy B-2 [N10665]).
[g] Solid plastic, glass, and impervious graphite (or other materials susceptible to mechanical breakage) should be safe-guarded as required for Class M materials in ASME/ANSI B31.3.[24] Chemical resistance alone is not a criterion in services in which pillage is intolerable.
[h] Standard FEP- or PVDF-lined; heavy PTFE-lined.
[i] Seal hard-facing of chromium oxide.
[j] Envelope PTFE/flexible graphite in United States.

Storage Tanks

Tanks used to store sulfuric acid must be designed and fabricated to specific requirements to cater for the following factors:

- Corrosive nature of the acid
- Velocity sensitivity
- Desire to limit iron pickup
- Possibility of low temperature fracture
- Hydrogen effects producing blisters and grooving

The details of design, fabrication, erection, inspection, and maintenance of carbon steel tanks for sulfuric acid are contained in a number of documents based on many years of experience in the field.[18,19] These documents provide recommendations for such things as supporting structures, design of inlet and outlet nozzles, weld details, radiography, and inspection. The main features are presented below, while issues of hydrogen blistering and grooving have been addressed in Chapter 8. Because specifications such as the referenced NACE documents are under regular review and update, the latest edition should always be consulted before manufacturing acid storage tanks.

Tanks should be provided with dikes to contain spills. Where other protection is not afforded, tanks (and their associated pumps, valves, and piping) should be protected against vehicular collision damage by appropriate physical barriers. Water must not be allowed to accumulate within the diked area, as dilute acid formed by a tank spill will cause severe corrosion of structures and concrete.

Desiccating vents are incorporated when the average absolute humidity of the ambient external atmosphere is consistently high, as in the Texas Gulf Coast and other industrial marine locations. Arid climates (e.g., high plateaus and desert areas) are not conducive to ingress of atmospheric moisture. In less well-defined geographical locations an individual assessment must be made.

There is an incremental effect of natural convection that increases corrosion of steel due to thermal effects on the motion of the stored acid. To minimize this effect, tanks may be painted white and may be provided with sunscreens, which are inexpensive and do not interfere with ultrasonic thickness measurements from the outside. Insulating a tank is effective but may interfere with inspection and can lead to external corrosion by wet insulation if the insulation is not properly applied over a coated surface and maintained.

The ingress of atmospheric humidity is less important with higher acid concentrations. In storing 70% acid in steel on the Texas Gulf Coast, expected corrosion rates are of the order of 40 mpy (1 mm/y) uninsulated and without vent-gas drying, but only about 10 mpy (0.25 mm/y) if insulated and provided with gas drying.[25]

It is prudent to provide vent-gas drying in any consistently humid geographical location. It should be noted that some systems for vent-gas drying are not effective for tanks with low usage rates. Other systems have proven ineffective even with high usage. Vent-gas drying units are high-maintenance devices requiring frequent inspection and service.

Design of Vertical Tanks

Vertical atmospheric storage tanks should be designed and built in accordance with API Standard 650,[26] with modifications to design and fabrication requirements as required by the specific recommendations of other standards, such as NACE RP 0294.[18] Vessel walls will be thicker than for oil storage because of the much higher density of sulfuric acid. A corrosion allowance (typically 3–6 mm [0.13–0.25 in]) should be added to the design thickness based on the desired life of the tank and expected corrosion rate. The corrosion allowance can be calculated based on published corrosion rates such as Figure 13.1[27] or Figure 13.2.[28] Such published data should only be used to provide preliminary information based on the anticipated range of acid strength and temperature to be encountered. Factors such as acid purity, velocity, presence of AP, and/or coatings all influence the appropriate allowance to be used.

Nozzles should be placed such as to avoid erosion-corrosion, drip, and erosive effects (grooving) due to hydrogen evolution.[29] Tanks should have a minimum shell-plate wall of 0.32 in (8 mm) and a minimum base-plate thickness of 0.5 in (13 mm) and be supported on elevated structural beams or built on a raised, sloped, and grooved-to-drain foundation. The base of the tank should drain to the outlet.

The roof should be self-supporting, with external support girders as required, although properly designed internal support columns may be required for larger tanks. The vent should be centered to prevent hydrogen buildup. (Special provisions must be made for venting oleum tanks because of SO_3 pressure.) Overflow and vent pipes should have a desiccating vent where atmospheric conditions require it (i.e.,

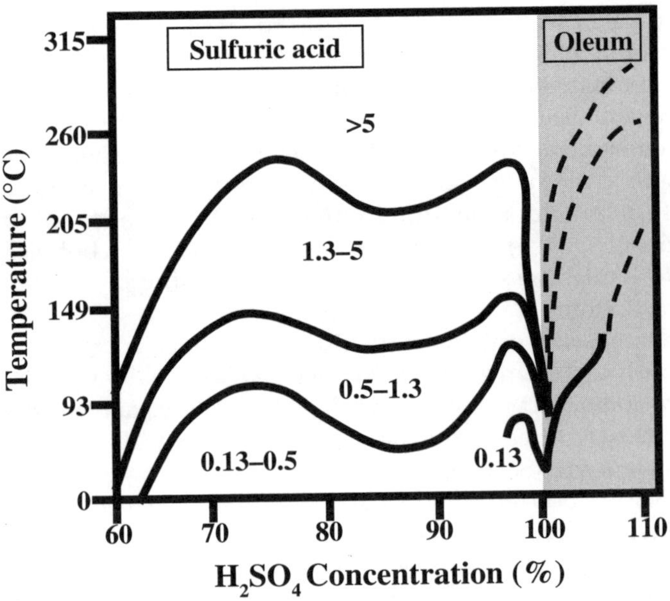

Figure 13.1 Isocorrosion Curves for Carbon Steel in Sulfuric Acid and Oleum

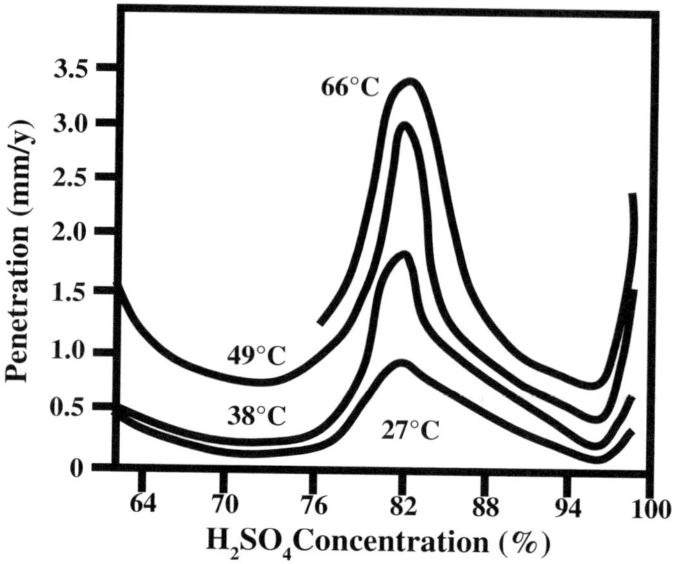

Figure 13.2 Effect of Temperature on the Rate of Corrosion of Carbon Steel in 60–100% Sulfuric Acid

with perennial or frequent high relative humidities). Acid inlet should be effected through a pipe section close to the center of the roof, about 10 ft (3 m) away from the tank wall and protruding 6 in (15 cm) or more into the tank. Alternatively, a stainless-steel dip-tube may be run about 6 in (15 cm) minimum away from the wall, extending under the acid level to about 3 ft (1 m) above the base and directed at a stainless-steel impingement plate or weld overlay. (Any dip-tube below the acid level must be provided with a siphon break.) All tanks should be provided with an appropriate level indicator (e.g., a remotely operated air-purged level gauge).

Horizontal manholes and nozzles should protrude about 1 in (3 cm) into the tank in the upper hemisphere, which should be lined or weld-overlaid over the 180-degree arc surface with a suitable stainless or high-performance alloy (e.g., type 304L [S30403] in 94–100% acid; type 316L [S31603] in 90–93%; alloy 20Cb-3 [N08020] or C-276 [N10276] below 90%). This protrusion directs hydrogen bubbles away from the tank wall. To facilitate iron sulfate removal, an API flush-type clean-out door may be provided in the bottom.

The bottom outlet connection should also be of lined, clad, weld-overlaid, or solid alloy construction to minimize erosion effects. An external valve, backed up with a remotely operated adjunct valve, should be of high-alloy stainless steel (e.g., CN-7M [N08007]) and closely fitted to the tank or installed on a long-radius outlet bend of type 304 (S30400) or 316 (S31600) construction.

For service below about 40°C (104°F), steels should be selected to avoid brittle fracture (see Chapter 8). The principle cause of brittle fracture is the use of steels with inadequate toughness. Modern analysis techniques can facilitate toughness consideration in structural design. A number of brittle fractures of sulfuric acid storage tanks in service have occurred, but it should be noted that their construction antedates modern practice. The latest edition of API 650[26] provides a high margin of safety. In practical applications, the design is predicated upon certain assumptions regarding wintertime temperatures, and the fill height for the tank is adjusted to provide a safety margin.

Fabrication of Vertical Tanks

The shielded metal arc welding (SMAW) process should be employed to assure full-penetration welds. Floor welds should be double-sided butt welds, without backing strips. The tank top may employ either butt or lap welds, fully welded from both sides. Side-to-bottom welds should have a minimum ¼-in (6-mm) fillet weld inside and out. The roof-to-shell weld should be fully welded internally, as should all nozzles and manholes. Welds below liquid level should be made with at least three passes to minimize leaks caused by through-weld slag or porosity, and ground smooth. All welds below the liquid level should undergo thorough nondestructive inspection, with any indications ground out and repaired as required. Clips attached for scaffolding, rigging, or other construction requirements must be removed prior to hydrostatic testing and service.

All welds should be tested to API 620.[30] Bottom welds should be tested with a vacuum box. The tank should be filled with water and allowed to stand 24 hours before draining for final inspection and before being placed in service.

Design of Horizontal Tanks

Horizontal tanks should be designed and built in accordance with Section VIII, Division 1 of the ASME Boiler and Pressure Vessel Code for Unfired Pressure Vessels[31] (or outside the United States, in accordance with equivalent codes or standards). Small day-tanks and similar containers should be of similar construction. Horizontal tanks should have a bottom outlet and valve. A second emergency drain valve should be provided if alternative means of emptying the tank are not available. A backup should be provided in the form of an internal, tank-top-operated plug valve. The tank should have a top-entry manhole and a top-entry inlet pipe protruding at least 6 in (15.24 cm) into the tank. A vent pipe should be installed to prevent fume or spray emission and, where atmospheric conditions require, a desiccating vent to prevent ingress of atmospheric moisture.

Fabrication of Horizontal Tanks

All welding is to be full-penetration, welded from both sides by the SMAW process. Fillet welds should have a minimum throat thickness equal to the thinnest plate joined. Branch nozzles and manholes should be fully welded internally. All welds below the liquid level should be made in three passes and inspected and repaired as previously discussed.

Tank Cleaning

Washing for decontamination purposes should be required before entry into a tank. Serious damage may occur if tank washing is not properly done, due to dilution and exotherm effects. The following procedures are recommended for tanks of up to 1,000-ton (900,000-kg) capacity:

- Empty tank of acid or oleum.
- Physically remove iron sulfate/sludge as much as possible.
- Using copious amounts of water, flush out remaining sludge. (Do not allow acidic solutions to stand for any length of time before neutralization, as severe corrosion with hydrogen evolution will result.)
- Charge tank with soda ash (Na_2CO_3), refit manhole covers, and fill with water. If necessary, add more soda ash to attain a pH > 8.3 (i.e., alkaline to phenolphthalein indicator). This neutralization step can be eliminated, provided the tank can be flushed clean and the sidewalls hosed down thoroughly. This step is not permissible if the tank is lined with a baked phenolic coating, which is attacked by alkalis.
- Air-dry or mop dry after ensuring that the tank is safe for entry (i.e., free of hydrogen as a flammable/explosive mixture, oxygen approximately 20–22%, CO_2 < 5,000 ppm). After drying, the tank should be kept sealed against inadvertent ingress of atmospheric moisture.

Inspection

Planning and carrying out internal and external inspection of acid tanks requires an understanding of the nature of problems that might be encountered. Likely locations of localized grooving or erosion should be identified and given special attention. Some cleaning and possibly surface preparation, such as grit blasting, may be necessary to permit accurate inspection. Welds are particularly prone to attack and should be carefully examined. Associated valves, piping, heaters, and vents can also be attacked and should be included in an overall inspection protocol. Frequency of inspections required will depend on many factors including tank size, strength and temperature of acid, iron content, presence of coating or AP, and previous history of the tank. Reference should be made to specialist guidance documents to aid planning and execution of this essential inspection process.[18,19]

Piping

Carbon-steel piping is suitable for concentrated acid at ambient temperature provided that the acid velocity is less than about 0.9 m/s (3 ft/s). Higher velocities, up to about 1.5 m/s (4.9 ft/s) can be acceptable if duration of pumping through the line is short, e.g., a few hours a day. Carbon-steel piping in intermittent service can suffer from hydrogen grooving. Draining the line and blowing it free of acid using dry air or inert gas might be considered. Ductile iron piping can also be used for handling ambient-temperature concentrated acid, and this piping is more tolerant of increased acid velocities. Iron and steel piping should be provided with means of pressure relief to prevent hydrogen pressure buildup.

Austenitic stainless steel is recommended for small diameter piping, <75 to 100 mm (<3 to 4 in). Type 304/304L (S30400, S30403) is suitable for 93 to 100% acid while 316/316L (S31600, S31603) is preferred for acid in the range of 90 to 100%. The standard steel can withstand higher velocities than steel or ductile iron. For velocities above about 1.8 m/s (5.9 ft/s), 20Cb-3 (N08020) has been successfully used in small-diameter piping. Where solar heating is extreme or when heat tracing is installed, type 316L (S31603), high-nickel stainless steels, or nickel alloys have been used.

Plastic-lined pipe has been generally satisfactory, with PTFE being used in up to 98% acid. Polypropylene-lined pipe has also been used, but there have been occasional failures.[1] The use of polypropylene-lined pipe, fitting, and valves in the 93–98% sulfuric acid concentration range is not recommended by one supplier because in his experience, polypropylene is susceptible to a liquid oxidative degradation mechanism. This mechanism leads to dehydrogenation, charring of the polypropylene liner, and brittle failure of the plastic liner.[32]

Pumps

Pumps associated with tanks should be physically located as close as possible to the tank itself. Long suction lines are to be avoided and must not penetrate the dike wall. The high velocities within the pump demand materials resistant to erosion-corrosion, abrasion, and cavitation.

A centrifugal, sealless, magnetic-drive pump is preferred for sulfuric acid transfer. The wetted parts should be 316 (S31600) stainless steel, alloy 20Cb-3 (N08020), or Teflon®-lined. CN-7M (J95150) is the standard material used for pumps in this service, although nickel-chromium-molybdenum pumps are also used. Glass-lined pumps or graphite pumps are also suitable, but they must be protected from mechanical damage.

Valves

Plug valves or full-port ball valves are generally recommended for sulfuric acid service. Shut-off valves are usually CF-3M (J92800) or CF-8M (J92900) at ambient temperatures, while higher alloys may be used for elevated temperatures or special conditions. The copper-bearing J93370 valves may be available at a similar cost and these are more resistant to weaker acid. Nonmetallic lined valves have been used in this service, as has J95150, and this material is recommended for acid strength >70%.

Throttling valves need to resist erosion-corrosion and abrasion and for this duty, higher alloys—especially Lewmet® and high-silicon cast irons or stainless steels, such as the silicon-containing grades—are preferred. The nickel-based cast alloys are also used, as are fluoroplastic-lined and glass-lined steel valves.

Gaskets

In the United States, compressed asbestos-type gaskets are not used. They are resistant to sulfuric acid, however, and their use continues in some countries. Spiral-wound stainless-steel/PTFE gaskets are used successfully.

Flexible graphite is a suitable gasket material up to 100% acid. Solid PTFE is not a good gasket material itself because of cold-flow characteristics, but it can be used as a multiple braid assembly with 25% glass or calcium fluoride filler. It may also be used when filled with alumina (aluminum oxide).

Envelope gaskets are employed as a styrene-butadiene rubber enclosed in a fluorinated thermoplastic. Solid thermoplastic gaskets may be used in PVDF/hexafluoropropylene or fluoroplastics formulations.

O-Rings and Packing

Materials suitable for use in O-ring seals have been summarized, Table 13.7.[33]

Table 13.7 O-Ring Materials Compatible with Concentrated Sulfuric Acid and Oleum

Material	Concentrated Sulfuric Acid	Oleum
Aflas® (fluoroelastomer, Asahi)	4	0
Buna-N® (Nitrile)	1	1
Butyl	0	1
Chemraz® (perfluoroelastomer Greene, Tweed)	0	4
Epichlorohydrin	1	1
Ethylene-propylene	0	1
Fluorocarbon	0	3
Fluorosilicone	0	1
Hypalon®	0	1
Kalrez®	4	4
Natural rubber	0	1
Neoprene	0	1
Nitrile, hydrogenated	0	3
Polyacrylate	0	1
Polysulfide	1	1
Polyurethane, millable	0	1
Silicone	0	1
Styrene Butadiene	0	1
Teflon®, virgin	4	4
Vamac® (Ethylene acrylic polymer, DuPont)	1	1

Key to compatibility
4—Good, both for static and dynamic seals
3—Fair, usually OK for static seals
2—Sometimes OK for static seals; not OK for dynamic seals
1—Poor
0—No data

Hoses

PTFE-lined hoses are acceptable for 93–98% sulfuric acid service. The hose must be dedicated to sulfuric acid service, be designed with a 200-psi minimum working pressure, and be full vacuum rated. The end fittings must be crimped or swaged; banding is not recommended. The hose end fittings should be type 316 (S31600) stainless steel with flanges or quick-connect fittings. The gaskets must be FKM, Viton®, or equivalent. The user should have a hose-management program in place to ensure the integrity of the hose.[34]

Instrumentation

All instrumentation should be fabricated from corrosion-resistant alloys or fluorinated plastic materials.

References

1. Anon, "Materials for the Handling and Storage of Concentrated (90 to 100%) Sulfuric Acid at Ambient Temperatures," RP 0391 (Houston, TX: NACE International, latest edition).
2. Anon, "Sulfuric Acid Chemical Backgrounder," NSC, National Safety Council (2003), http://www.nsc.org/library/chemical/sulfuric.htm.
3. Anon, "Shipping Container Spec.," 49 CFR, Parts 179–198 (Washington, DC: DOT, latest revision).
4. T. R. Dickey, "Coast Guard Requirements for Carriage of Sulfuric Acid," AIChE Session 77, PETRO EXPO 86, New Orleans, LA, 1986.
5. R. E. Phillips, "Tank Car Design—Past, Present, and Future," PETRO EXPO 86, New Orleans, LA.
6. DOT Specification 111A100A2, 49 CFR Part 179.200, U.S. Code of Federal Regulations (Washington, DC: Government Printing Office). Also: AAR Manual of Standards and Recommended Practices, Section C, Part III, Specifications for Tank Cars, Specification M-1002 (Washington, DC: Association of American Railroads).
7. ASTM A 516/A 516M-90, "Standard Specification for Pressure Vessel Plates, Carbon Steel, for Moderate- and Lower-Temperature Service" (West Conshohocken, PA: ASTM, 2001).
8. ASTM A 515/A 515M-92, "Standard Specification for Pressure Vessel Plates, Carbon Steel, for Intermediate- and Higher-Temperature Service" (West Conshohocken, PA: ASTM, 1997).
9. A808/A808M-00a, "Standard Specification for High-Strength, Low-Alloy Carbon, Manganese, Niobium, Vanadium Steel of Structural Quality with Improved Notch Toughness" (West Conshohocken, PA: ASTM, 2000).
10. S. W. Dean, G. D. Grab, "Corrosion of Carbon Steel Tanks in Concentrated Sulfuric Acid Service," Corrosion 41, 4 (1985).

11. S. L. Pohlman, S. N. Anderssen, "Acid Car Corrosion—Mechanism, Monitoring and Protection," CORROSION/87, paper no. 24 (Houston, TX: NACE International).
12. Anon, "Application of a Coating System to Interior Surfaces of New and Used Rail Tank Cars," RP 0295 (Houston, TX: NACE International, latest edition).
13. Heresite Protective Coatings, private communication (July 1987) in C. P. Dillon, ed., "Concentrated Sulfuric Acid and Oleum," vol. MS-1, Materials Selector for Hazardous Chemicals (St. Louis, MO: MTI Inc., 1997): 212 pp.
14. Anon, "Ships Carrying Bulk Liquid, Liquefied Gas, or Compressed Gas Hazardous Materials," CFR Title 46, Part 153 (amended 1983).
15. Anon, "Multicargo Tankers Heating Coils, Cargo Pumps and Piping," Nickel Magazine 4, 2 (Toronto, ON, Canada: Nickel Development Institute, 1988): p. 10.
16. B. Leffler, "Alloy 2205 for Marine Chemical Tankers," MP 30, 4 (1990): pp. 60–63.
17. ASTM A 212, discontinued specification, replaced by A 515 and A 516, Annual Book of ASTM Standards (West Conshohocken, PA: ASTM).
18. Anon, "Design, Fabrication, and Inspection of Tanks for the Storage of Concentrated Sulfuric Acid and Oleum at Ambient Temperatures," RP 0294 (Houston, TX: NACE International, latest edition).
19. M. Tiivel, F. J. McGlynn, A. A. Trickett, "Carbon Steel Sulfuric Acid Storage Tank: Inspection Guidelines" (North York, ON, Canada: MARSULEX Inc., 1986): 46 pp.
20. Anon, "Corrosion in Sulfuric Acid," Proceedings of 1985 NACE Symposium (Houston, TX: NACE International, 1985).
21. Anon, "Recommended Safe Practices and Emergency Procedures for Sulfur Trioxide, Oleum and Chlorosulphonic Acid," The Soap and Detergent Industry Association, April (1979): 39 pp.
22. Anon, "Brink® Oleum Vent Package" (St Louis, MO: EnviroChem Systems, Inc., 2004), http://www.enviro-chem.com/plant-tech/3rdtier/brinktop.html.
23. Anon, "Oleum Vent Sytems," Koch-OttoYork® Separation Technology (2004), http://www.koch-ottoyork.com/industry/sulfuricacid.htm#ov.
24. ASME/ANSI B31-3, "Chemical Plant and Petroleum Refinery Piping," ASME Code for Pressure Piping (1990).
25. S. W. Dean, G. D. Grab, "Corrosion Of Carbon Steel Tanks In Concentrated Sulfuric Acid Service," Corrosion/85 paper no. 298 (Houston, TX: NACE International, 1985): 9 pp.
26. API 650, "Welded Steel Tanks for Oil Storage" (Washington, DC: American Petroleum Institute, latest edition).
27. M. G. Fontana, "Corrosion," Industrial and Engineering Chemistry 43, August (1951): p. 65A.
28. D. W. McDowell, "Handling Mineral Acids—Sulfuric Acid," Chem Eng 81, 24 (1974): pp. 118–128.
29. D. Fyfe, R. Vanderland, J. Rodda, "Corrosion in Sulfuric Acid Storage Tanks," Chemical Engineering Progress, March (1977): pp. 65–69.
30. API 620, "Recommended Rules for Design and Construction of Large Welded, Low-Pressure Storage Tanks" (Washington, DC: American Petroleum Institute).
31. "ASME Boiler and Pressure Vessel Code" (New York: ASME International).

32. Anon, "Handling Sulfuric Acid," Resistoflex Plastic Lined Pipe and Fittings (2003), http://www.resistoflex.com/sulfuric.htm.
33. Anon, "O-Ring Compatibilities," Engineering Fundamentals, Efunda (2002), http://www.efunda.com/DesignStandards/oring.
34. Anon, "Equipment" (Wilmington, DE: DuPont Sulfur Products, 2003), http://www.dupont.com/sulfurproducts/techdata/equipment.html.

Appendix A. Nominal Composition of Alloys

Common name	UNS No.	Nominal Composition %
Alloy Steels		
1.25 Cr, 0.5 Mo	K11597	1.25 Cr, 0.5 Mo, 0.45 Mn, 0.75 Si, <0.15 C
2.25 Cr, 1 Mo	K21590	2.25 Cr, 1 Mo, 0.45 Mn, 0.5 Si, <0.15 C
5 Cr, 0.5 Mo	K41545	5 Cr, 0.5 Mo, 0.45 Mn, 0.5 Si, <0.15 C
NiResist® Alloy Cast Irons		
Type 1	F41000	15.5 Ni, 2 Cr, 6.5 Cu, 1 Mn, 2 Si, <3 C
Type 2	F41002	20 Ni, 2 Cr, >0.5 Cu, 1 Mn, 2 Si, <3 C
Type 3	F41002	20 Ni, 4.5 Cr, >0.5 Cu, 0.8 Mn, <3 C
Type 5	F41006	35 Ni, <0.1 Cr, <0.5 Cu, 1 Mn, 1.5 Si, <2.4 C
Type D2	F43000	20 Ni, 2.25 Cr, 1 Mn, 2.25 Si, <3 C
Type D5	F43006	35 Ni, <0.1 Cr, <1 Mn, 2 Si, <2.4 C
Austenitic Stainless Steels		
22-13-5	S20910	22 Cr, 12.5 Ni, 5 Mn, 2.25 Mo, <0.06 C
302	S30200	18 Cr, 9 Ni, <0.15 C
304	S30400	18 Cr, 8 Ni, <0.08 C
304L	S30403	19 Cr, 10 Ni, <0.03 C
304H	S30409	19 Cr, 9.5 Ni, 0.04–0.1 C
309	S30900	22 Cr, 12 Ni, <0.20 C
309S	S30908	22 Cr, 12 Ni, <0.08 C
310	S31000	25 Cr, 20 Ni, <0.25 C
310L	S31050	25 Cr, 20 Ni, <0.02 C
310S	S31008	25 Cr, 20 Ni, <0.08 C
316	S31600	17 Cr, 12 Ni, 2.7 Mo, <0.08 C
316L	S31603	17 Cr, 12 Ni, 2.7 Mo, <0.03 C
316Ti	S31635	17 Cr, 12 Ni, 2.7 Mo, 0.04 C, Ti
317L	S31703	19 Cr, 13.5 Ni, 3.5 Mo, <0.03 C
321	S32100	18 Cr, 10.5 Ni, 0.05 C, 0.4 Ti
347	S34700	18 Cr, 11 Ni, 0.05 C, 0.04 Nb

Common name	UNS No.	Nominal Composition %
Iron-Based and Iron-Rich Cast Alloys		
CD-4MCu	J93370	25 Cr, 5 Ni, 2 Mo, 3 Cu, 0.03 C
CF-3	J93500	18 Cr, 12 Ni, 1 Si, 0.02 C
CF-3M	J92800	19 Cr, 10 Ni, 2 Mo, 1 Si, 0.02 C
CF-8	J92600	19 Cr, 9 Ni, 1 Si, 0.05 C
CF-8M	J92900	19 Cr, 10 Ni, 2 Mo, 1 Si, 0.05 C
CH-20	J93402	22 Cr, 12 Ni, 1 Si, 0.1 C
CK-20	J94202	23 Cr, 19 Ni, 1 Si, 0.1 C
CN-7M	J95150	20 Cr, 29 Ni, 2.2 Mo, 3.4 Cu, 1 Si, 0.05 C
Silicon iron	F47003	14.5 Si, <1.5 Mn, <0.5 Cr, <0.5 Mo, 0.9 C
High-Performance Alloys		
904L	N08904	21 Cr, 25.5 Ni, 4.5 Mo, 1.5 Cu, <0.02 C
2RK65	N08904	20 Cr, 25 Ni, 4.5 Mo, 1.5 Cu, <0.025 C
254SMO	S31254	20 Cr, 18 Ni, 6.2 Mo, 0.7 Cu, 0.2 N, <0.02 C
25-6MO	N08926	20 Cr, 25 Ni, 6.5 Mo, 0.9 Cu, 0.2 N
1925hMo	N08926	20 Cr, 25 Ni, 6.2 Mo, 0.8 Cu, 0.2 N, <0.02 C
Alloy 926	N08926	20 Cr, 25 Ni, 6.2 Mo, 0.8 Cu, 0.2 N, <0.02 C
27-7MO	S31277	21.5 Cr, 27 Ni, 7.2 Mo, 1 Cu, 0.35 N
AL-6X	N08366	21 Cr, 24.5 Ni, 6.5 Mo, <0.035 C
AL-6XN	N08367	20.5 Cr, 24 Ni, 6.3 Mo, 0.2 Cu, 0.2 N, <0.02 C
654SMO	S32654	24 Cr, 22 Ni, 7.3 Mo, 3 Mn, 0.5 Cu, 0.5 N, <0.01 C
Alloy 18	S38100	18 Cr, 18 Ni, 2 Mo
20Cb-3	N08020	20 Cr, 35 Ni, 2.5 Mo, 3.5 Cu, 0.07 Cb, <2 Mn, <1 Si, <0.02 C
20 Mod	N08320	22 Cr, 26 Ni, 5 Mo, <2.5 Mn, <1 Si, <0.05 C, Ti
20Mo-6	N08026	24 Cr, 35 Ni, 5.8 Mo, 3 Cu, 0.03 C
Worthite®	—	20 Cr, 24 Ni, 3 Mo, 1.75 Cu, 3.3 Si, 0.07 C
Nitronic® 50	S20910	22 Cr, 12.5 Ni, 2.2 Mo, <1 Si, 0.3 N, 0.2 Cb, 0.2 V
Nitronic® 60	S21800	17 Cr, 8.5 Ni, 4 Si, 8 Mn, 0.13 N, <0.1 C
800	N08800	20 Cr, 31 Ni, <0.08 C, 0.4 Si, 0.3 Al, 0.4 Ti
825	N08825	21.5 Cr, 42 Ni, 3 Mo, 2.3 Cu, <0.05 C, 0.9 Ti, <0.2 Al
Ni-o-nel	—	22 Cr, 42 Ni, 3 Mo, 1.8 Cu, 1 Ti
3127hMo	N08031	27 Cr, 31 Ni, 6.5 Mo, 1.2 Cu, 0.2 N, <0.015 C
Alloy 28	N08028	27 Cr, 32 Ni, 3.5 Mo, 1.0 Cu, <0.03 C
Alloy 31	N08031	27 Cr, 31 Ni, 6.5 Mo, 1.2 Cu, 0.2 N, <0.02 C
Alloy 33	R20033	33 Cr, Bal Ni, 32 Fe, 1.6 Mo, 0.6 Cu, 0.4 N
Lewmet® 66	—	31 Cr, Bal Ni, 16 Fe, 3 Cu, 6 Co, 3 Si, 3 Mn, <0.08 C
G	N06007	22.5 Cr, Bal Ni, 6.5 Mo, 2 Cu, <0.03 C, 19.5 Fe, 2 Cb
G-3	N06985	22.5 Cr, Bal Ni, 7 Mo, 2 Cu, <0.015 C, 19.5 Fe, <0.5 Cb + Ta
G-30	N06030	30 Cr, Bal Ni, 5 Mo, 1.7 Cu, <0.03 C, 15 Fe, 2.7 W, <5 Co, 0.9 Cb + Ta
Illium® P	—	28 Cr, 8 Ni, 2 Mo, 3 Cu, 0.2 C

Common name	UNS No.	Nominal Composition %
Silicon-Containing Austenitic Stainless Steels		
A610/1815 LCSi	S30600	17.5 Cr, 15.2 Ni, 4.0 Si, <0.018 C
A611	S30601	17.5 Cr, 17.5 Ni, 5.3 Si, <0.015 C
ZeCor®	S38815	14 Cr, 15 Ni, 1.0 Mo, 1.0 Cu, 6.0 Si
SX®	S32615	18 Cr, 20 Ni, 1.0 Mo, 2.0 Cu, 5.5 Si, 1.5 Mn, <0.04 C
SARAMET® 23	S30601	17.5 Cr, 17.5 Ni, 5.3 Si, <0.02 C
Bohler 614	S32615	18 Cr, 20 Ni, 1.0 Mo, 2.0 Cu, 5.5 Si
AL 388	S38815	14 Cr, 15 Ni, 1.0 Mo, 1.0 Cu, 6.0 Si
2509Si7	S70003	9 Cr, 25 Ni, 7 Si, <0.02 C
8204	—	20 Cr, 11 Ni, 5 Si, 12 Co, 2 Cb, <0.06 C
Durcomet® 5	—	21 Cr, 16 Ni, 5 Si, 0.025 C
Duplex and Superduplex Stainless Steels		
329	S32900	25.5 Cr, 3.5 Ni, 1.5 Mo, 0.05 N, <0.02 C
3RE60	S31500	18.5 Cr, 4.7 Ni, 2.7 Mo, 1.6 Mn, 0.08 N, <0.03 C
2205	S31803	22 Cr, 5.5 Ni, 3.0 Mo, 0.14 N, <0.03 C
2304	S32304	23 Cr, 4 Ni, 0.1 N, < 0.03 C
2507	S32750	25 Cr, 7 Ni, 4.0 Mo, 0.3 N, <0.03 C
—	S32906	29 Cr, 6 Ni, 2 Mo, 0.4 N, <0.03 C
Alloy 255	S32550	25.5 Cr, 5.5 Ni, 3.4 Mo, 2.0 Cu, <0.04 C, <1.5 Mn
Zeron® 100	S32760	25 Cr, 7 Ni, 3.5 Mo, 0.25 N, <0.03 C, 0.75 Cu, 0.75 W
7-Mo	—	27.5 Cr, 4.5 Ni, 1.5 Mo, <2 Mn, <0.1 C
7-Mo PLUS®	S32950	27.5 Cr, 4.5 Ni, 2 Mo, <2 Mn, <0.03 C
Lewmet® 55	—	32 Cr, Bal Ni, 4 Mo, 16 Fe, 3 Cu, 3.5 Si, 6 Co, 3 Mn, 0.03 B, <0.08 C
Ferritic and Superferritic Stainless Steels		
409	S40900	11 Cr, 0.5 Ni, <0.08 C, Ti
430	S43000	17 Cr, <0.12 C
434	S43400	17 Cr, 1 Mo, <0.12 C
444	S44400	18 Cr, 2 Mo, <0.02 C, Cb/Ti
446	S44600	25 Cr, <0.2 C
E-Brite®/26-1	S44627	26 Cr, 1 Mo, 0.002 C, Cb
XM-27®	S44627	26 Cr, 1 Mo, 0.002 C, Cb
Sea-Cure®	S44660	27.5 Cr, 1.2 Ni, 3.5 Mo, 0.5 Ti, 0.3 Si
Monit®	S44635	25 Cr, 4.0 Ni, 4.0 Mo, 0.5 Ti, 0.35 Si
29-4C	S44735	29 Cr, 0.3 Ni, 4.0 Mo, 0.5 Ti, 0.35 Si
29-4-2	S44800	29 Cr, 2.1 Ni, 4.0 Mo, 0.1 Si
Precipitation-Hardening Steels		
15-5 PH	S15500	5 Ni, 15 Cr, 3.5 Cu, <1 Mn, <1 Si, <0.07 C
17-4 PH	S17400	4 Ni, 17 Cr, 4 Cu, <1 Mn, <1 Si, <0.07 C

Common name	UNS No.	Nominal Composition %
Nickel-Based Alloys		
Balance is Ni unless Ni content is stated.		
200	N02200	>99 Ni, <0.4 Fe, <0.25 Cu, <0.35 Mn, <0.02 C
400	N04400	66.5 Ni, Bal Cu, <0.3 C, <2.5 Fe, <2 Mn, <0.5 Si, <0.3 C
600	N06600	>72 Ni, 16 Cr, <0.5 Cu, <0.15 C, 8 Fe
625	N06625	61 Ni, 22 Cr, 9 Mo, <0.1 C, <5 Fe, 3.6 Nb
686	N06686	21 Cr, 16 Mo, 3.7 W, <5 Fe
B-2	N10665	68 Ni, <1.0 Cr, 28 Mo, <0.02 C, <1 Co, 1.8 Fe
B-3	N10675	1.5 Cr, 28.5 Mo, 1.5 Fe, 0.003 C
B-4	N10629	1.3 Cr, 28 Mo, 3 Fe, 0.005 C
C-4	N06455	54 Ni, 16 Cr, 15.5 Mo, <0.015 C, <3 Fe, 0.7 Ti, <2 Co
C-22	N06022	21 Cr, 13 Mo, 3 W, 4 Fe, 0.2 V, 1.7 Co, 0.003 C
C-276	N10276	54 Ni, 15.5 Cr, 16 Mo, <0.02 C, <2.5 Co, 5.5 Fe, 4 W
C-2000	N06200	23 Cr, 16 Mo, 1.6 Cu, <0.01 C, <0.08 Si
Allcorr	N06110	31 Cr, 10 Mo, 2 W, 0.02 C
Alloy 59	N06059	23 Cr, 16 Mo, 1 Fe
Illium® G	—	22 Cr, 6 Mo, 5 Fe, 0.2 Si
Illium® 98	—	28 Cr, 8 Mo, 5 Cu, 0.05 C
Illium® B	—	28 Cr, 8 Mo, 5 Cu, 3.5 Si, 0.05 C
Nickel-Based Cast Alloys		
CW-2M	N26455	1 Fe, 15 Cr, 69 Ni, 15 Mo, 0.5 Si, 0.01 C
CW-6M	N30107	1 Fe, 17 Cr, 65 Ni, 17 Mo, 0.5 Si, 0.05 C
CW-6MC	N26625	3 Fe, 20 Cr, 65 Ni, 8 Mo, 3.5 Cb, 0.5 Si, 0.05 C
CW-12MW*	N30002	6 Fe, 16 Cr, 56 Ni, 17 Mo, 0.5 Si, 0.07 C, 4.5 W, 0.3V
N-7M	N30007	1 Fe, 1 Cr, 67 Ni, 31 Mo, 0.5 Si, 0.04 C
N-12MV*	N30012	4 Fe, 1 Cr, 69 Ni, 26 Mo, 0.5 Si, 0.1 C, 0.2 V
Titanium Alloys		
Grade 2	R50400	<0.3 Fe, Bal Ti
Grade 7	R52400	<0.3 Fe, Bal Ti, 0.15 Pd
Grade 12	R53400	<0.3 Fe, Bal Ti, 0.3 Mo, 0.8 Ni
Grade 26	R52404	<0.3 Fe, Bal Ti, 0.11 Ru, 0.25 O_2
Grade 27	R52254	<0.2 Fe, Bal Ti, 0.11 Ru, 0.18 O_2
Zirconium Alloys		
Zirconium 702	R60702	<0.2 Fe + Cr, 99.2 Zr + Hf, <4.5Hf
Zirconium 704	R60704	0.3 Fe + Cr, 97.5 Zr + Hf, <4.5 Hf, 1.5 Sn
Zirconium 705	R60705	0.2 Fe + Cr, 95.5 Zr + Hf, <4.5 Hf, 2.0–3.0 Nb
Aluminum Alloys		
1100	A91100	0.1 Cu, <0.05 Mn, <0.1 Zn
3003	A93003	0.1 Cu, <0.07 Fe, 1.2 Mn, <0.6 Si, <0.1 Zn

* Class 1 is post-weld heat treated; Class 2 is not.

Common name	UNS No.	Nominal Composition %
Copper Alloys		
Silicon bronze	C65500	Rem Cu, 3.2 Si, Mn 0.9, Zn <1.5, Fe <0.8, Ni <0.6
Lead Alloys		
Chemical Pb	L51120	>99.9 Pb, 0.06 Cu
Antimonial Pb	L52901	96 nom Pb, 4 nom Sb
Cobalt Alloys		
ULTIMET®	R31233	26 Cr, Bal Co, 9 Ni, 5 Mo, 2 W, 3 Fe
Stellite 1	R30001	30 Cr, Bal Co, >3 Fe, 1.5 Ni, 0.5 Mn, 1.3 Si, 13 W, 2.5 C
Havar	R30004	42 Co, 20 Cr, 13 Ni, 2.4 Mo, 2.8 W, 1.5 Mn, 0.04 Be
Stellite 21	R30021	27 Cr, Bal Co, <3 Fe, 2.75 Ni, <1 Mn, 5.5 Mo, <1 Si, 0.25 C

Appendix B. Approximate Equivalent Grade of Some Cast and Wrought Alloys

Structure	Alloy Name	Cast (ACI)	Cast UNS	Wrought
Austenitic SS	304L	CF3, CF3A	J92500	S30403
	304	CF8	J92600	S30400
	316	CF8M	J92900	S31600
	316L	CF3M, CF3MA	J92800	S31603
	310	CK20	J94202	S31000
	309	CH20	J93402	S30900
	Alloy 20	CN7M	N08007 J95150	N08020
Duplex SS	Alloy 2205	CD3MN	J92205	S31803 S32205
	Zeron® 100	CD3MWCuN	J93380	S32760
	—	CD4MCu	J93370	—
Martensitic or Ferritic SS	Alloy 410	CA15	J91150	S41000
	Alloy 420	CA40	J91153	S42000
Nickel-Based Alloys	Alloy 825	Cu5MCuC	N08826	N08825
	Alloy 600	CY40	N06040	N06600
	Alloy 625	CW6MC	N26625	N06625
	Alloy 400	M35-2	N04020	N04400
	Alloy B-2	N7M	N30007	N10665
		N12MV	N30012	
	Alloy C-4	CW2M	N26455	N06455
	Alloy C-22	CX2MW	N26022	N06022
	Alloy C-276	CW6M	N30107	N10276

Appendix C. Glossary of Corrosion and Materials Terms

active—(1) the negative direction of electrode potential; (2) a state of a metal that is corroding without significant influence of reaction product

amphoteric—a metal that is susceptible to corrosion in both acid and alkaline environments

anion—a negatively charged ion that migrates through the electrolyte

anode—the electrode of an electrochemical cell at which oxidation occurs. Electrons flow away from the anode in the external circuit. Corrosion usually occurs and metal ions enter the solution at the anode

anodic protection—polarization to a more oxidizing potential to achieve a reduced corrosion rate by the promotion of passivity

anodizing—oxide coating formed on a metal surface (generally aluminum) by an electrolytic process

austenite—the face centered cubic structure of iron-based alloys

austenitic—a steel in which the predominant structure at room temperature is austenite

brittle fracture—fracture with little or no plastic deformation

carbon steel—alloy of carbon and iron containing up to 0.5% carbon, manganese, and residual quantities of other elements, except those intentionally added in specific quantities for deoxidation (usually silicon and/or aluminum)

casting (cast component)—metal that is obtained at or near its finished shape by the solidification of molten metal in a mold

cast iron—iron-carbon alloy containing approximately 2 to 4% carbon

cathode—the electrode of an electrochemical cell at which reduction is the principal reaction. Electrons flow toward the cathode in the external circuit

cathodic corrosion—corrosion resulting from a cathodic condition of a structure, usually caused by the reaction of an amphoteric metal with the alkaline products of electrolysis

cathodic protection—a technique to reduce the corrosion of a metal surface by making that surface the cathode of an electrochemical cell

cation—a positively charged ion that migrates through the electrolyte toward the cathode under the influence of a potential gradient

cavitation—the formation and rapid collapse of cavities or bubbles within a liquid, which often results in damage to a material at the solid/liquid interface under conditions of severe turbulent flow

corrosion—the deterioration of a material, usually a metal, that results from a reaction with its environment

corrosion fatigue—fatigue-type cracking of metal caused by repeated or fluctuating stresses in a corrosive environment characterized by shorter life than would be encountered as a result of either the repeated or fluctuating stress alone or the corrosive environment alone

corrosion inhibitor—a chemical substance or combination of substances that, when present in the environment, prevents or reduces corrosion

corrosion potential (Ecorr)—the potential of a corroding surface in an electrolyte relative to a reference electrode under open-circuit conditions (also known as rest potential, open-circuit potential, or freely corroding potential)

corrosion rate—the rate at which corrosion proceeds

corrosion resistance—ability of a material, usually a metal, to withstand corrosion in a given system

corrosion-resistant alloy (CRA)—alloy intended to be resistant to general and localized corrosion of oilfield environments that are corrosive to carbon steels

corrosiveness—the tendency of an environment to cause corrosion

creep—time-dependent strain occurring under stress

crevice corrosion—localized attack of a metal at or near an area that is shielded from the bulk environment

dealloying—the selective corrosion of one or more components of a solid solution alloy (also known as parting or selective dissolution)

dezincification—a corrosion phenomenon resulting in the selective removal of zinc from copper-zinc alloys (this phenomenon is one of the more common forms of dealloying)

ductile (nodular) cast iron—cast iron that has been treated while molten with an element (usually magnesium or cerium) that spheroidizes the graphite

electrochemical cell—a system consisting of an anode and a cathode immersed in an electrolyte so as to create an electrical circuit. The anode and cathode may be different metals or dissimilar areas on the same metal surface

electrolyte—a chemical substance containing ions that migrate in an electric field

embrittlement—loss of ductility of a material resulting from a chemical or physical change

environment—the surroundings or conditions (physical, chemical, mechanical) in which a material exists

environmental cracking—brittle fracture of a normally ductile material in which the corrosive effect of the environment is a causative factor. Environmental cracking is a general term that includes corrosion fatigue, hydrogen embrittlement, hydrogen-induced cracking (stepwise cracking), hydrogen stress cracking, liquid metal cracking, stress corrosion cracking, and sulfide stress cracking

erosion—the progressive loss of material from a solid surface due to mechanical interaction between that surface and a fluid, a multicomponent fluid, or solid particles carried with the fluid

erosion-corrosion—a conjoint action involving corrosion and erosion in the presence of a moving corrosive fluid or a material moving through the fluid, leading to accelerated loss of material

ferrite—body-centered cubic crystalline phase of iron-based alloys

ferritic steel—steel whose microstructure at room temperature consists predominantly of ferrite

fretting corrosion—deterioration at the interface of two contacting surfaces under load that is accelerated by their relative motion

galvanic corrosion—accelerated corrosion of a metal because of an electrical contact with a more noble metal or nonmetallic conductor in a corrosive electrolyte

graphitic corrosion—deterioration of gray cast iron in which the metallic constituents are selectively leached or converted to corrosion products, leaving the graphite intact

graphitization—the formation of graphite in iron or steel, usually from decomposition of iron carbide at elevated temperatures (should not be used as a term to describe graphitic corrosion)

heat-affected zone (HAZ)—that portion of the base metal that is not melted during brazing, cutting, or welding, but whose microstructure and properties are altered by the heat of these processes

heat treatment—heating and cooling a solid metal or alloy in such a way as to obtain desired properties. NOTE: Heating for the sole purpose of hot working is not considered heat treatment

hydrogen blistering—the formation of subsurface planar cavities, called hydrogen blisters, in a metal resulting from excessive internal hydrogen pressure. Growth of near-surface blisters in low-strength metals usually results in surface bulges

hydrogen embrittlement—a loss of ductility of a metal resulting from absorption of hydrogen

hydrogen-induced cracking—stepwise internal cracks that connect adjacent hydrogen blisters on different planes in the metal, or to the metal surface (also known as stepwise cracking)

inhibit—to retard or slow the rate of corrosion, usually by the addition of other chemicals to the system

intergranular corrosion (IGC)—preferential corrosion at or near the grain boundaries of a metal

iron rot—deterioration of wood in contact with iron-based alloys

knife-line attack (KLA)—local corrosion along a line adjacent to a weld after heating into the sensitization temperature range

liquid metal cracking (LMC)—cracking of a metal caused by contact with a liquid metal

low-alloy steel—steel with a total alloying element content of less than about 8%, but more than specified for carbon steel

metallizing—the coating of a surface with a thin metal layer by spraying, hot dipping, or vacuum deposition

oxidation—(1) loss of electrons by a constituent of a chemical reaction; (2) corrosion of a metal that is exposed to an oxidizing gas at elevated temperatures

passivation—a reduction in the anodic reaction rate of an electrode involved in a corrosion process

passive—(1) the positive direction of electrode potential; (2) a state of a metal in which a surface reaction product causes a marked decrease in the corrosion rate relative to that in the absence of the product

pH—the negative logarithm of the hydrogen ion activity written as: pH = –log10(aH+), where aH+ = hydrogen ion activity = the molar concentration of hydrogen ions multiplied by the mean ion-activity coefficient

pitting—localized corrosion of a metal surface that is confined to a small area and takes the form of cavities called pits

pitting factor—the ratio of the depth of the deepest pit resulting from corrosion divided by the average penetration as calculated from mass loss

rust—corrosion product consisting of various iron oxides and hydrated iron oxides (this term properly applies only to iron and ferrous alloys)

sensitization—precipitation of constituents (usually carbides) in a structure as a result of heating and cooling through a certain temperature range. Can lead to intergranular corrosion

stress corrosion cracking (SCC)—cracking of metal involving anodic processes of localized corrosion and tensile stress (residual and/or applied)

sulfidation—the reaction of a metal or alloy with a sulfur-containing species to produce a sulfur compound that forms on or beneath the surface of the metal or alloy

transpassive—the noble region of potential where an electrode exhibits a higher-than-passive current density

weld (verb)—join two or more pieces of metal by applying heat and/or pressure with or without filler metal to produce a union through localized fusion of the substrates and solidification across the interfaces

weld decay—intergranular corrosion associated with sensitization due to welding

weldment—that portion of a component on which welding has been performed, including the weld metal, the heat-affected zone (HAZ), and the base metal

weld metal—that portion of a weldment that has been molten during welding

wrought metal—metal in the solid condition that is formed to a desired shape by working (rolling, extruding, forging, etc.), usually at an elevated temperature

yield strength—stress at which a material exhibits a specified deviation from the proportionality of stress to strain

Appendix D. Glossary of Acronyms and Abbreviations

AAR—Association of American Railroads
ACGIH—American Conference of Governmental Industrial Hygienists
ANSI—American National Standards Institute
AP—Anodic Protection
API—American Petroleum Institute
ASME—American Society of Mechanical Engineers
ASTM—American Society for Testing and Materials
AWA – Alkylation Waste Acid
BP—Boiling Point
BSI—British Standards Institution
CAF—Compressed Asbestos Fiber
CAS—Chemical Abstracting Service
CBM—Constant Boiling Mixture
CFR—Code of Federal Regulations
CI—Corrosion Index; also, Cast Iron
CMA—Chemical Manufacturers Association
CP—Cathodic Protection
C.P.—Chemically Pure
CPVC—Chlorinated Polyvinyl Chloride
CSA—Contact Sulfuric Acid
CSCC—Chemical Stress Corrosion Cracking (plastics)
°Bé—Degrees Baumé
DBT—Ductile-Brittle Transition
DI – Ductile Iron
DIN—Deutsches Institut für Normung
DO—Dissolved Oxygen
DOT—Department of Transportation (U.S.)
ECTFE—Ethylene Chlorotrifluoroethylene
EFMA—European Fertilizer Manufacturers Association
EPA—Environmental Protection Agency
ERPG—Emergency Response Planning Guideline
ESCC—Environmental Stress Corrosion Cracking (plastics)
FEP—Fluorinated Ethylene Propylene
FP—Freezing Point
FRP—Fiber-Reinforced Plastic

GRP—Glass Fiber-Reinforced Plastic
HAC—Hydrogen Assisted Cracking
HAZ—Heat-Affected Zone
HBN—Hardness Brinell Number
HDPE—High-Density Polyethylene
HRC—Hardness Rockwell C
HSC—High Stage Concentrator (Make)
IDLH—Immediately Dangerous to Life or Health
IGA—Intergranular Attack
IGC—Intergranular Corrosion
IMO—International Maritime Organization
LMC—Liquid Metal Cracking
MIC—Microbiologically Influenced Corrosion
MSDS—Material Safety Data Sheet
NACE—National Association of Corrosion Engineers
NDT—Nil Ductility Transition
NDTT—Nil Ductility Transition Temperature
NIOSH—National Institute for Occupational Safety and Health
NPS—Nominal Pipe Size
OSHA—Occupational Safety and Health Agency
PE—Polyethylene
PEL—Permissible Exposure Limit
PFA—Perfluoroalkoxy
PP—Polypropylene
PTFE—Polytetrafluoroethylene
PVC—Polyvinyl Chloride
PVDC—Polyvinylidene Chloride
PVDF—Polyvinylidene Fluoride
PVF—Polyvinyl Fluoride
Re—Reynolds Number
REL—Recommended Exposure Limit
RTP—Reinforced Thermoset Plastic
SAC—Sulfuric Acid Concentration
SBR—Styrene Butyl Rubber
SCC—Stress Corrosion Cracking
Sp. Gr.—Specific Gravity
SS—Stainless Steel
STEL—Short Term Exposure Limit
TLV—Threshold Limit Value
TWA—Time Waited Average
UV—Ultraviolet

Index